Microalgae in Waste Water Remediation

T0199563

Arun Kumar

Department of Environmental Science
Guru Nanak Khalsa College
Yamuna Nagar, Haryana, India

Jay Shankar Singh

Department of Environmental Microbiology
Babasaheb Bhimrao Ambedkar University
Lucknow, Uttar Pradesh, India

CRC Press
Taylor & Francis Group
Boca Raton London New York

CRC Press is an imprint of the
Taylor & Francis Group, an **informa** business

A SCIENCE PUBLISHERS BOOK

Cover illustrations provided by one of the authors of the book, Dr. Jay Shankar Singh.

First edition published 2021
by CRC Press
6000 Broken Sound Parkway NW, Suite 300, Boca Raton, FL 33487-2742

and by CRC Press
2 Park Square, Milton Park, Abingdon, Oxon, OX14 4RN

Library of Congress Cataloging-in-Publication Data

Names: Kumar, Arun, 1991- author. | Singh, Jay Shankar, author.
Title: Microalgae in waste water remediation / Arun Kumar, Department of
    Environmental Science, Guru Nanak Khalsa College, Yamuna Nagar, Haryana,
    India, Jay Shankar Singh, Department of Environmental Microbiology,
    Babasaheb Bhimrao Ambedkar University, Lucknow, Uttar Pradesh, India.
Other titles: Microalgae in wastewater remediation
Description: First edition. | Boca Raton : CRC Press, Taylor & Francis
    Group, 2021. | Includes bibliographical references and index.
Identifiers: LCCN 2020051491 | ISBN 9780367276034 (hardcover)
Subjects: LCSH: Sewage--Purification--Biological treatment. |
    Microalgae--Industrial applications. | Microalgae--Utilization.
Classification: LCC TD755 .K8154 2021 | DDC 628.3/5--dc23
LC record available at https://lccn.loc.gov/2020051491

ISBN: 978-0-367-27603-4 (hbk)
ISBN: 978-0-367-70737-8 (pbk)
ISBN: 978-0-429-29808-0 (ebk)

Typeset in Palatino Roman
by Innovative Processors

Preface

During recent years, the intensity of agro-industrial and urbanization activities has been generating huge amounts of wastewaters. There are still constant municipal and rural domestic wastewater that is discharged directly into water ways, often without proper treatment. The indiscriminate disposal of wastewaters without an effective treatment to water sources has been causing severe water pollution hazards, particularly in developing countries. In general physico-chemical methods have been developed to remove nutrients from wastewater, but these are expensive and generate high sludge contents. Moreover, the occurrence of emerging organic and inorganic pollutants, such as micro-plastics, pharmaceuticals, flame retardants, personal care products, hazardous and noxious substances have been increasing since conventional wastewater treatment plants are not yet equipped and suitable to remove these new contaminants. Therefore, finding a solution for the treatment and safe discharge of wastewater is urgently required because the integrated treatment processes entail technical, economic and financial challenges . It is important to ensure that appropriate treatment standards should be selected to meet local conditions, and alternative innovative technologies along with conventional methods for treating wastewater should be evaluated. It has been reported that due to the requirement of large amounts of nutrients, especially nitrogen and phosphorus for growth, microalgae species have been proposed as an effective biological treatment to remove nitrogen and phosphorus from a different source of wastewater for a long time. Recent research investigations indicated that the formation of Microalgae-Bacteria Flocculation (MBF) and Fungi-Assisted Microalgae Pellets (FAMP) can be a new tool to enhance nutrient removal from wastewater. However, the impact of environmental drivers on these new tools to clean wastewaters is not adequately understood yet. Therefore, it is vital to pay more attention to these new biological-mediated tools to enhance its application.

The need to reduce costs associated with the production of microalgae biomass has encouraged pairing the process with wastewater treatment. The application of microalgae on water bioremediation is an evolving research field that currently focuses on developing efficient and cost-effective treatments methods known as phycoremediation. Designing the cultivation system is a critical parameter for microalgae production in order to achieve optimal growth rates and minimize costs. Microalgae are typically cultivated in open or closed culture systems. Furthermore, diverse strategies, namely in batches, continuous and semi-continuous, are applied around the world and which differ mainly on nutrient supply and operational modes. New methods of microalgae cultivation, like the immobilization method on alginate beads and the development of bioreactors that use microalgae biofilm, have emerged and have been coupled with the treatment of wastewater. Interest in this field of research is increasing, related to the potential of microalgae cultivation for wastewater remediation. Nevertheless, cost efficacy of commercialization is still an issue and essential to attract investors to up-scale production, so technology is still being developed in terms of research and development. The present book *Microalgae in Waste Water Remediation* aims to point out trends and hot topics concerning the use of microalgae in wastewater treatment and to identify potential paths for future research regarding microalgae-based bioremediation. In order to achieve this goal, available literature has been updated to assess and analyze the topics that have attracted more attention among the scientific community and their evolution through time. We are sure that this book will be useful for students, scientists and policy makers concerned with microalgae-mediated management of wastewater effluents and its applications in the overall future of sustainable development.

Arun Kumar
Jay Shankar Singh

Contents

Microalgae I: Origin, Distribution and Morphology

Introduction

Microalgae are considered as a large group of microscopic, phototrophic organisms that include cyanobacteria, prochlorophytes and eukaryotic algae. They are found in diverse ranges of terrestrial, freshwater and marine habitats; and also exist in extreme environments, such as snow, sea ice, deserts, hot springs and salt lakes (Rothschild and Mancinelli 2001, Vincent 2010, Singh 2014, Hopes and Mock 2015, Singh et al. 2016). They are ubiquitous in distribution, survive in most habitats that have moisture and sufficient light conditions, and are quite common in freshwater lakes and oceans as part of the phytoplankton community.

Organisms like cyanobacteria originated about 2.7-2.6 billion years ago in the Archean eon of the Precambrian era; which later may have been responsible for a radical transformation involving global oxygenation in the atmosphere of the Earth (Canfield 2000, Holland 2006, Kulasooriya 2011). The primary endosymbiotic event between cyanobacteria and the unicellular heterotrophic eukaryotic host gave rise to chlorophytes, glaucophytes and rhodophytes containing membrane-bound organelles known as plastids (Falkowski et al. 2004, Miyagishima 2011). Later a secondary endosymbiotic event involving green and red algae with new heterotrophic eukaryotes gave rise to euglenoids and chlorarachniophyte, cryptophytes, dinoflagellates and chromophytes (diatoms) having green or red plastids with additional membranes (Cavalier-Smith 1999).

In oceans, cyanobacteria and eukaryotic microalgae are together responsible for just over 45% of global primary production (Field et al. 1998), which play a large role in nutrient cycling by providing nutrition to other organisms and later on death, sinking their biomass (organic matter) to the interior of oceans. The regular sinking of algal biomass also enriches the ocean's interior with fixed atmospheric CO_2, resulting in maintaining global CO_2 levels. This phenomenon is more closely observed during seasonal excess microalgal growth (often a single or a few species of microalgae) or simply algal blooms, occurring due to excess availability of nutrients such as nitrate and iron (Boyd et al. 2000). On depletion of nutrients, the microalgal community begins to sink, which are responsible for cycling organic and inorganic products. Some microalgal groups i.e., diatoms and coccolithophores have geological contributions such as diatomite and chalk deposits, through sinking of their intricate shells of silica and calcium carbonate. It was also indicated that sinking and depositions of microalgal lipids was linked to marine oil reserves.

Origin and Evolution

The phylogeny of microalgae and related organisms are understood or interpreted through ultrastructural and molecular evidence of conserved features like plastid and mitochondrial structures, flagellar hairs and roots, plastid mitotic apparatus and ribosomal RNA gene sequencing (Fig. 1.1). There are several problems such as the dynamic nature of research and discussions related to the phylogeny of eukaryotes, current doubts about the placement of a number of groups, requiring further research in ecology and distribution, biochemical diversity and applied phycology; which limits the scope of a classification scheme based on a traditional approach. The traditional approach, primarily considers features like pigmentation, thylakoid organization and other ultrastructural features of the chloroplast, the chemical nature of the photosynthetic storage product, structure and chemistry of the cell wall, and the presence of flagella (if any) and their number, arrangement and ultrastructure; to classify the algae into major groups in to 'divisions' (which are equivalent to zoological 'phylum').

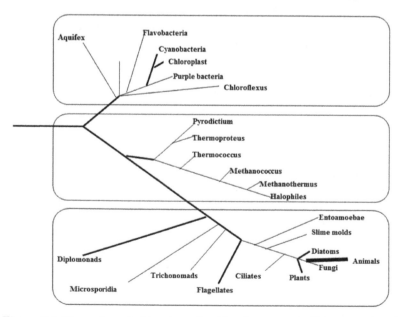

Figure 1.1: Three domain kingdom classification systems (Woese et al. 1972).

Based on fossil records, it was established that the first cyanobacteria originated about 2.7–2.6 billion years ago in the Archean era of Precambrian period that created small oxygen 'oases' within the anoxygenic environment (Andersen 1996, Buick 2008, Blank and Sanches-Baracaldo 2010, Shestakova and Karbysheva 2017). Due to these small oxygen oases, global oxygenation of the atmosphere occurred between 2.45 and 2.23 billion years ago, often known as the Great Oxidation Event (GOE). Sergeev et al. (2002) and Zavarzin (2010) suggested that cyanobacteria gradually replaced methane from the anoxygenic environment, leading to the transformation of global geochemical conditions that significantly affected the development of interdependent biogeochemical cycles i.e., carbon, nitrogen, phosphorus and sulfur. Due to the alternation in methane concentration and some lithospheric processes, it induced the cooling of the Earth's surface and later glaciation activities in the early Proterozoic period. These alternative changes in climatic conditions, further paved the way for the evolution and diversification of cyanobacteria (Garcia-Pichel 1998, Sorokhtin 2005, Kopp et al. 2005).

In recent times, the eukaryotic domain is categorized in to six major super-groups: Archaeplastida, Chromalveolata, Excavata, Rhizaria, Amoebozoa and Opisthokonta (fungi and animals) (Fig. 1.2) (Adl et al. 2005, Keeling et al. 2005, Reyes-Prieto et al. 2007, Gould et al. 2008, Archibald 2009), of which the first four super-groups mostly includes the photosynthetic members. The chloroplasts originated more than 1 billion years ago, through an endosymbiotic incident between a free-living cyanobacterium and an endosymbiont in a eukaryotic host cell. It is also indicated that all chloroplasts and their non-photosynthetic relatives (plastids) are directly or indirectly evolved through a single endosymbiotic event (Reyes-Prieto et al. 2007, Gould et al. 2008, Archibald 2009). Through this primary (original) endosymbiosis, cyanobacterium invokes the development of the plastids (primary plastids) in divisions of Glaucophyta, Rhodophyta (red algae) and Viridiplantae (green algae and land plants) that are placed in the Archaeplastida.

Then a later subsequent endosymbiotic incident invoking the integration of eukaryotic algae into other eukaryotes, led to the development of all other plastids; that existed in Chromalveolata (kelps, dinoflagellates and malaria parasites), Excavata (euglenids) and Rhizaria (chlorarachniophytes) (Archibald 2012, Ball et al. 2011). The super group Archaeaplastida comprises only of the organisms with primary plastids, while secondary and tertiary plastids primarily exist in the members of the Chromalveolata (Cryptophyta, Stramenopiles, Haptophyta, Apicomplexa, Chromerida and certain Dinoflagellata), Excavata (Eugleonophyta) and Rhizaria (Chlorarachniophyta) (Fig. 1.2). The presence of plastid-lacking organisms is more surprising and confusing, it could be possible they never had plastids or lost their plastids.

Based on the recent phylogenetic analyses on the host level, the group Stramenopiles is placed together with the Alveolata (includes Chromerida, Apicomplexa and Dinoflagellata) and Rhizaria (includes Chlorarachniophyta that have green-plastids), these groups are collectively abbreviated as SAR.

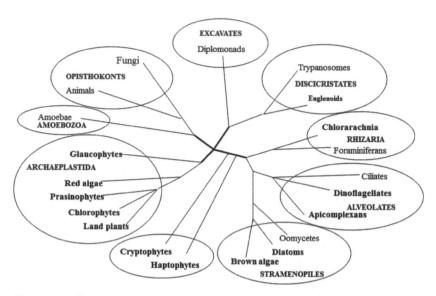

Figure 1.2: Current reorganization of super-groups in the domain Eukarya
(Based on Gould et al. 2008; Achibald 2009).

Haptophyta (group Chromalveolata) which is considered as a sister group to the SAR; and Cryptophyta was found to be more close to the group Viridiplantae (Burki et al. 2012), while the Euglenophyta (group Excavata) was distantly related to the above mentioned groups.

Based on the phylogenetic analyses on the plastid level, there are evidences of secondary endosymbiotic events from either a Chlorophyta (green lineage) or a Rhodophyta (red lineage), but there are no reports of endosymbiotic events from a Glaucophyta. Chlorophyta, the green lineage undergoes two independent endosymbiotic events, where members of core families UTC (Ulvophyceae,Trebuxiophyceae, Chlorophyceae) and Prasinophyceae family leads to the origin of the Chlorarachniophyta (Rhizaria) and Euglenophyta (Excavata), respectively (Rogers et al. 2007, Turmel et al. 2009). While it is correct that no close relation was found between the host of Rhizaria and Excavata, from which they originated, but there is much closeness in genes of their individual chloroplasts (Archibald 2009, Keeling 2010). The Rhodophyta, the red lineage was further diversified and found in various groups including the Cryptophyta, Haptophyta, Stramenopiles, Apicomplexa and Chromerida. Based on the genetic and phenotypic evidence, it was established that all plastids of these groups arose from the same ancestral red lineage (Bodył et al. 2009, Keeling 2010); but there are still unresolved issues about the exact circumstances of this event.

The group Pyrrophyta (dinoflagellates) has the presence of different plastids that originated from various lineages such as Chlorophyta, Cryptophyta, Stramenopiles (heterokontophyta) and Haptophyta. It is suggested that there may be two explanations for this acquisition: (a) either a tertiary endosymbiotic event occurs between a heterotrophic eukaryote and a secondary plastid-containing alga, leading to the origin of these different plastids; (b) or a sequential secondary endosymbiotic event occurs, in which acquired secondary plastids were lost out and they are replaced by a primary plastid-containing alga endosymbiotically.

Phylogeny and Classification

It is now understood that algae are not evolved as a cohesive, natural assemblage of organisms and their positions are usually dependent on problematic interpretations of the data related to groups of algae. There are placed in as many as seven (Andersen 1992), eight (O'Kelley 1993) or more eukaryotic and one or more eubacterial lines within the domains Eukarya and Bacteria (Woese et al.1990). Van den Hock et al (1994) recognized about 10 of the divisions which represent almost all the groups of algae existing or are known to the scientific communities (Table 1.1).

In relation to eukaryotic algae, some scientists emphasized on the available data which was sufficient to represent a minimum of four of five phyletic assemblages of algae: green lineage (including green algae and land plants), red algae, euglenoids, chromophyte algae and dinoflagellates. There is no clear consensus about the groups such as, glucophytes, prymnesiophytes, chloroacrachniophytes, which might be because of unavailability of information or incorrect interpretation of available evidences (Fig. 1.3).

Figure 1.3: Phylogeny of major eukaryotic microalgae.

Table 1.1: Major micro-algal supergroups (divisions), sub-groups and their distribution and representative genera

Division (common name) and distinguishing features	Major groups (common name and/or features)	Distribution of microscopic species	Representative genera
Cyanophyta (Blue-green algae) Chlorophyll a,d; phyobilins; β-carotene; zeaxanthin; echinenone; canthaxanthin; myxoxanthophyll, 3 minor xanthophylls, oscillaxanthin,; prokaryotic; Gram-negative cell wall	Chroococcales (Unicellular cocci and rods; Binary fission and budding	Terrestrial, freshwater, marine, symbiotic	Microcystis, Synechococcus, Synechocystis
	Pleurocapsales (Unicellular and aggregate-forming; multiple fission)	Freshwater, marine	Chroococcidiopsis, Pleurocapsa
	Oscillatoriales (Filamentous; binary fission; no specialized cells)	Terrestrial, freshwater, marine	Lyngbya, Leptolyngbya, Microcoleus, Oscillatoria, Phormidium, Planktothrix
	Nostocales (Filamentous; binary fission; specialized cells; heterocysts, akinetes)	Terrestrial, freshwater, marine, symbiotic	Anabaena, Calothrix, Nostoc, Scytonema, Tolypothrix
	Stigonematales (Filamentous; branched; cell division in 3 planes;	Terrestrial, freshwater	Mastigocladus (Fischerella), Stigonema
Prochlorophyta (prochlorophytes) Chlorophyll a, b; phyobilins; β-carotene; several xanthophylls, prokaryotic; Gram-negative cell walls	Prochlorales (Coccoid and filamentous; binary fission	Marine	Prochlorococcus, Prochlorothrix
Glaucophyta (Glaucophytes) Chlorophyll a, b; phyobilins; β-carotene; zeaxanthin; β-cryptoxanthin; starch; cellulose wall	Cyanophorales Gloeochaetales Glaucocystales	Freshwater, but rarely observed	Cyanophora Gloeochaete Glaucocystis

Rhodophyta (Red algae) Chlorophyll a; phyobilins; α-, β-carotenes; zeaxanthin; 5 minor xanthophylls; floridean starch; cellulose and other cell wall materials	Porphyridales Bangiophycidae Floridophycidae (sea weeds only)	Mostly marine; very rare in freshwater	Porphyridium, Rhodella, Bangia, Polysiphonia
Cryptophyta (cryptophytes) Chlorophyll a, c_2 phyobilins; α-, β-ε-carotenes; alloxanthin, zeaxanthin, α-1,4 glucan starch; exterior organic periplast		Freshwater, marine	Cryptomonas
Chlorophyta (green algae) Chlorophyll a, b; α-, β-, γ- carotenes; zeaxanthin, lutein, vioalxanthin, neoxanthin; several minor xanthophylls; starch; diverse cell wall materials including cellulose	Prasinophyceae Charophyceae (stonewarts, filamentous microalgae) Pleurastrophyceae Ulvophyceae (seaweeds, filamentous microalgae) Trebouxiophyceae Chlorophyceae		Micromonas Spirogyra Tetraselmis Chlorella, Perietochloris, Botryococcus, Chlamydomonas, Dunaliella, Haemotococcus
Euglenophyta (euglenoids) Chlorophyll a, b; α-, γ- carotenes; neoxanthin, diadinoxanthin, diatoxanthin; paramylon; wall-like organic pellicle internal to cell membrane	Euglenales		Euglena
Chlorarachiniophytes Chlorophyll a, b; unknown carotenoids; amoeboid with zoospores	Minorisida Chlorarachniales		Minorisa

(Contd.)

Table 1.1: *(Contd.)*

Division (common name) and distinguishing features	*Major groups (common name and/or features)*	*Distribution of microscopic species*	*Representative genera*
Pyrrhophyta (dinoflagellates) Chlorophyll a, c; β-carotene; diadinoxanthin, diatoxanthin, fucoxanthin, peridinin, 5 minor xanthophylls; α-1,4 starch; cellulose walls			*Noctiluca, Crypthecodinium*
Chromophyta (heterokont algae) Chlorophyll a, c_1, c_2, c_3; α-, β-, ε-carotenes; several major and minor xanthophylls; chrysolaminarin, other glucans, oils; variety of cell wall materials	Chrysophyceae (Golden-brown algae)	Predominately freshwater	*Ochromonas*
	Bacillariophyceae (diatoms)	Freshwater, marine	*Skeletonema, Chaetoceros, Thalassiosira*
	Xanthophyceae (yellow-green algae)	Freshwater	*Tribonema and Vaucheria*
	Eustigmatophyceae (eustgmatophytes)	Terrestrial, freshwater, marine	*Nannochloropsis*
	Raphidophyceae (raphidophytes, chloromonads)	Freshwater, marine	*Chattonella, Gonyostomum*
	Prymnesiophyceae (Haptophytes or prymnesiophytes, including coccolithophorids)	Marine	*Isochrysis,Pavlova*
	Dictyochophyceae (silicoflagellates)	Predominately marine	*Dictyocha*
	Pheophyceae (brown seaweeds, no microalgae)	Marine	

Cyanobacteria (Blue-Green-Algae)

Cyanobacteria are gram-negative bacteria which are currently placed in the phylum cyanobacteria (division cyanophyta) in the domain Bacteria; while earlier they were considered as microscopic algae (Cox 1980, Cox et al. 1989). There are approximately 2500 known species of cyanobacteria, which can be unicellular coccoid, colonial and filamentous forms; they could be with or without branching. Some species also have specialized cells like heterocysts (site for anaerobic N_2-fixation) and akinetes (such as, spores in bacteria and fungi). The photosynthetic membranes (thylakoids) are not arranged in stacks, as compared to prochlorophytes and most other algae. They comprise the pigments including chlorophyll a and d, β-carotene, blue and red phycobilins (Glazer, 1999) and several xanthophylls. The storage material in cyanobacteria is cyanobacterian starch (α-1-4 linked glucan), which is synthesized as granules between the thylakoids (Van den Hoek et al. 1994). In addition to many secondary metabolites, cyanobacteria are known to produce at least 50 toxins that are harmful to many aquatic organisms like vertebrates (Carmichael 1992).

Prochlorophytes

Prochlorophytes are also gram-negative bacteria like cyanobacteria and placed in the same phylum as cyanobacteria but in a different group of prochlorophyta (Cox 1980, Cox et al. 1989); till their discovery less than 50 years ago as extracellular symbionts within subtropical ascidians (sea squirts); but later free-living freshwater and marine planktonic communities were also found (Palenik and Haselkorn 1992). They are different from the cyanobacteria in many features: (a) possessing both the chlorophylls a and b or divinyl derivatives, (b) photosynthetic membranes (thylakoids) are arranged in stacks; both are characteristic features of green algae, euglenoids and land plants.

Glaucophytes (Glucophyta)

Glucophytes comprises naturally occurring, unicellular and biflagellate microalgae that have only three genera. Like in cyanobacteria, the chloroplasts are strongly coccoid and the thylakoids are not stacked and are surrounded by a thin peptidoglycan layer. They have chlorophyll a and phycobilins, but have a smaller range of xanthophyll pigments (Van den Hoek et al. 1994).

Green algae (Chlorophyta)

Chlorophyta is the largest, most diverse and widely distributed group of algae; found in freshwater, marine or even terrestrial environments. They include unicellular, colonial, filamentous and pseudo parenchymatous uni- or multinucleate forms. The chloroplasts are very diverse in number and arrangement which often have tightly stacked thylakoids that are made up of two to six lamellae within the plastids; usually surrounded by two envelope membranes like in red algae and glucophytes. They contain the pigments chlorophyll a and b and the storage *Perietochloris*) and Chlorophyceae (*Dunaliella, Haemotococcus*) are characterized with the absence of multilayered structures, the presence of cruciate flagellar microtubular root system, phycoplast formation-assisted cytokinesis and carry out closed mitosis.

Red algae (Rhodophyta)

Rhodophyta comprises primarily seaweeds and includes two classes: Bangiophycaee that shows a relatively simple structure; and Floridophyceae that show a complex structure. They exhibit chlorophyll a, phycobilins (phycoerythrin which is responsible for their characteristic red color), β-carotene, zeaxanthin with five other known xanthophylls. They are distinct from other algal divisions by these features: (1) lack of any flagellate stages; (2) the phycobilin pigments as accessory pigments (such as in cyanobacteria, glucophytes and cryptomonads); (3) chloroplasts with no stacked thylakoids; (4) photosynthate storage product is floridean starch (an α-(1-4-glucan); (5) with an incidence of oogamy in species with sexual reproduction (Van den Hoek et al. 1994).

Cryptophytes (Cryptophyta)

Cryptophyta is a small division that includes mostly unicellular biflagellate microalgae. They consist of pigments chlorophyll a and c2 and phycobilins. Cryptophytes have a single chloroplast with a thylakoid structure which shows an intermediate connection between red algae and other plants. The storage product is starch presented in distinct granules; some species also have oils as a storage product (Van den Hoek et al. 1994). They have a stiff proteinaceous periplast made of rectangular or polygonal plates that enclose the cell. Cryptophytes are distinct from other algae due to their chloroplast's ultrastructural properties and unique cellular inclusions, but Cavelier-Smith (1989) placed the cryptophytes with the chromophyte algae.

Chlorarachniophytes

Chlorarachniophytes is a rare group that includes two orders Minorisida and Chlorarachinales that seems have originated from endosymbiosis of eukarotic organisms that are derived from a euglenoid or chlorophyte with an amoeboid, plasmodial, phagotropic host (O'Kelley 1993). They contain pigments of chlorophyll a and b and are known for their association with marine siphonous green algae (Van den Hoek et al. 1994).

Euglenoids (Euglenophyta)

Euglenophyta have the same pigments like in prochlorophytes, green algae and land plants, but some unique cellular and biochemical characteristics along with ultrastructural and molecular evidence shows closeness to trypanosomes rather than to any algal group (O'Kelley 1993). They are microscopic, unicellular or colonial; the presence of single nucleus and flagella, are some characteristic features of euglenoids (Van den Hoek et al. 1994).The common storage compound is paramylon, a β-1,3 linked glucan that is located freely in the cytoplasm; this is also found in prymesiophytes (a class of chromophyte algae). Regardless of a defined cell wall, they possess a pellicle, a proteinaceous layer which protects the cell membrane. The common nutrition modes are phagotrophic, osmotrophic, facultative and obligate heterotrophic.

Chromophyte algae (Chromophyta)

Chromophyta are a highly diverse group, which includes golden, golden brown, brown or yellow-green colored algae rather than the grass-green, blue-green or red coloration typical of other major algal divisions. This division also includes diatoms, coccolithophorids and silicoflagellates, which are quite distinct microalgal classes in an ecological and morphological manner (Vanden Hoek et al. 1994). They have some characteristic features: (a) the presence of a common storage product chrysolaminarin (a β-linked glucan); (b) preponderance of carotenes over chlorophylls. Some chromophytes and prymnesiophytes also store fats and oils to synthesize paramylon, which is quite often present in euglenoids.

Dinoflagellates (Pyrrhophyta)

Dinoflagellates are important freshwater and marine planktonic communities which include a diverse grouping of coccoid, filamentous, amoeboid and palmelloid microalgae. In this group, there are multiple encounters of evolutionary acquisition of plastids and phototrophy (O'Kelley 1993). Dinoflagellates also show many unique and diverse ultrastructural aspects of flagellar and nuclear organization (Van den Hoek et al. 1994). Dinoflagellates contain the pigments chlorophyll a, c1, and c2 and some carotenoids like in chromophyte algae, but have a cellulose wall and their carotenoids are especially distinct in nature. The characteristic feature of most donoflagellates is the presence of two flagella that are uniquely positioned. Besides the photosynthetic ability, several species also show a number of heterotrophic modes like symbiotic, parasitic, phagotrophic and saprophytic.

Ecology and Distribution

Microalgae are known as the principal photosynthetic planktonic communities in marine and freshwater environments, and it is estimated that alone are responsible for 40-50% of the total global photosynthetic primary production by all the algae (Harlin and Darley, 1988). Based on their size, microalgae are divided in microplankton (20-200 μm), nanoplankton (2-20 μm) and picoplankton (0.2-2 μm) (Sieburth et al. 1978). It is well established that ultraplankton with heterotrophic plankton (plankton having size of <3-5 μm in diameter), contributes most of the primary production up to 80% of the pelagic biomass (Andersen 1992, Shapiro and Guillard 1986). In comparison to aquatic i.e., pelagic, coastal and freshwater microalgal communities, the role of terrestrial microalgae are not so precisely known, but it may be significant in a local or regional level due to their abundance in exposed land surfaces in arid and semi-arid steppes and deserts (Metting 1991).

In the aquatic arena, phytoplankton inhabits the photic zone that exists from the surface layer (neuston) to 250 m or in deeper in clear oligotrophic lakes and it is also depends upon the latitude of the region. It is calculated that the world's oceans contain as many as 3.6×10^{25} individual phytoplanktons at any given time and they might contribute over 5×10^{13} kg annual primary production which makes the base of the marine food web. The diurnal, seasonal, vertical and geographic variation in nutrient availability, temperature, light and other factors affect the distribution and metabolic activities of groups of microalgae.

Cyanobacteria are omnipresent microalgal communities and found in freshwater, marine and terrestrial ecosystems as free living or symbiotic with lichens, bryophytes, cycad roots and marine animals (Castenholz and Waterbury 1989, Castenholz 2001). Ultraplanktonic cyanobacterial species like *Synechococcus* proved to have a more potential role in oceanic primary production in comparison to any individual group of microalgae (Fogg 1987, Waterbury et al. 1979). There are filamentous cyanobacteria such as *Nostoc, Anabaena, Oscillatoria, Microcoleus* and *Mastigocladus*, known for their dominance in microbial mats or cyanobacterial mats; existing in brackish waters, hot springs, deserts and paddy fields (Metting 1981, Roger and Kulasooriya 1980). Prochlorophytes (e.g., *Prochloron*) were first discovered five decades ago and considered as symbionts with marine didemnids and holothurians. But now it is clearer that they are free living communities in the pelagic naoplankton; which includes their distribution and importance in this habitat (Chishom et al. 1992).

Red algae are mostly comprised of seaweeds which are found from polar waters to the tropics, while common in tide pools and coral reefs, freshwater or unicellular species are rare. The red algae of the order corallinales or coralline algae play an important role in coral reefs stabilizing through providing food and patching the gaps in coral reefs. Some unicellular species of order porphyridales like *Porphyridiumcruentum* and *Rhodellareticulata* are mainly exploited for commercial production of phycobilins.

In earlier times, the position of cryptophytes was controversial as they were treated as algae by botanists or flagellate protozoa by zoologists under the order cryptomonadina. Later they were assigned as algae in the division of cryptophyta due to presence of plastids in different species and their close relation with dinoflagellates. They are common in freshwater and also occur in marine and brackish water. Species such as *Chroomonas* are found to exist in estuaries and salt marshes; having a tolerance to a range of salt concentrations (Thomsen 1986).

Green algae (Chlorophyta) are diverse and dominant phytoplankton communities in freshwater and terrestrial habitats. They could be unicellular (e.g., *Chlamydomonas, Phacotus*), coccoid non-flagellate (e.g., *Chlorella, Pediastrum, Ankistrodesmus, Scenedesmus*), colonial flagellates (e.g.,*Volvox, Pyrobotrys*) and filamentous forms (e.g., *Chaetophora, Oedogonium*); mat forming (e.g., *Spirogyra, Hydrodictyon*) or attached (e.g., *Cladophora, Ulothrix, Stigeoclonium*) (Bold and Wynne 1985, Lembi et al. 1988, Van den Hoek et al. 1994). The representative genera such as uniflagellate (e.g., *Pedinomonas, Micromonas*), biflagellate prasinophytes (e.g., *Mamiella, Mantoniella*) and chlorophytes (e.g., *Pyramimonas, Tetraselmis*) are widely distributed in marine environments.

Green algae also exists in terrestrial habitats like soil, rocks; could be coccoid (*Chlorococcum*), colonial palmelloid (e.g., *Palmella, Gloeococcus*), colonial sarconoid (e.g., *Tetracystis, Chlorosarcina*) and filamentous (e.g., *Stichococcus, Klebshormidium*) (Metting 1981). Being predominantly photosynthetic, there are a number of facultative and obligate heterotrophic communities existing in green algae (Van den Hoek et al. 1994).The genera *Trentepholia* are known to be inhabitants of tree trunks and *Trebouxia* are found to be symbiotic with lichens.

Euglenoids such as *Euglena, Colacium* and *Trachelomonas* are common freshwater and edaphic species; rare in marine environments. They are found to be heterotrophic and/or phagotrophic. There is also a common presence of *Perenema*, a colorless phagotrophic euglenoids (Van den Hoek et al. 1994).

Chromophyte algae (Chromophyta) comprise a large number of microalgae that are widely distributed in most freshwater and marine environments. They are divided into many families: Chrysophyceae (golden-brown algae), Dictyochophyceae (silicoflagellates) Bacillariophyceae (diatoms), Xanthophyceae (yellow-green algae) and Prymnesiophyceae/Haptophyceae. Golden-brown algae are important freshwater phytoplanktonic communities and are often responsible for algal blooms in lakes and ponds; rare in marine environments except of *Ochromonas* and *Chromulina*; but there are no reports in the soil. Silicoflagellates (e.g., *Dichtyota*) are common oceanic phtoplanktonic where they are responsible for dense blooms; rare in freshwater and with no reports from the terrestrial habitats. *Pedinella* is one of the pelagic silicoflagellates that occurs below ice, can give the sea a yellow color (Thomsen 1986). Diatoms are the largest group of microalgae (up to 10 million species), and found everywhere in freshwater, terrestrial and marine environments (John 1994, Round et al. 1990). They are primarily unicellular, but diverse and abundant presence of filamentous species are also reported; found in all latitudes, from temperate coastal and pelagic (in and below ice) to freshwater (in still or moving) to terrestrial (some soils) environments (Round et al. 1990). Based on the bilateral or radial symmetry, they are divided in to three classes: (a) Bacillarophyceae (bilateral symmetry with raphes e.g., *Navicula, Nitzschia, Phacodactylum*); (b) Fragilariophyceae (bilateral symmetry without raphes e.g., *Diatoma, Synedra, Fragilaria*); (c) Coscinodiscophyceae (radial symmetry e.g., *Chaetoceros, Biddulphia, Thalassiosira*). Yellow-green microalgae (Xanthophyceae) primarily exist in freshwater (e.g., *Vaucheria, Bumilleria*), but are also common in terrestrial habitats in unicellular, colonial and filamentous forms. Eustigmatophytes e.g., *Nanochloris* and *Nanochloropsis* (Eustigmatophyceae) were discovered five decades ago, and found to be an important contributor to primary production in salt marshes, estuaries and brackish environments; but not common in open oceans (Andersen, 1996).

Prymnesiophytes or haptophytes (Prymnesiophyceae or haptophyceae) are the dominating phytoplankton in marine environments in terms of numbers and biomass (e.g., *Phaeocystis*); less common in freshwater and not reported from terrestrial habitats. They are very diverse in warm oligotrophic regions like the Red Sea and Gulf of California, but also cause blooms in eutrophic environments. The genera like *Emiliana*, *Discosphaera*, *Rhabdosphaera* abundantly exist in subtropical and tropical marine waters, while *Coccolithus* and *Emiliana* are common in open oceans at Arctic latitudes (Winter et al. 1994).

Dinoflagellates (Pyrrhophyta) e.g., *Dinophysis, Gymnodinium, Aphidinium, Peridinium, Ceratium* and *Prorocentrum* are important microplanktonic communities in freshwater and marine waters at most latitudes (Thomsen 1986). Some species of dinoflagellates are found to be endo- and acto-parasites (e.g., *Blastodinium* and *Haplozoon*) or photosynthetic partners of many corals (known as zooxanthellae) (Van den Hoek et al. 1994). Some genera e.g., *Gonyaulax* and *Pyrocystis* are capable of bioluminescence and also responsible for 'red tides'.

Morphology, Colony Features and Adaptations

Morphology

Microalgae are unicellular, colonial and filamentous in their cell organization. Unicellular microalgae could be non-motile or not, but cyanobacteria are primarily non-motile due to the absence of flagella. However some motility occurs through gliding and swimming in unicellular cyanobacteria; like swimming motility in *Synechococcus* sp. and gliding motility in baeocytes (earlier known as endospore), reproductive cells formed through multiple fission of a parental cell. The motile forms of microalgae have the presence of flagella; their reproductive structures like gametes and zoospores are usually flagellate and motile. Red algae *Porphyridium*, some pennate diatoms and also some green algae show a type of gliding motility.

Unlike unicellular forms, colonial forms of cyanobacteria also develop non-motile colonies like in *Gloeocapsa*; while non-motile microalgae might be arranged into coenobic forms with a fixed number of cells (e.g., *Scenedesmus*), or into non-coenobic forms with a variable number of cells (e.g., *Pediastrum*). Motile flagellate microalgal cells might be arranged in to motile (e.g., *Volvox*) or nonmotile colonies (e.g., *Gloeocystis*).

Several filamentous cyanobacteria might show a gliding motility, often facilitated through rotation and creeping motions (e.g., *Oscillatoria*), but other filamentous cyanobacteria may have motility at the stage of hormogonia being formed in asexual reproduction (e.g., *Nostoc*). In branched or unbranched microalgae, mostly microalgae are found non-motile, but reproductive structures like zoospores and gametes are motile as an exception. Unlike the presence of siphonaceous and parenchymatous cells in most of the macroalgae, microalgae generally lack these types of cells.

Colony characteristics

The colonies could be flat, spherical, cubic, palmelloid and dendroid shapes; flagellate and non-flagellate. To keep the cells in a colony intact, a polysaccharide envelope of an amorphous (e.g., *Microcystis*) or microfibrillar (e.g., *Gloeothece*) structure, which are known as a sheath, glycocalyx, capsule or slime; depending up on their consistency. The cells might be orderly (e.g., *Pediastrum*) or irregularly e.g., (*Palmella*) arranged and embedded in mucilage inside the colony (e.g., *Microcystis*). Moreover, the colonies could be non-motile (e.g., *Coelastrum*) and motile consisting flagellate cells (e.g., *Gonium*). There is the presence of some pigments in cyanobacterial sheaths which may act as sun-screen compounds or UV-A/B-absorbing mycosporine-like amino acids (Garcia-Pichel et al. 1992, Ehling-Schulz et al. 1997).

Morphological adaptations

Cyanobacteria have some specialized cells such as heterocysts, akinetes, hormogonia and flagella, that are adapted for various purposes like nitrogen fixation, reproduction and locomotion.

Heterocysts

Heterocysts are spherical, thick walled, less pigmented and larger cells than vegetative cells; and usually have cyanophycin granules at the pole position neighboring to vegetative cells. The thickness of the cell wall of heterocysts provides the protection to the enzyme nitrogenase from the toxic effect of oxygen. In mature heterocysts, an essential microoxic environment facilitates nitrogen fixation, which provides a separate space for this process without being concerned by the toxic effect of oxygen evolved by vegetative cells during photosynthesis.

Akinetes or resting cells

In many cyanobacterial species of the order Nostocales and Stigonematales, there is the presence of resting cells or akinetes (Kaplan-Levy et al., 2010). They are spore-like, thick-walled and non-motile cells that serve a reproductive structure. In comparison to vegetative cells, akinetes are thick walled and large in size (sometimes by up to tenfold); and contain large reserves of food resources and DNA. The akinete have different shapes which could sphere to oblate spheroid; this variety in position and distribution within a filament (trichome) is considered as a taxonomic feature. In Nostocales, akinetes have some similar features with other prokaryotes like endospore in *Bacillus* and eukaryotes like a spore in yeast, cyst in protists in various developmental stage i.e., dormancy, germination, differentiation and maturation (Errington 2003, Corliss 2001).

Hormogonia

Phycologists initially use the term 'hormogonium' to differentiate short motile trichomes produced from immotile unsheathed parental trichomes. Later this term extended to describe motile or non-motile trichomes that often contain gas-vacuoles and lack heterocysts, but are distinguishable due to the difference of their cell shape and size in comparison to vegetative cells (Rippka et al. 1979). There are many differences in structure and mobility of hormogonia; the hormogonia of many cyanbacterial genera like *Lyngbya* or *Leptolyngbya* and *Scytonema* are non-motile, not gas-vacuolated and morphologically similar to vegetative trichomes but their length is relatively short; while the hormogonia of *Chlorogloeopsis* sp. and *Fischerella* sp. are found to be highly motile and not gas-vacuolated, but morphologically distinct vegetative trichomes through the presence of small cylindrical cells (Rippka et al. 1979, De Marsac 1994).

Flagella

Many eukaryotic microalgae have flagella for locomotion purposes. They might be smooth or hairy, and usually embedded in the outer layer of the cytoplasm via a basal body. Flagella are very complex in structure, usually comprising of an axoneme which is made up of two central microtubules surrounded by nine peripheral double microtubules. The whole structure of the flagella is covered by the plasma membrane.

Conclusion

The origin and evolution of diverse microalgal groups is still debatable, there are many ways to establish the link between the groups such as fossil records, phylogeny, study of plastids and flagella, ecology and distribution. There are many doubts about the origin of cyanobacteria, but it is well established that all eukaryotic microalgal groups acquired their photosynthetic activity by endosymbiotic events. Current phylogenetic studies have helped in significant progress related to the understanding of the mechanism and regulation of plastid division, but a number of queries still remain unresolved and many new questions are still arising.

Despite many differences and missing links, the microalgal world is currently categorized in to multiple groups most likely on the basis of their plastid's pigment and other cellular characteristics. To resolve this problem, many scientists attempted to reclassify the microalgal world thus far, which resulted in the use of different classification schemes; but as yet no attempt has been able to present a satisfactory classification system. The repeated instances of reclassification and further transfer of the particular microalgal species from one group to another group have not provided much help in resolving the issue. Along with phylogentic information, there is also a need to adopt a readily defined strategy based on the existing classification schemes. Thus it is more appropriate to adapt such classification systems in which groupings are created on the basis of more readily refined and available information.

Most of the microalgae acquired their genes from photosynthetic organisms, heterotrophic eukaryotes and bacteria, which provide them a unique and diverse range of abilities. To survive in diverse habitats, microalgae are found to develop many temporary or permanent adaptations. Their difference in strategies related to processes such as photosynthesis, nutrient uptake and acquisition of specific genes from other organisms; facilitate microalgae to have a better ability to manage a range of conditions. Microalgal diversity and their abundance have a remarkable influence on the aquatic environments particularly oceans; including primary production, nutrient cycling, initiating a food chain and geochemical deposits.

References

Adl, S.M., A.G.B. Simpson, M.A. Farmer, R.A. Andersen, O.R. Anderson, J.R. Barta, et al. 2005. The new higher level classification of eukaryotes with emphasis on the taxonomy of protists. J. Eukaryot. Microbiol. 52: 399–451.

Andersen, R.A. 1992. Diversity of eukaryotic algae. Biodivers. Conserv. 1: 267–292.

Andersen, R.A. 1996. Algae. pp. 29–64. *In*: J.C. Hunter-Cevera and A. Belt (eds.). Maintaining Cultures for Biotechnology and Industry. Academic Press, San Diego, USA.

Archibald, J. 2012. The evolution of algae by secondary and tertiary endosymbiosis. Adv. Bot. Res. 64: 87–118.

Archibald, J.M. 2009. The puzzle of plastid evolution. Curr. Biol. 19: R81–R88.

Ball, S., C. Colleoni, U. Cenci, J.N. Raj and C. Tirtiaux. 2011. The evolution of glycogen and starch metabolism in eukaryotes gives molecular clues to understand the establishment of plastid endosymbiosis. J. Exp. Bot. 62: 1775–1801.

Blank, C.E. and P. Sanchez-Baracaldo. 2010. Timing of morphological and ecological innovations in the cyanobacteria—a key to understanding the rise in atmospheric oxygen. Geology 8: 1–23.

Bodył, A., J.W. Stiller and P. Mackiewicz. 2009. Chromalveolate plastids: direct descent or multiple endosymbioses. Trends Ecol. Evol. 24: 119–121.

Bold, H.C. and M.J. Wynne. 1985. Introduction to the Algae, 2nd edition. Prentice-Hall, Englewood Cliffs, New Jersey.

Boyd, P.W., A.J. Watson, C.S. Law, E.R. Abraham, T. Trull, R. Murdoch, et al. 2000. A mesoscale phytoplankton bloom in the polar Southern Ocean stimulated by iron fertilization. Nature 407(6805): 695–702.

Buick, R. 2008. When did oxygenic photosynthesis evolve? Phil. Trans. R. Soc. B 263: 2731–2743.

Burki, F., N. Okamoto, J.-F. Pombert and P.J. Keeling. 2012. The evolutionary history of haptophytes and cryptophytes: phylogenomic evidence for separate origins. Proceed. Roy. Soc. B Biol. Sci. 279: 2246–2254.

Canfield, D.E., K.S. Habicht and B. Thamdrup. 2000. The Archean sulfur cycle and the early history of atmospheric oxygen. Sci. 288: 658–661.

Carmichael, W.W. 1992. Cyanobacteria secondary metabolites—the cyanotoxins. J. Appl. Bacteriol. 72: 445–459.

Castenholz, R.W. 2001. Phylum BX. Oxygenic photosynthetic bacteria. pp. 473–599. *In*: D.R. Boone and R.W. Castenholz (eds.). Bergey's Manual of Systematic Bacteriology, 2nd edition. Springer, Berlin.

Castenholz, R.W. and J.M. Waterbury. 1989. Oxygenic photosynthetic bacteria, Group I. Cyanobacteria. pp. 1710–1727. *In*: W.R. Hensyl (ed.). Bergey's Manual of Systematic Bacteriology. Williams and Wilkins, Baltimore, MD.

Cavalier-Smith, T. 1989. The kingdom Chromista. pp. 381–407. *In*: J.C. Green, B.S.C. Leadbetter and W.L. Diver (eds.). The Chromophyte Algae: Problems and Perspectives. Clarendon Press, Oxford.

Cavalier-Smith, T. 1999. Principles of protein and lipid targeting in secondary symbiogenesis: euglenoid, dinoflagellate, and sporozoan plastid origins and the eukaryote family tree. J. Eukaryot. Microbiol. 46 (4): 347–366.

Chishom, S.W., S.L. Frankel, R. Goericke, R.J. Olson, B. Palenik, J.B. Waterbury, et al. 1992. *Prochlorococcus marinus* nov gen sp: an oxyphototrophic marine prokaryote containing divinyl chlorophyll a and b. Archive fur Mikrobiologie 157: 297–300.

Corliss, J.O. 2001. Protozoan cysts and spores. pp. 1–8. *In*: Encyclopedia of Life Sciences (eLS). John Wiley & Sons, Chichester, UK.

Cox, E.R. 1980. Phytoflagellates: Developments in Marine Biology 2. Elsevier, North Holland Publishing Company, New York.

Cox, J., H. Chen, C. Kabacoff, J. Singer, S. Hoeksema and D. Kyle. 1989. The production of ^{2}H-, ^{13}C-, and ^{15}N-labelled biochemicals using microalgae. *In*: T. Chapman (ed.). Stable isotopers in pediatric, nutritional, and metabolic research. Intercept Ltd Press, Andover, USA.

De Marsac, N.T. 1994. Differentiation of hormogonia and relationships with other biological processes. pp. 825–842. *In*: D.A. Bryant (ed.). The Molecular Biology of Cyanobacteria. Kluwer Academic Publishers, Dordrecht, Netherlands.

Ehling-Schulz, M., W. Bilger and S. Scherer. 1997. UV-B-induced synthesis of photoprotective pigments and extracellular polysaccharides in the terrestrial cyanobacterium *Nostoccommune*. J. Bacteriol. 179: 1940–1945.

Errington, J. 2003. Regulation of endospore formation in *Bacillussubtilis*. Nat. Rev. Microbiol. 1: 117–125.

Falkowski, P.G., M.E. Katz, A.H. Knoll, A. Quigg, J.A. Raven, O. Schofield, et al. 2004. The evolution of modern eukaryotic phytoplankton. Sci. 305(5682): 354–360.

Field, C.B., M.J. Behrenfeld, J.T. Randerson and P. Falkowski. 1998. Primary production of the biosphere: integrating terrestrial and oceanic components. Sci. 281(5374): 237–240.

Fogg, G.E. 1987. Marine planktonic cyanobacteria. pp. 393–414. *In*: P. Fay and C. Van Baalen, (eds.). The Cyanobacteria. Elsevier, Amsterdam.

Garcia-Pichel, F. 1998. Solar ultraviolet and the evolutionary history of cyanobacteria. Origins Life Evol. Biospheres 28: 321–347.

Garcia-Pichel, F., N.D. Sherry and R.W. Castenholz. 1992. Evidence for an ultraviolet sunscreen role of the extracellular pigment scytonemin in the terrestrial cyanobacterium *Chlorogloeopsis* sp. Photochem. Photobiol. 56: 17–23.

Glazer, A.N. 1999. Phycobiliproteins. pp. 281–280. *In*: Z. Cohen (ed.). Chemicals from Microalgae. Taylor & Francis, London.

Gould, S.B., R.F. Waller and G.I. McFadden. 2008. Plastid evolution. Annu. Rev. Plant. Biol. 59: 491–517

Harlin, M.M. and W.M. Darley. 1988. The algae: An overview. pp. 3–27. *In*: C.A. Lembi and R.A. Waaland (eds.). Algae and Human Affairs. Cambridge University Press, Cambridge.

Holland, H.D. 2006. The oxygenation of the atmosphere and oceans. Philos. Trans. R. Soc. B 361: 903–915.

Hopes, A. and T. Mock. 2015. Evolution of microalgae and their adaptations in different marine ecosystems. *In*: eLS. John Wiley & Sons Ltd, Chichester, England.

John, D.M. 1994. Biodiversity and conservation: an algal perspective. Phycologist 38: 3–15.

Kaplan-Levy, R.N., O. Hadas, M.L. Summers, J. Rücker and A. Sukenik. 2010. Akinetes: dormant cells of cyanobacteria. pp. 5–27. *In*: E. Lubzens et al. (eds.). Dormancy and Resistance in Harsh Environments. Springer-Verlag, Berlin, Heidelberg.

Keeling, P.J. 2010. The endosymbiotic origin, diversification and fate of plastids. Philosoph. Transact. Roy. Soc. B Biol. Sci. 365: 729–748.

Keeling, P.J., G. Burger, D.G. Durnford, B.F. Lang, R.W. Lee, R.E. Pearlman, et al. 2005. The tree of eukaryotes. Trends Ecol. Evol. 20: 670–676.

Kopp, R.E., J.L. Kirschvink, I.A. Hilbum and C.Z. Nash. 2005. The Paleoproterozoic snowball Earth: a climate disaster triggered by the evolution of oxygenic photosynthesis. Proc. Natl. Acad. Sci. U.S.A. 102: 11131–11136.

Kulasooriya, S.A. 2011. Cyanobacteria: pioneers of planet Earth. Ceylon J. Sci. (Bio. Sci.) 40(2): 71–88.

Lembi, C.A., S.W. O'Neil and D.F. Spencer. 1988. Algae as weeds: economic impact, ecology, and management alternatives, pp. 455–481. *In*: C.A. Lembi and J.R. Waaland (eds.). Algae and Human Affairs. Cambridge University Press, Cambridge.

Metting, F.B. 1981. The systematics and ecology of soil algae. Bot. Rev. 47: 195–312.

Metting, F.B. 1991. Biological surface features of semi-arid lands and deserts. pp. 257–293. *In*: J. Skujins (ed.). Semi-arid Lands and Deserts: Soil Resource and Reclamation. Marcel Dekker, New York.

Miyagishima, S. 2011. Mechanism of plastid division: from a bacterium to an organelle. Plant Physiol. 155: 1533–1544.

O'Kelley, C.J. 1993. Relationships of eukaryotic algal groups to other protists. pp. 269–293. *In*: T. Berner (ed.). Ultrastructure of Microalgae. CRC Press, Boca Raton.

Palenik B. and R. Haselkorn. 1992. Multiple evolutionary origins of prochlorophytes, the chlorophyll-b containing prokaryotes. Nature 355: 265–267.

Reyes-Prieto, A., A.P. Weber and D. Bhattacharya. 2007. The origin and establishment of the plastid in algae and plants. Annu. Rev. Genet. 41: 147–168.

Rippka, R., J. Deruelles, J.B. Waterbury, M. Herdman and R.Y. Stanier. 1979 Generic assignments, strain histories and properties of pure cultures of cyanobacteria. J. Gen. Microbiol. 111: 1–61.

Roger, P.A. and S.A. Kulasooriya. 1980. Blue-Green Algae and Rice. International Rice Research Institute, Los Baños, Philippines.

Rogers, M.B., P.R. Gilson, V. Su, G.I. McFadden and P.J. Keeling. 2007. The complete chloroplast genome of the chlorarachniophyte *Bigelowiellanatans*: evidence for independent origins of chlorarachniophyte and euglenid secondary endosymbionts. Mol. Biol. Evol. 24: 54–62.

Rothschild, L.J. and R.L. Mancinelli. 2001. Life in extreme environments. Nature 409(6823): 1092–1101.

Round, F.E., R.M. Crawford and D.G. Mann. 1990. The Diatoms. Cambridge University Press, Cambridge.

Sergeev, V.N., L.M. Gerasimenko and G.A. Zavarzin. 2002. The Proterozoic history and present state of cyanobacteria. Microbiol. (Moscow) 71(6): 623–637.

Shapiro, L.P. and R.R.L. Guillard. 1986. Physiology and ecology of the marine eukaryotic ultraplankton. Can. Bull. Fish Aquat. Sci. 214: 371–389.

Shestakova, S.V. and E.A. Karbysheva. 2017. The origin and evolution of cyanobacteria. Biol. Bull. Rev. 7(4): 259–272.

Sieburth, J.McN., V. Smetacek and J. Lang. 1978. Pelagic ecosystem structure: heterotrophic compartments of the plankton and their relation to plankton size fractions. Limnol. Oceanogr. 23: 1256–1263.

Singh, J.S. 2014. Cyanobacteria: a vital bio-agent in eco-restoration of degraded lands and sustainable agriculture. Clim. Change Environ. Sustain. 2: 133–137.

Singh, J.S., A. Kumar, A.N. Rai and D.P. Singh. 2016. Cyanobacteria: a precious bio-resource in agriculture, ecosystem, and environmental sustainability. Front. Microbiol. 7: 529.

Sorokhtin, O.G. 2005. The bacterial nature of the glaciation of the Earth. Herald Russ. Acad. Sci. 75(6): 571–578.

Thomsen, H.A. 1986. A survey of the smallest eukaryotic organisms of the marine phytoplankton. Can. Bull. Fish Aquat. Sci. 214: 121–158.

Turmel, M., M.-C. Gagnon, C.J. O'Kelly, C. Otis and C. Lemieux. 2009. The chloroplast genomes of the green algae *Pyramimonas*, *Monomastix*, and *Pycnococcus* shed new light on the evolutionary history of prasinophytes and the origin of the secondary chloroplasts of euglenids. Mol. Biol. Evol. 26: 631–648.

Van den Hoek, C., D. Mann and H.M. Jahns. 1994. An Introduction to Phycology. Cambridge University Press, Cambridge.

Vincent, W.F. 2010. Cyanobacteria. pp. 226–232. *In*: G.E. Likens (ed.). Protists, Bacteria and Fungi: Planktonic and Attached. Academic Press, Cambridge, US.

Waterbury, J.B., S.W. Watson, R.R.L. Guillard and L.E. Brand. 1979. Widespread occurrence of a unicellular marine, planktonic cyanobacterium. Nature 277: 293–294.

Winter, A., R.W. Jordan and P.H. Roth. 1994. Biogeography of living coccolithophores in ocean waters. pp. 161–176. *In*: A. Winter and W.G. Siesser (eds.). Coccolithophores. Cambridge University Press, Cambridge.

Woese, C.R., G.E. Fox, L. Zablen, T. Uchida, L. Bonen, K. Pechman, B.J. Lewis and D. Stahl. 1975. Nature 254: 83–86.

Woese, C.R., O. Kandler and M.L. Wheelis. 1990. Towards a natural system of organisms: proposal for the domains Archaea, Bacteria, and Eukarya. Proc. Natl. Acad. Sci. USA 87: 4576–4579.

Zavarzin, G.A. 2010. Initial stages of biosphere evolution. Herald Russ. Acad. Sci. 80(6): 522–533.

CHAPTER

2

Microalgae II: Cell Structure, Nutrition and Metabolism

Introduction

Microalgae include prokaryotic groups—cyanobacteria and prochlorophytes; and eukaryotic groups—red algae, green algae, diatoms and dinoflagellates, etc. Prokaryotic cells have a thin plasma membrane, the peptidoglycan cell wall could have sheaths, capsules and slime as an outer covering of the cell wall. They lack membrane-bound organelles like nucleus and the genetic material suspended freely in the cytoplasm. The cyanobacterial cell contains unstacked thylakoids in their chloroplasts as photosynthetic apparatus and also contains phycobilioproteins accessory pigments. They often perform asexual reproduction through binary fission or multiple fission and special structures like hormogonia and akinetes. The presence of nitrogen fixing cells i.e., heterocyst makes them cyanobacteria, a distinctive and important group of the microalgal world.

Eukaryotic cells are characterized with a cellulose cell wall and might lack sheaths or capsule-like structures but often have flagella for locomotion in many species. They have well organized genetic material in the membrane-bound nucleus. They have stacked thylakoids in their chloroplasts, chlorophylls and carotenoids as their photosynthetic pigments but lack phycobilioproteins. Eukaryotic microalgae have different reproduction strategies—vegetative through fragmentation, asexual through spores and sexual reproduction through production of gametes.

Most of the microalgae either prokaryotic or eukaryotic are found to be obligate photoautotrophs, that harness light energy to convert inorganic materials in to organic compounds, the process is known as photosynthesis. Other modes such as auxotrophy or mixotrophy are also observed in many microalgal species; auxotrophy could be associated with the requirement of additional organic compounds like vitamins and amino acids for growth, while mixotrophy is associated with the necessary requirement of glucose and CO_2 for growth. Although there is a presence of both autotrophic and heterotrophic modes of nutrition, there is no clear distinction related to conditions which induce switching the various nutritional modes.

Photosynthesis is the main process that is responsible for fixation of carbon that is required as the main building block element of organic compounds. Other essential macronutrients such as nitrogen and phosphorous uptake in acceptable forms like nitrogen absorbed as urea, nitrite-, nitrate- and ammonium-ions, which are finally converted into ammonium before assimilation; while phosphorous uptake in orthophosphate- and phosphate-ions. Micronutrients like iron and other trace metals are also required for biochemical and metabolic processes required for growth. Some nutrients like phosphorous are also stored for unfavorable conditions in the form of polyphosphate granules.

Cell Structure: Prokaryotic Microalgae

Cell wall and plasma membrane

Cyanobacteria and prochlorophytes have a Gram-negative type cell wall which is composed of many layers (2-3): the inner layer lies between the plasma membrane and the outer layer, composed of peptidoglycan (murein); the outer layer outside the inner layer is composed of lipo-polysaccharide. Due to the absence of cellulose, cyanobacteria cells have high digestibility, making them suitable for human consumption as healthy food (e.g., *Spirulina*). The cell wall is further surrounded by mucilaginous envelopes, known as sheaths, glycocalix, capsules or slime. The cell wall might have small perforations or pores and appendages such as fimbriae and pili for reproduction. The plasma membrane or plasmalemma found beneath the cell wall, is a thin unit membrane of about 8 nm thickness.

Cytoplasm, nucleus and other organelles

The genetic material (DNA) of cyanobacteria and prochlorophytes is not arranged in chromosomes, and found freely in the cytoplasm without being covered by the membrane. The photosynthetic membranes i.e., thylakoids are also suspended in the cytoplasm. Thus the prokaryotic cell lacks the membrane-bounded organelles.

Photosynthetic apparatus

Cyanobacterial cells have flattened sacs membrane-bound structures, known as thylakoids; found freely suspended in the cytoplasm. They might be found dispersed, in parallel bundles or arranged in concentric rings etc. They contain photosynthetic pigments chlorophyll to absorb light and biochemical reactions. Phycobilisomes, the aggregates of light-harvesting proteins are attached to the surface of the thylakoid membrane in regularly spaced rows. The phycobilisomes contain phycobiliproteins that have a characteristic color: phycocyanin, allophycocyanin, phycoerythrin, which are widely used as fluorescent tags (Glazer 1999). In prochlorophytes, thylakoids are arranged in stacks as in other microalgae, but phycobilisomes are absent.

Cell inclusions

The cell inclusion in cyanobacterial cells commonly include cyanophycin granules, glycogen granules, polyphosphate granules, carboxysomes, lipid droplets, ribosomes and gas vacuoles. The glycogen granules (a-1,4-linked glucan) act as reserve material and are found in between the thylakoids. Besides glycogen granules, other reserve material such as cyanophycin granules, polymer of arginine and aspartic acid are also found. Carboxysomes are located centrally in the cytoplasm, which contain the RuBisCO (ribulose 1,5-bisphosphate carboxylase-oxygenase) enzyme. Ribosomes are randomly dispersed all through the cytoplasm. There are also unusual inclusions of poly-hydroxybutyrate (PHB) granules that appear as empty holes; which could be a potential source of natural biodegradable thermoplastic polymers (Suzuki et al. 1996).

Cell division and reproduction

Cell division could also take place via binary fission or multiple fission. Binary fission involves the constriction of the cell wall layers to grow inward, or invagination of the plasma membrane to form two cells; while in multiple fission, a number of small cells are formed, known as baeocytes. There is also the presence of budding-like cell division occurring in *Chamaesiphon*. Cyanobacteria that undergo mainly asexual reproduction normally through the fragmentation (hormogonia) or producing akinetes, it occurs in many filamentous cyanobacteria. It is evident that cyanobacteria do not perform sexual reproduction, but there might be the presence of genetic recombination by transformation or conjugation.

Cell Structure: Eukaryotic Microalgae

Cell wall, plasma membrane and outer investments

The eukaryotic microalgal cell wall is a biphasic structure consisting of cellulose (β1,4-glucan) microfibril embedded in an amorphous gel-like non-cellulosic matrix. In algae and plants, non-cellulosic material of the cell wall is usually assembled and exported by the Golgi apparatus; it might be silicified or calcified, and further strengthened with plates and scales. There might be an occurrence of a laminated polysaccharide investment outside the outer amorphous layer; which supports the production of polysaccharides (alginates, agar and carrageenans) in various macro algae as well as in microalgae *Porphyridium* (Arad 1999). The cytoplasm is covered by a thin plasma membrane which follows the cytoplasm. Some microalgal species especially members of euglenophyta lack a cell wall, but have a proteinaceous outer covering, known as pellicle. Like the euglenophyta, chryptophyta also lack the cell wall but they possess an outer covering for the protection of the cytoplasm, known as periplast.

Cytoplasm, nucleus and other organelles

There are a number of cell organelles in the cytoplasm: nucleus, chloroplast, Golgi apparatus, mitochondria, endoplasmic reticulum, plastids, ribosomes, lipid globules, vacuoles, contractile vacuoles, microtubules and flagella; that are formed by invagination of the plasma membrane and endoplasmic reticulum. The nucleus is bound by a dependable membrane, which have a nucleolus and contains the genetic material (genome) in the form of chromosomes. Some microalgae are uninucleate, while others are multinucleate which have coenocytes cellular organization.

Photosynthetic apparatus

The eukaryotic microalgal cell wall has a special photosynthetic organelle, the chloroplast which contains two parts: grana, composed of stacked thylakoids, a chlorophyll containing a membrane system and stroma, the surrounding matrix. Chloroplast is usually a double membrane bound organelle; while some algal division also has one or two membranes of endoplasmic reticulum besides the double membrane. There is also the presence of phycobiliproteins in thylakoids either in phycobilisomes (in Rhodophyta) or dispersed within the thylakoids (in the Cryptophyta). Pyrenoids could be present in the chloroplasts of some algal species. Some motile forms of algae have an orange-red eyespot or stigma, which is made up of lipid globules.

Cell division and reproduction

There are many types of reproduction observed in eukaryotic microalgae: (a) vegetative reproduction through cell division (an increase in cell or colony size) and fragmentation, which is widespread in the algae; (b) asexual reproduction by the production of spores, which are called zoospores (if flagellate), aplanospores or hypnospores (if non-flagellate). There is also the presence of autospores which are like aplanospores but lack the ontogenetic capacity for motility; (c) sexual reproduction, involves the production of gametes, which have a different morphology and dimension; based on the size of the two opposite gametes of the same species, the arrangement is called isogamy (same size), anisogamy (different size) or oogamy (one gamete is too small and the other is too large). The life-histories can be classified into five categories:

(i)   Most part of the life cycle has a diploid life form which produces the gametes that are haploid life form;

(ii) Most part of the life cycle has a haploid life form which have the zygote that are diploid life form;

(iii) Isomorphic alternation of generation that have similar a haploid gametophytic form and diploid sporophytic forms;

(iv) Heteromorphic alternation of generation that have either a small haploid gemetophytic form and a large diploid sporophytic form, or large gametophytic form and small diploid sporophytic form;

(v) Triphasic life cycle, includes a haploid gametophytic form, diploid carposporophytic form and diploid tetrasporophytic form, found in the red algae.

Nutrition Mode

There are several nutritional modes that are possible for algae (Devi et al. 2012, Kong et al. 2012, Farooq et al. 2013). It is well known that autotrophic organisms get their energy from light (photoautotrophic)or chemicals (chemoautotrophic) and an electron donor usually water or H_2S, sulfur, ammonium and ferrous ion to reduce the CO_2 in to the organic compounds and released O_2 as a byproduct (Qiao and Wang 2009, Yoo et al. 2010, Lee et al. 2013). The photoautotrophic organisms are also categorized in to two categories: (a) facultative, (b) obligate; facultative photoautotrophs can survive with or without light but obligate photoautotrophs cannot grow in the dark. Most of the algae belong to the obligate photoautotrophs category, which only need inorganic mineral ions to synthesize organic compounds.

In comparison to autotrophic, heterotrophic organisms are known to get their energy needs from organic compounds that are synthesized by other organisms. Some algal species are found to grow exclusively on organic compounds, but they still use light as an energy source, known as photoheterotrophic organisms. The need of organic compounds could be large enough to satisfy all the energy requirements or only need it in small quantities such as essential organic compounds like vitamins and amino acids; this is known as auxotrophy. Further the use of both modes autotrophic or heterotrophic for the energy requirements known as mixotrophic or amphitrophic growth, where both organic compounds like glucose and CO_2 are necessary for growth of algae. There is no specific determination about switching between autotrophy and heterotrophy; it might be either both the nutritional modes or an individual mode, except in total darkness. It is observed that *Chlorella sorokiniana* uses glucose at night showing heterotrophy, while in the day time they use both the glucose and CO_2 showing mixotrophy (Lee et al. 1996).

Except the obligate photoautotrophic nutrition, there is no clear distinction between the other types of nutritional modes that is possible for microalgae. It is understood that some interchange between the various trophic possibilities depends on the growth conditions under which they increase. It is also unclear how the excreted organics by the algae play a role in their trophic (nutrition) choices. From a long time, algae are known to release glycolic acid, but as yet there is not much information about its function and role (Fogg 1966). Grobbelaar (1983) stated that algae are well known to secrete alkaline phosphatase enzymes in the condition of phosphorus deficiency; this enzyme is known for its role in mobilization of adsorbed organic-P to be available for the algae. These exudates are reported to inhibit either the algal growth to limit the nutrient competition between the species (known as autoinhibitors) or as a predator to defense itself. It is emphasized that autoinhibitors are usually produced in ultrahigh-density mass algal communities and play an important role to prevent attaining the maximal productivity potential, leading to the reduction in net growth. It is also suggested that excreted organics could act as an energy source for the algae, especially during the night, while in daylight algae are capable of mixotrophic growth through photosynthesis related growth (Grobbelaar 1985, Garcia et al. 2011, Mohan et al. 2011).

Photosynthesis: Process, Components and Reactions

Photosynthesis is considered a unique process, in which inorganic compounds are converted in to organic matter using the energy from sunlight, the organisms are known as photoautotrophs. On the planet Earth, the photosynthesis process is directly or indirectly essential for the metabolism and growth of all forms of life. It was estimated that about 3.5 billion years ago, the earliest photoautotrophic organisms, anoxygenic photosynthetic bacteria originated, which used light energy, protons and electrons from different donor molecules, such as H_2S, to synthesize organic molecules and release CO_2 as a byproduct. But here the emphasis is on oxygen evolving photosynthetic microalgae, which emerged later and was responsible for the evolution of oxygenous atmosphere on planet Earth.

In prokaryotic microalgae e.g., cyanobacteria and prochlorophytes, the nucleoplasm region containing the DNA is located in the center; surrounded by a peripheral chromoplast region comprising of photosynthetic membranes.The prokaryotes have simpler photosynthetic membranes that are usually arranged in parallel sheets close to the cell surface, while in eukaryotic microalgae, the photosynthetic membranes are well organized in special organelles, known as the chloroplasts; which is composed of many thylakoids, the stacked lipoprotein membranes and stroma, the surrounding matrix-like structure (Staehelin 1986). Based on their light-harvesting photosynthetic pigments, eukaryotic microalgae are divided in to: Chlorophyta (green algae), Rhodophyta (red algae), Phaeophyceae (brown algae) and Chrysophyceae (golden algae).

Oxygenic photosynthesis can be considered as a redox reaction, which starts with the harvesting of light energy (by chlorophyll molecules); then leads to the conversion of carbon dioxide and water in to carbohydrates and oxygen released as abyproduct. This conversion is usually categorized into two stages: (a) light reactions that occur in photosynthetic membranes when the conversion of light energy in to chemical energy through formation of $NADPH_2$ and ATP; (b) dark reactions that happens in the stroma, whereas $NADPH_2$ and ATP are used in the synthesis of carbohydrates from carbon dioxide.

The photosynthesis can be measured as the amount of oxygen evolved in proportion to light intensity, but some amount of oxygen is also used in the process of respiration in the dark. The gross photosynthesis is calculated as the sum of net O_2 evolution (photosynthesis), and O_2 consumption (respiration).

It has been observed that the rate of photosynthesis is linearly proportional to the light intensity at low irradiance conditions (light-limited region), while at high irradiance conditions, the rate of photosynthesis is not much greater with increasing light intensity. Finally the photosynthesis reaches a peak at a particular light intensity called light saturated value, beyond which light intensity becomes the limiting factor for the photosynthesis rate; this leads to the temporary halting of the enzymatic reactions that utilize fixed energy. If this situation of supra-optimal irradiance persists, photosynthetic rates usually start to decline from the light-saturated value; the phenomenon is known as photoinhibition.

Nature of light

The visible light wavelengths i.e., 400-750 nm used in photosynthesis, known as Photosynthetically Active Radiation (PAR). The incidence of light energy on a particular surface is measured as radiant flux energy or irradiance, and the units are Wm^{-2} or mmol photon m^{-2} s^{-1}. The average solar irradiance is estimated about 1000 Wm^{-2} that directly reaches the Earth's surface on a sunny day, from which about 40% i.e., 400 Wm^{-2} or 1800 mmol photon $m^{-2}s^{-1}$ is considered as photosynthetically active radiation.

Photosynthetic pigments

Most of the microalgae contain chlorophylls (Chl a, b, c, d) and carotenoids (carotenes and xanthophylls) as their photosynthetic pigments, but cyanobacteria and red algae also have phycobilins (phycoerythrobilin, phycocyanobilin and phycourobilin). Chlorophylls have two absorption ranges: (a) blue or blue-green (450–475 nm) and (b) red (630–675 nm); while carotenoids and phycobilins have an absorption range of 400–550 nm and 500–650 nm respectively. Chl a is the integral component of all photosynthetic machinery in all oxygenic photoautotrophs, and makes the maximum portion of the core and reaction center (pigment–protein complexes), while Chl b or Chl c in light harvesting antennae, as an accessory to widen light absorption range. Besides the chlorophylls, carotenoids are functions as accessory light-harvesting pigments in photosynthetic apparatus to transfer excitation energy to Chl a. They also make the structural components of the pigment–protein complexes of the light-harvesting antenna and reaction center. The most important role of carotenoids is the protection of photosynthetic machinery from the harmful effect of excess irradiance, chlorophyll triplets and reactive oxygen species.

Chl–proteins and carotenoid–proteins are water insoluble pigment–protein complexes, while phycobili–proteins are water-soluble, and are found to be covalently bound to the apoprotein. Some carotenoids particularly xanthophylls i.e., astaxanthin and canthaxanthin are not used for the transfer of excitation energy, known as secondary carotenoids; often accumulated in the cytoplasm of some microalgal species e.g., *Haematococcus pluvialis* in response to certain unfavorable conditions that combine nutrient deficiency, high irradiance and temperature extremes.

The photosynthetic membrane system

The photosynthetic membrane system orthylakoids is a closed, interconnected flat vesicles structure which includes the thylakoid membrane surrounded by an intrathylakoidal space, lumen. The thylakoid membrane is composed of bilayer of mono- and di-galactosylglycerol lipids, which are interspersed with large proteins molecules (Singer and Nicholson 1972). In cyanobacteria and red algae, photosynthetic membranes are found in the form of single lamellae (sheet-like membrane system in comparison to the thylakoid's disk-like structure) that are extended parallel from one end to another of the chloroplast; which have phycobilisomes, light harvesting protein complexes attached on the stroma side of the thylakoid membrane. In contrast to this, other eukaryotic microalgae contain chloroplast with stacked thylakoids (generally in pairs or stacks of three) called grana that are connected by single thylakoids known as stromal lamellae. The thylakoid membranes have five major complexes:

(a) light-harvesting antennae—harvesting of light energy and transfer to the reaction center,
(b) photosystem II (PS II)—have a reaction center,
(c) cytochrome b6/f—electron transfer between PS II and PS I,
(d) photosystem I (PS I)—have a reaction center,
(e) ATP synthase—facilitates energy for electron transport and photophosphorylation.

Light-harvesting antennae

The outer light-harvesting antennae harvest the light energy and transfer this energy to the reaction centers of the photosystems, and some amount of heat is also released during the energy transfer. The photosynthetic pigments i.e., chlorophylls, carotenoids and phycobilins) are coupled with proteins, called pigment-protein complexes, which are responsible for various specific functions in light harvesting and electron transfer. Two types of pigment–protein complexes are present: (i) hydrophilic phycobiliproteins, that occur in cyanobacteria and red algae, and (ii) hydrophobic pigment–protein complexes, such as LHC II and LHC I that comprises of Chl a, Chl b and carotenoids.

Cyanobacteria and red algae have phycobiliproteins containing multimeric particles called phycobilisomes that are attached to the protoplasmic side of the thylakoid membrane. In phycobilisomes, there is a central allophycocyanobilin-containing core that is attached to the reaction center of the PS II. The central allophycocyanobilin-containing core extends outward to another core containing phycocyanobilin. There is a further presence of a more distant core that either contains phycoerythrobilin or phycourobilin, which specifically depends on the species. Based on the presence or absence of phycobiliproteins in the division Cyanophyta (cyanobacteria), a separate group Prochlorophyceae is created into Cyanophyta, that contain Chl a and Chl b but lacks phycobiliproteins (Bryant 1994).

In other eukaryotic microalgae (and higher plants), pigment–protein complexes e.g., LHC II and LHC I, which are attached to the reaction center contain Chl a and b as well as xanthophylls. LHC II and LHC I are quite genetically and biochemically different, and serves PS II and PS I respectively. In diatoms, there is a slight difference in the outer light-harvesting complex, which imparts Chl a and c, and fucoxanthin as the major carotenoid.

Photosystem II (P_{680})

Photosystem II is a membrane-embedded-protein-complex that comprises of more than 20 subunits and a molecular mass of about 300 kDa. It has a reaction center, inner light-harvesting antennae and oxygen-evolving complex. The PS II reaction center comprises of proteins D1 and D2; and two subunits cyt b559 i.e., a and b subunits. The D1 and D2 proteins hold all essential prosthetic groups that are the integral part of the primary electron donor, P_{680}, tyrosine Z, pheophytin and the quinone acceptors, QA and QB; and are also required for the charge separation and its stabilization. The inner core antennae contains the intrinsic Chl a-proteins i.e., CP43 and CP47 that are located on the opposite sides of D1-D2 reaction center (Hankamer et al. 2001); and is responsible for the transfer of excitation energy from the outer antennae to the reaction center.

Plastoquinone, the cytochrome b6/f complex and plastocyanin

For the electron transport between PS II and PS I, the need of cytochrome b6/f complex and two kinds of mobile carriers such as plastoquinone and plastocyanin are required to pick off on or pick up from cyt b6/f complex (Gross 1996).

Photosystem I (P_{700})

Photosystem I is an intermembrane complex that is made up of about 10 proteins, 100 chlorophylls with molecular mass of about 360 kDa. It has a reaction center that comprises of centrally located two large PsaA and PsaB proteins, which keeps the prosthetic cofactors for the reaction center. The complex have a Chl dimer P_{700} (where primary charge separation is initiated), electron carriers A_0 (Chl a) and A_1 (phylloquinone); and FX (4Fe–4S). The primary function of PS I is to facilitate the photochemical reactions that produce low redox potential (about 1 V); which is essential for the reduction of ferredoxin and subsequent production of $NADPH_2$.

ATP synthase/ATPase

ATP synthase is a membrane-bound enzyme complex that has two oligomeric subunits: CF_0 and CF_1 of relative molecular masses of about 110–160 kDa and 400 kDa respectively. The CF_0 subunit is hydrophobic in nature located in the thylakoid membrane, while CF_1 is hydrophilic in nature and attached to the CF_0 on the stromal side of the membrane. This complex facilitates the synthesis of ATP by creating the pH gradient. The subunit CF_0 functions as a membrane proton channel and facilitates the movement of protons to compel the subunit CF_1, which is composed of a ring catalytic core containing catalytic sites for ATP synthesis. For the synthesis of one ATP molecule, about four protons are required (Kramer et al. 1999).

Electron transport and phosphorylation

The last step of light reactions involves the use of energy molecules $NADPH_2$ and ATP (generated in light reactions) in the assimilation of inorganic carbon. The light energy capturing photosystems i.e., PS I and PS II are functions in a series, which are connected through a chain of electron carriers usually known as the 'Z' scheme (Hill and Bendall 1960). In this Z scheme, redox components are associated with their equilibrium mid-point potentials; and the movement of electrons starts downhill, from a more negative redox potential to a more positive redox potential. This process starts with the illumination, which is induced by the extraction of two electrons from dissociation of water (and O_2 is evolved); then these electrons move through a chain of electron carriers to generate one molecule of $NADPH_2$. Simultaneously, a pH gradient is created through protons transportation from an external space (stroma) into the intra-thylakoid space (lumen), which leads to ATP synthesis, that is catalyzed by the protein complex called ATPase or ATP synthase. This process is known as photophosphorylation and could be follows as:

$$2NADP + 3H_2O + 2ADP + 2Pi \xrightarrow{\text{light energy}} 2NADPH_2 + 3ATP + O_2$$

The dark reactions: Carbon assimilation

The fixed CO_2 in light reactions assimilate in the dark reaction using the $NADPH_2$ and ATP which is generated in the light reaction of photosynthesis. The reaction can be followed as:

$$CO_2 + 4H^+ + 4e^- \xrightarrow[\text{enzymes}]{2NADPH_2, 3ATP} (CH_2O) + H_2O$$

For the fixation of one molecule of CO_2, two molecules of $NADPH_2$ and three molecules of ATP are required. It is also reported that a minimum 10 quanta of absorbed light are required for each molecule of CO_2 fixed or O_2 evolved; this is the quantum efficiency of CO_2 fixation. In 1940, Calvin and Benson discovered the reaction mechanism of carbon fixation using ^{14}C radio-labeling technique and were awarded the Nobel Prize in 1961. The conversion of CO_2 to sugar (or other compounds), known as Calvin–Benson cycle, involves four phases:

1. Carboxylation phase: The CO_2 molecule combines with the ribulosebisphosphate (RuBP) to form two molecules of 3-phospho-glycerate (3PGA); by the help of the enzyme ribulosebisphospate carboxylase/oxygenase (RuBisCo).

2. Reduction phase: 3-Phosphoglycerate (3PGA) is reduced to intermediate 1,3-bisphosphoglyceric acid (1,3-BPGA) and finally in to Glyceraldehydes-3-Phosphate (G3P) and the energy is used either as ATP in conversion of 3PGA TO 1,3BPGA or as $NADPH_2$ in conversion of 1,3BPGA to G3P.

3. Regeneration phase: Most of the G3P molecules are used to regenerate Ribulose phosphate (RuPB) molecules for further CO_2 fixation. The enzymes transketolase and aldolase are required in the conversion of 6-C and 3-C sugars in to a 5-C compound.

4. Production phase: Some G3P molecules turn in to sugars which are considered primary end products of the Calvin cycle, but some G3P molecules are also used in the synthesis of fatty acids, amino acids and organic acids through various mechanisms.

Photorespiration

Photorespiration is a similar process like respiration in other organisms, in which fixed organic carbon is consumed for their growth and CO_2 is generated as a byproduct in place of O_2 in photosynthesis. This process starts with the reaction of O_2 with ribulosebisphosphate to form phosphoglycolate by the help of the enzyme RuBisCo (which functions as an oxygenase). Then phosphoglycolate phosphorylates in to glycolate, which is later transported in to peroxisome to convert to intermediate compound glyoxylate to glycine; and again conveyed to mitochondrion to carry out the final conversion in serine, ammonia and CO_2. In this process, no metabolic gain is achieved. The photorespiration process primarily depends on the relative ratio of O_2 and CO_2 concentrations; it has been observed that a high O_2/CO_2 ratio (i.e., high concentration of O_2 and low concentration of CO_2) encourages this process, while a low O_2/CO_2 ratio prefers carboxylation. Due to low half saturation value i.e., km (which is roughly equal to the level of CO_2 in air), RuBisCo shows low affinity to CO_2, which leads to shifting of the reaction equilibrium to photorespiration, on facing high irradiance, high oxygen level and reduced CO_2 conditions.

Uptake and Assimilation of Nutrients

Carbon fixation and assimilation

Microalgae including cyanobacteria obtain their carbon requirement through reducing atmospheric CO_2 into organic matter that are needed for cell growth. The pathway of the Calvin cycle or reductive pentose phosphate cycle is used for this reductive biosynthesis of organic matter (Pelroy and Bassham 1972, Jansz and Maclean 1973). In the Calvin cycle, the first step involves the carboxylation of ribulose 1,5-bisphosphate (RuBP) into a 6-carbon carboxylated intermediate (2-carboxy-3-keto-Dribitol 1,5-bisphosphate) (Siegel and Lane 1973, Sjodin and Vestermark 1973); then it subsequently breaks down into two molecules of 3-phosphoglyceric acid. This phosphoglyceric acid then reduces to produce glyceraldehydes-3-phosphate which is later transported to cytosol for glucose synthesis (Fig. 2.1).

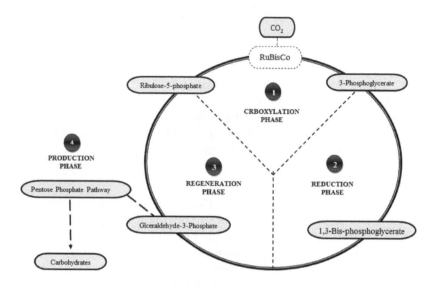

Figure 2.1: Carbon fixation and synthesis of carbohydrates (Calvin-Benson Cycle).

Nitrogen uptake and assimilation

Nitrogen-containing compounds are frequently found at low concentrations in the environment; and the incorporation of this nitrogen into the cyanobacterial cell is facilitated through permeases that are found in the cytoplasmic membrane. These permeases could be ABC (ATP-Binding Cassette)-type or nitrate-nitrite transporters; which are involved in the uptake of nitrate and nitrite (Omata et al. 1993, Luque et al. 1994, Sakamoto et al. 1999) or urea (Valladares et al. 2002) in a number of cyanobacteria. ABC-type permeases utilizes ATP to facilitate an active and concentrative transport of nitrogen-containing compounds; and is also required for arginine and glutamine transport (Quintero et al. 2001). For the transport of ammonium, secondary permeases Amt type (ammonium transporters) permeases are involved in the transport of an ammonium form of nitrogen. To examine the function of these permeases, Montesinos et al. (1998) and Vázquez-Bermúdez et al. (2002) incorporated the ^{14}C methyl ammonium in to the medium in which *Synechocystis* sp. strain PCC6803 was growing. It was observed that ^{14}C methyl ammonium is found to be accumulated into the cells of *Synechocystis* sp. strain PCC6803 up to a level that could be considered a membrane potential-driven transport.

Luque et al. (1993) and Rubio et al. (1996) suggested that the enzymes nitrate reductase (narB gene) and nitrite reductase (nir gene) are responsible for the sequential reduction of nitrate into nitrite and then nitrite into ammonium respectively. The organic sources of nitrogen e.g., urea and arginine is assimilated by a different path, urea goes in to the urea cycle to render ammonium by a standard Ni^{2+}-dependent urease enzyme and CO_2 is released, while the arginine is incorporated through a unique pathway (combined form of the urea cycle and the arginase pathway) to give ammonium and glutamate. All the nitrogen forms either inorganic (nitrate, nitrite, ammonium) or organic (urea, arginine) uptake by the cell, first converted in to ammonium, and then ammonium is incorporated into carbon skeletons through the glutamine synthetase–glutamate synthase pathway. Although cyanobacteria lack 2-oxoglutarate dehydrogenase (OGDH), an enzyme responsible for the substrate level phosphorylation and nitrogen incorporation; could deprive the cyanobacteria for the above mentioned processes (Smith et al. 1967, Pearce and Carr 1967, Hodges 2002). But cyanobacteria also uses two other enzymes i.e., 2-OG decarboxylase and succinic semialdehyde dehydrogenase to substitute the OGDH and succinyl-CoA ligase to generate reducing equivalents for completion of the entire cycle (Fig. 2.2) (Zhang and Bryant 2011).

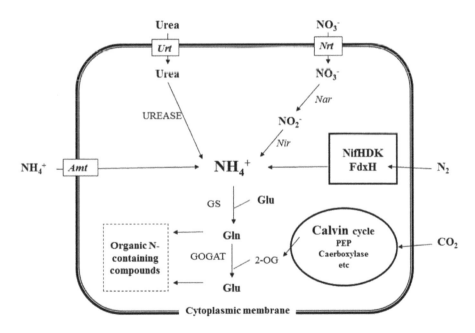

Figure 2.2: Nitrogen uptake and their incorporation.

Nitrogen assimilation in eukaryotic microalgae involves four steps in which two are transport steps and two are reduction steps (Guerrero et al. 1981, Fernandez and Galvan 2007, 2008): (a) in the first step, transport of nitrate in to the cell; (b) reduction of nitrate in to nitrite through cytosolic enzyme nitrate reductase (narB); (c) subsequent transport of nitrite in to the chloroplast; (d) reduction of nitrite in to ammonium through the enzyme nitrite reductase (nir) (Fig. 2.3). The incorporation of this reduced ammonium takes place as an amide group of glutamine in a reaction involving glutamate and ATP in the presence of enzyme glutamine synthetase, and then the amide group is transferred reductively to α-oxoglutarate to form two molecules of glutamate, this whole process was earlier known as glutamate synthase cycle (Miflin and Lea 1975). By investigating *Chlamydomonas* microalgae, three families of proteins NRT1 (NPF), NRT2, and NAR1 are found to be involved in nitrate and/or nitrite transport (Crawford and Glass 1998, Forde 2000, Galvan and Fernández 2001, Forde and Cole 2003).

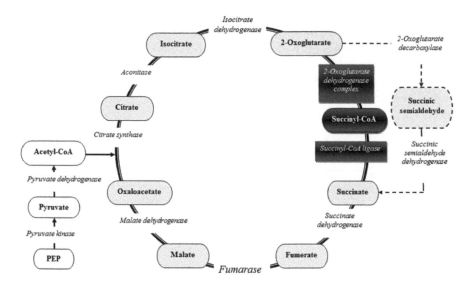

Figure 2.3: TCA Cycle and their variants in cyanobacteria. (Products and enzymes in blackened boxes are absent in cyanobacteria, in place of this a broken arrow shows the alternative product and enzymes).

Phosphorous uptake and storage

Cyanobacteria and microalgae primarily take phosphorous in the form of orthophosphate, preferably as $H_2PO_4^-$ or HPO_4^{2-}. They are also able to take organic phosphates, when facing a short supply of orthophosphates, but phosphates are converted in to orthophosphates by phosphatases at the cell surface. The process of phosphorous uptake is active; it needs energy for transport (Becker 1994). Beside active uptake, microalgae are able to perform excess uptake of the phosphorus, which is stored as polyphosphate in polyphosphate (volutin) granules within the cells. It is suggested that polyphosphate synthetase (polyphosphate kinase) is responsible for the synthesis of polyphosphates in cyanobacteria, and is known for their role as catalysts in the formation of polyphosphate from ATP. Grillo and Gibson (1979) stated that there is no requirement of polyphosphate as a primer but the enzyme polyphosphate synthetase requires magnesium for their activation.

These reserved polyphosphate granules are used to cope with phosphorous starvation and unfavorable conditions in the environment (Achbergerová and Nahalka 2011). On need, these polyphosphates are metabolized in simple phosphates through the action of enzyme alkaline polyphosphatase (Grillo and Gibson 1979, Doonan and Jensen 1977, 1980a, 1980b, Zevenboom 1980).

Cembella et al. (1984) observed that both cyanobacteria and eukaryotic microalgae are able to store phosphorous as polyphosphates, except certain species which show the inability to form polyphosphates. Cembella et al. (1984) also suggested that some eukaryotic microalgal species like *Chlamydomonas* sp. show the capability of storing phosphate in a dissolved state in the vacuole. Tillberg et al. (1984) investigated the formation of polyphosphate bodies in *Scenedesmus* sp. and observed these bodies in cisternae of the vacuoles. Kulaev and Vagabov (1983) and Sianoudis et al. (1986) suggested that the long chain polyphosphates can also be present in the cell periphery, outside the cytoplasmic membrane in the eukaryotes.

Sulfur uptake and assimilation

Sulfur is mainly present as sulfate and found bounded to the organic substances e.g., humic acids as part of organic soil biomass (Eriksen et al. 1998). To separate the sulfate from the humic acids, microalgae secrete arylsulfatases to cleave ester bonds between sulfate and organic substances, e.g., humic acids. On facing sulfate deficiency, encoding ARS genes are induced to produce extracellular enzymes to make sulfate available for transferring into the cell. Sulfate transporters are considered as integral membrane proteins that endorse the cotransport of sulfate/H^+ ions. These high-affinity transporters are specialized for the missing ions and are usually inducted by the conditions of sulfate or phosphate deprivation (Smith et al. 1997, Raghothama 2000, Takahashi et al. 2000), this is the reverse of nitrate transporter genes which are induced by the presence of nitrate (Hell and Hillebrand 2001). Yildiz et al. (1994) studied the microalgal sulfate transporters in *Chlamydomonas reinhardtii*, and observed that the mechanism of transport is not light-dependent; it is facilitated through an ATPase mediated proton gradient which is positioned in the plasmalemma.

After uptake of sulfate ions into the cell, it is either transported to the plastids or stored in the vacuole (only if present in excess). In the plastids, the sulfate ion is first activated by the ATP to produce 5'- adenylsulfate (APS) through the enyme ATP sulfurylase (ATP-S); and then APS is reduced in to sulfite by receiving two electrons from glutathione and the reaction is catalyzed by enzyme APS reductase (Bick and Leustek 1998). Gao et al. (2000) and Koprivova et al. (2000) suggested that APS reductase act as a primary regulation site for the sulfate assimilation pathway in plants and algae. Gao et al. (2000) found remarkable APS reductase activity in some microalgae which may be 400 times more than in the plants; and it mainly depends on microalgal growth rate and the N availability (Gao et al. 2000). Bork et al. (1998) observed that sulfite is further reduced in to sulfide through the enzyme sulfite reductase, which shows structural and functional similarity to nitrite reductase. In the end, this free sulfide is incorporated into cysteine.

Conclusion

Besides the presence of different pigments and storage products, most microalgal groups have similar photosynthetic apparatus for photosynthesis. Some microalgal groups have simple cell characteristics like gram negative bacteria in the cell wall, naked genetic material and absence of membrane-bound organelles like chloroplast, mitochondrion and Golgi apparatus; making them true members of prokaryotic world. But their photosynthetic apparatus such as the presence of thylakoids and two photosystems (like in eukaryotic microalgae) andoxygenic photosynthesis, makes them an important member of the microalgal world. Later these similarities and differences in cellular characteristics establishes the ancestral links and evolution between the prokaryotic and eukaryotic members of the microalgal world.

The oxygenic photosynthesis process started by the earliest cyanobacteria such as organisms and the subsequent evolution of true cyanobacteria responsible for changing the anoxygenic environment to the oxygenic one in the Precambrian era. Most of microalgal groups are obligate photoautotrophs and play an important role in nutrient cycling and productivity in the aquatic world especially in oceans. But there is the presence of photoheterotrophic mode of nutrition mode in both prokaryotic and eukaryotic microalgal species, with some microalgal in light-limited and dark conditions. From the simplest photosynthetic structure in cyanobacteria to complex chloroplast in eukaryotic microalgae use the same mechanisms to fix the inorganic carbon into organic compounds.

Apart from carbon fixation, there might be some alternations and adaptations in microalgal strategies related to uptake, storage and metabolism of the other nutrients i.e., nitrogen, phosphorous, iron and others. Some cyanobacterial members have specialized cells heterocysts for nitrogen fixation, some are able to store nitrogen in the form of cyanophycin and phycocyanin; but all microalgal groups including cyanobacteria benefit from the uptake of phosphorous and store it as polyphosphate granules. These differences in cell structure, nutrition modes and metabolism provide a wider range adaptability to successfully thrive in different ecological conditions.

References

Achbergerová, L. and J. Nahálka. 2011. Polyphosphate: an ancient energy source and active metabolic regulator. Microb. Cell Fact. 10: 63.

Arad, S.M. 1999. Polysaccharides of red microalgae. pp. 282–291. *In*: Z. Cohen (ed.). Chemicals from Microalgae. Taylor & Francis, London.

Becker, E.W. 1994. Microalgae: Biotechnology and Microbiology. Cambridge University Press, Cambridge.

Bick, J.A. and T. Leustek. 1998. Plant sulfur metabolism – the reduction of sulfate to sulfite. Curr. Opin. Plant Biol. 1: 240–244.

Bork, C., J.-D. Schwenn and R. Hell. 1998. Isolation and characterization of a gene for assimilatory sulfite reductase from *Arabidopsis thaliana*. Gene 212: 147–153.

Bryant, D.A. (ed.). 1994. The Molecular Biology of Cyanobacteria. Kluwer Scientific Publishers, Dordrecht, Netherlands.

Cembella, A.D., N.J. Antia and P.J. Harrison. 1984. The utilization of inorganic and organic phosphorus compounds as nutrients by eukaryotic microalgae: a multidisciplinary perspective. Part 2. Crit. Rev. Microbiol. 11(1): 13–81.

Crawford, N.M. and A. Glass. 1998. Molecular and physiological aspects of nitrate uptake in plants. Trends Plant Sci. 3: 389–395.

Devi, M.P., G.V. Subhash and S.V. Mohan. 2012. Heterotrophic cultivation of mixed microalgae for lipid accumulation and wastewater treatment during sequential growth and starvation phases. Effect of nutrient supplementation. J. Renew. Energ. 43: 276–283.

Doonan, B.B. and T.E. Jensen. 1977. Ultrastructural localization of alkaline phosphatase in the blue-green bacterium *Plectonema boryanum*. J. Bacteriol 132: 967–973.

Doonan, B.B. and T.E. Jensen.1980a. Ultrastructural localization of alkaline phosphatase in the cyanobacteria *Coccochloris peniocytis* and *Anabaena cylindrica*. Protoplasma 102: 189–197.

Doonan, B.B. and T.E. Jensen.1980b. Physiological aspects of alkaline phosphatase in selected cyanobacteria. Microbios. 29: 185–207.

Eriksen, J., M.D. Murphy and E. Schnug. 1998. The soil sulfur cycle. pp. 39–73. *In*: E. Schnug (ed.). Sulfur in Agroecosystems. Kluwer Academic Publishers, Dordrecht, Netherlands.

Farooq, W., Y.-C. Lee, B.-G. Ryu, B.-H. Kim, H.-S. Kim, Y.-E. Choi, et al. 2013. Two-stage cultivation of two *Chlorella* sp. strains by simultaneous treatment of brewery wastewater and maximizing lipid productivity. Bioresour. Technol. 132: 230–238.

Fernandez, E. and A. Galvan. 2007. Inorganic nitrogen assimilation in *Chlamydomonas*. J. Exp. Bot. 58: 2279–2287.

Fernandez, E. and A. Galvan. 2008. Nitrate assimilation in *Chlamydomonas*. Eukaryotic Cell 7: 555–559.

Fogg, G.E. 1966. Algal Cultures and Phytoplankton Ecology. University of Wisconsin Press, Madison, Wisconsin, USA.

Forde, B.G. 2000. Nitrate transporters in plants: structure, function and regulation. Biochem. Biophys. Acta 1465: 219–235.

Forde, B.G. and J.A. Cole. 2003. Nitrate finds a place in the sun. Plant Physiol. 131: 395–400.

Galvan, A. and E. Fernández. 2001. Eukaryotic nitrate and nitrite transporters. Cell. Mol. Life Sci. 58: 225–233.

Gao, Y., O.M.E. Schofield and T. Leustek. 2000. Characterization of sulfate assimilation in marine algae focusing on the enzyme 5'-adenylylsulfate reductase. Plant Physiol. 123: 1087–1096.

Garcia, O.P., M.E. Froylan-Escalante, L.E. de-Bashan and Y. Bashan. 2011. Heterotrophic cultures of microalgae: metabolism and potential products. Water Res. 45: 11–36.

Glazer, A.N. 1999. Phycobiliproteins. pp. 261–276. *In*: Z. Cohen (ed.). Chemicals from Microalgae. Taylor & Francis, London.

Grillo, J.F. and J. Gibson. 1979. Regulation of phosphate accumulation in the unicellular cyanobacterium *Synechococcus*. J. Bacteriol. 140: 508–517.

Grobbelaar, J.U. 1983. Availability to algae of N and P adsorbed on suspended solids in turbid waters of the Amazon River. Arch. Hydrobiol. 96(3): 302–316.

Grobbelaar, J.U. 1985. Carbon flow in the pelagic zone of a shallow turbid impoundment. Wuras Dam. Arch. Hydrobiol. 103(1): 1–24.

Gross, E.L. 1996. Plastocyanin: structure, location, diffusion and electron transfer mechanisms. pp. 213–247. *In*: D.R. Ort and C.F. Yocum (eds.). Advances in Photosynthesis, Vol. 4, Oxygenic Photosynthesis: The Light Reactions. Kluwer Academic Publishers, Dordrecht, Netherlands.

Guerrero, M.G., J.M. Vega and M. Losada. 1981. The assimilatory nitrate-reducing system and its regulation. Annu. Rev. Plant Physiol. Plant Mol. Biol. 32: 169–204.

Hankamer, B., E. Morris, J. Nield, C. Gerle and J. Barber. 2001. Three-dimensional structure of the photosystem II core dimer of higher plants determined by electron microscopy. J. Struct. Biol. 135: 262–269.

Hell, R. and H. Hillebrand. 2001. Plant concepts for mineral acquisition and assimilation. Curr. Opin. Biotechnol. 12: 161–168.

Hill, R. and R. Bendall. 1960. Function of the two cytochrome components in chloroplasts: a working hypothesis. Nature 186: 136–37.

Hodges, M. 2002. Enzyme redundancy and the importance of 2-oxoglutarate in plant ammonium assimilation. J. Exp. Bot. 53: 905–916.

Jansz, E.R. and F.I. Maclean. 1973. CO_2 fixation by the blue-green alga *Anacystis nidulans*. Can. J. Microbiol. 19: 497–504.

Kong, W.B., H. Song, S.F. Hua, H. Yang, Qi. Yang and C.G. Xia. 2012. Enhancement of biomass and hydrocarbon productivities of *Botryococcus braunii* by mixotrophic cultivation and its application in brewery wastewater treatment. Afr. J. Microbiol. Res. 6: 1489–1496.

Koprivova, A., M. Suter, R.O. den Camp, C. Brunold and S. Kopriva. 2000. Regulation of sulfate assimilation by nitrogen in *Arabidopsis*. Plant Physiol. 122: 737–746.

Kramer, D.M., C.A. Sacksteder and J.A. Cruz. 1999. How acidic is the lumen? Photosyth. Res., 60: 151–163.

Kulaev, I.S. and V.M. Vagabov. 1983. Polyphosphate metabolism in micro-organisms. Adv. Microb. Physiol. 24: 83–171.

Lee, Y.K., S.Y. Ding, C.H. Hoe and C.S. Low. 1996. Mixotrophic growth of *Chlorella sorokiniana* in outdoor enclosed photobioreactors. J. Appl. Phycol. 8: 163–169.

Lee, Y-C., B. Kim, W. Farooq, J. Chung, J.-I. Han, H.-J. Shin, et al. 2013. Harvesting of oleaginous *Chlorella* sp. by organoclays. Bioresour. Technol. 132: 440–445.

Luque, I., E. Flores and A. Herrero. 1993. Nitrite reductase gene from *Synechococcus* sp. PCC 7942: homology between cyanobacterial and higher-plant nitrite reductases. Plant Mol. Biol. 21: 1201–1205.

Luque, I., E. Flores and A. Herrero. 1994a. Nitrate and nitrite transport in the cyanobacterium *Synechococcus* sp. PCC 7942 are mediated by the same permease. Biochim. Biophys. Acta 1184: 296–298.

Miflin, B.J. and P.J. Lea. 1975. Glutamine and asparagines as nitrogen donors for reductant-dependent glutamate synthesis in pea roots. Biochem. J. 149: 403–409.

Mohan, S.V., M.P. Devi, G. Mohanakrishna, N. Amarnath, M.L. Babu and P.N. Sarma. 2011. Potential of mixed microalgae to harness biodiesel from ecological water-bodies with simultaneous treatment. Bioresour. Technol. 102: 1109–1117.

Montesinos, M.L., A.M. Muro-Pastor, A. Herrero and E. Flores. 1998. Ammonium/ methylammonium permeases of a cyanobacterium identification and analysis of three nitrogen-regulated Amt genes in Synechocystis sp. PCC 6803. J. Biol. Chem. 273: 31463–31470.

Omata, T., X. Andriesse and A. Hirano. 1993. Identification and characterization of a gene cluster involved in nitrate transport in the cyanobacterium *Synechococcus* sp. PCC7942. Mol. Gen. Genet. 236: 193–202.

Pearce, J. and N.G. Carr. 1967. The metabolism of acetate by the blue-green algae, *Anabaenavariabilis* and *Anacystisnidulans*. J. Gen. Microbiol. 49: 301–313.

Pelroy, R.A. and J.A. Bassham. 1972. Photosynthetic and dark carbon metabolism in unicellular blue-green algae. Arch. Mikrobiol. 86: 25–38.

Qiao, H. and G. Wang. 2009. Effect of carbon sources on growth and lipid accumulation in *Chlorella sorokiniana* GXNN01. Chin. J. Oceanol. Limnol. 27: 762–768.

Quintero, M.J., M.L. Montesinos, A. Herrero and E. Flores. 2001. Identification of genes encoding amino acid permeases by inactivation of selected ORFs from the *Synechocystis* genomic sequence. Genome Res. 11: 2034–2040.

Raghothama, K.G. 2000. Phosphate transport and signaling. Curr. Opin. Plant Biol. 3: 182–187.

Richmond, A. 2004. Handbook of Microalgal Culture: Biotechnology and Applied Phycology. Blackwell Science Ltd. Iowa, USA.

Rubio, L.M., A. Herrero and E. Flores. 1996. A cyanobacterialnar B gene encodes a ferredoxin-dependent nitrate reductase. Plant Mol. Biol. 30: 845–850.

Sakamoto, T., K. Inoue-Sakamoto and D.A. Bryant. 1999. A novel nitrate/nitrite permease in the marine cyanobacterium *Synechococcus* sp. Strain PCC 7002. J. Bacteriol. 181: 7363–7372.

Sianoudis, J., A.C. Kusel, A. Mayer, L.H. Grimme and D. Leibfritz. 1986. Distribution of polyphosphates in cell-compartments of *Chlorella fusca* as measured by P-NMR-spectroscopy. Arch. Microbial. 144: 48–54.

Siegel, M.I. and M.D. Lane. 1973. Chemical and enzymatic evidence for the participation of a 2-carboxy-3-ketoribitol 1,5-diphosphate intermediate in the carboxylation of ribulose 1,5-diphosphate. J. Biol. Chem. 248: 5486–5498.

Singer, S.J. and G.L. Nicholson. 1972. The fluid mosaic model of the structure of cell membranes. Science 175: 720–731.

Sjodin, B. and A. Vestermark. 1973. The enzymatic formation of a compound with the expected properties of a carboxylatedribulose 1,5-diphosphate. Biochem. Biophys. Acta 297: 165–173.

Smith, A.J., J. London and R.Y. Stanier. 1967. Biochemical basis of obligate autotrophy in blue-green algae and *Thiobacilli*. J. Bacteriol. 94: 972–983.

Smith, F.W., M.J. Hawkesford, P.M. Ealing, D.T. Clarkson, P.J.V. Berg, A.R. Belcher, et al. 1997. Regulation of expression of a cDNA from barley roots encoding a high affinity sulfate transporter. Plant J. 12: 875–884.

Staehelin, A. 1986. Chloroplast structure and supramolecular organization of photosynthetic membranes. pp. 1–84. *In*: L.A. Staehelin and C.A. Arntzen (eds.). Photosynthesis III: Photosynthetic Membranes and Light-Harvesting Systems. Springer-Verlag, New York.

Suzuki, T., M. Miyake, Y. Tokiwa, H. Saegusa, T. Saito and Y. Asada. 1996. A recombinant cyanobacterium that accumulates poly-(hydroxybutrate). Biotechnol. Lett. 18: 1047–1050.

Takahashi, H., A. Watanabe-Takahashi, F.W. Smith, M. Blake-Kalff, M.J. Hawkesford and K. Saito. 2000. The roles of three functional sulfate transporters involved in uptake and translocation of sulfate in *Arabidopsis thaliana*. Plant J. 23: 171–182.

Tillberg, J., T. Barnard and J.R. Rowly. 1984. Phosphorus status and cytoplasmic structures in *Scenedesmus* (Chlorophyceae) under different metabolic regimes. J. Phycol. 20: 124-136.

Vázquez-Bermúdez, M.F., J. Paz-Yepes, A. Herrero and E. Flores. 2002. The NtcA-activated amt1 gene encodes a permease required for uptake of low concentrations of ammonium in the cyanobacterium *Synechococcus* sp. PCC 7942. Microbiol. 148: 861–869.

Valladares, A., M.L. Montesinos, A. Herrero and E. Flores. 2002. An ABC-type, high-affinity urea permease identified in cyanobacteria. Mol. Microbiol. 43: 703–715.

Yildiz, F.H., J.P. Davies and A.R. Grossman. 1994. Characterization of sulfate transport in *Chlamydomonas reinhardtii* during sulfur-limited and sulfur-sufficient growth. Plant Physiol. 104: 981–987.

Yoo, C., S.Y. Jun, J.Y. Lee, C.Y. Ahn and H.M. Oh. 2010. Selection of microalgae for lipid production under high level of carbon dioxide. Bioresour. Technol. 101: 71–74.

Zevenboom, W. 1980. Growth and nutrient uptake kinetics of *Oscillatoria agardhii*. PhD thesis. University of Amsterdam, Netherlands.

Zhang, S. and D.A. Bryant. 2011. The tricarboxylic acid cycle in cyanobacteria. Sci. 334: 1551–1553.

Microalgae III: Stress Response and Wastewater Remediation

Introduction

Microalgae are most successful photoautotrophic communities that not only survive the primitive anoxygenic to oxygenic, extreme environments like deserts, hot springs, snow and polar ice (Hu et al. 2008, Gimpel et al. 2015, Wan et al. 2019); but are also the best adapted to various physical, chemical and nutritional stresses involving carbon, nitrogen, phosphorous, iron and others. It is notable that stresses induce various biochemical changes, often leading to alternation in cell composition through accumulation of either lipids or polysaccharides or other compounds to resist the negative impact of the stress.

Physical stresses like light intensity, temperature and turbulence mixing generally does not affect the growth much, but after a certain level they become a growth limiting factor for microalgal growth. High light intensity beyond the saturated value limits the photosynthesis and further starts to decline to temporarily suspended carbon fixation, known as photoinhibition. To protect the deleterious effect of UV light, some microalgae increase the accumulation of carotenoids and phycobiliproteins. Microalgae prefer medium range temperature, thus high temperature induces the accumulation of lipids, polyols or amino acids in their cell to overcome the stress. Turbulence does not directly affect the microalgal cells, but alters the availability of light intensity and nutrient status in a particular region of oceans.

The carbon source CO_2 is abundant in the environment, it is not growth limiting, but physical stresses often affect CO_2 fixation, ultimately affecting the uptake and assimilation of other nutrients. Nitrogen and phosphorous are limited, so their further scarcity or deficiency limit the growth of the microalgal cell, thus microalgae either uptake variants of these elements or those stored in the cell for nutrient-scarce conditions. Cyanobacteria are known to develop such modified cell 'heterocysts' that fix the atmospheric nitrogen, which might be an indication for future development of nitrogen fixing organelle like carbon fixing organelle. The primary response of microalgal cells to nutrient stress is the accumulation of lipids, polyols, amino acids and secondary carotenoids.

The phototrophic nature, minimum nutritive requirement and further different strategies and responses of microalgae to various stresses, makes them a potential candidate in wastewater remediation. Microalgae mediated wastewater remediation paves the way for affordable, economical and eco-friendly remediation approach that could be a game changer in the arena of wastewater treatment. They sequestrate a huge amount of nutrients like inorganic N & P form the municipal wastewater in an efficient manner, which often needs a lot of energy and modern techniques. Their survival in extreme conditions even in toxic industrial wastewater and their synergy with heterotrophic bacteria could be useful in adequate treatment of these wastewaters.

Physical Stress

Microalgae are well adapted to sequestrate the maximum resources from their environments; this could be through some structural changes, storage of the resources or enhancing resource utilization efficiency. In response to stresses, there are some biochemical and physiological changes that occur inside the cell, which sometimes in quite resilient scenarios could produce various extracellular compounds to render maximum nutrients that are available or limit the growth of competitors (Table 3.1). Physical stresses such as light, temperature, turbulence affect the nutrient uptake and change the strategy to further absorption of these short supplies of nutrients.

Light intensity

Light intensity can affect the biochemical composition of microalgal cells, and in response to this, the process known as photoacclimation or photoadaptation is adapted by the microalgae. The photoacclimation or photoadaptation induces some changes in the ultrastructure of the cell, and can also affect the biophysical and physiological properties, which leads to dynamic changes in cell composition of microalgae to support photosynthesis and growth (Dubinsky et al. 1995). It has been observed that in low light intensity, there is an increase in chlorophylls (a, b, c), phycobiliproteins and primary carotenoids) that are directly involved in photosynthesis. In response to high light intensity, chlorophylls (a, b, c), phycobiliproteins and primary carotenoids decrease, but increase in the secondary carotenoids (e.g., zeaxanthin, β-carotene, astaxanthin), which act as photoprotective pigments. These photoprotective carotenoids often start to gather in some special structures like cytoplasmic lipid bodies or plastoglobuli of plastids (Ben-Amotz et al. 1982, Vechtel et al. 1992); and filter out excess light intensity to protect the photosynthetic pigments. Besides high light intensity, carotenoid accumulation in the cells might occur in the case of change in C and N supply during various stresses.

It is reported that high light intensities induce the production of polysaccharide in microalgal cells. Friedman et al. (1991) and Tredici et al.

Table 3.1: Physical and nutritional stress and responses by the microalgae

Stress	Conditions	Microalgae	Responses	References
Light intensity	Dark	*Dunaliella virdis*	Increase in total lipid content Decrease in free fatty acids, alcohol, sterol	Gordillo et al. 1998
	Limited	*Nannochloropsi ssp.*	Increase in lipid content. Increase in EPA proportions Enhanced	Sukenik et al. 1989
	Red	*Porphyridium cruentum*	Photosystem II relative to	Cunningham et al. 1990
	Blue	*Chlorella vulgaris*	Photosystem I and phycobilisome	Miyachi and Kamiya 1978
		Chlorella vulgaris	Increase in sucrose and starch formation Increase in lipid fraction and alcohol-water insoluble non carbohydrate fraction	Miyachi and Kamiya 1978
Temperature	High	*Botryccoccus braunii*	Decrease in intracellular lipid content from 22% to 5% wt. Accumulation of polysaccharides	Kalacheva et al. 2002
		Chlorella vulgaris	Decrease in starch resulting in increase in sucrose	Nakamura and Miyachi 1982; Nakamura and Imamura 1983
		Chlorococcum sp. *Haematococcuspluvialis* *Nitellamucronata Miquel*	Transformation of L starch (high molecular weight) to S starch (low molecular weight) Reversible with temperature Two fold increase in total carotenoid content 3-fold increase in astaxanthin formation Increase in velocity of cytoplasmic streamin	Liu and Lee 2000 Tjahjono et al. 1994 Raven and Geider 1988

(Contd.)

Table 3.1: (*Contd.*)

Stress	Conditions	Microalgae	Responses	References
pH	Low	Chlamydomonasacidophila Coccochlorispeniocystis	Denaturation of V-lysin Decrease in total accumulated carbon and oxygen evolution	Visviki and Palladino 2001 Coleman and Colman 1981
Carbon	High pH	Chlamydomonasreinhardtii	Inefficient accumulation of carbon. High supply of carbonates required to maintain photosynthetic activity	Moroney and Tolbert 1985
	Elevated	Dunaliellasalina Spirulinaplatensis	increase in amount of fatty acid (dry weight basis) Increase in carbohydrate content; Decrease in proteins and pigments	Muradyan et al. 2004 Gordillo et al. 1998
Nitrogen	Limitation	Phaeodactylumtricornutum Haematococcuspluvialis	Increase in lipid synthesis from 7.90% to 15.31% [80] Increase in lipid synthesis; Decrease in protein content	Morris et al. 1974 Borowitzka et al. 1991
	Starvation	Nannochloropsisoculata Chlorella vulgaris	Increase in lipid synthesis from 5.90% to 16.41% [80] Increase in carotenoid formation (13% w/w)	Converti et al. 2009 Converti et al. 2009
Phosphorous	Limited	Chlamydomonasreinhardtii Ankistrodesmusfalcatus	Decrease in phosphatidylglycerol Decrease in chl a and protein; Increase in carbohydrate and lipids	Sato et al. 2000 Kilham et al. 1997; Healey, 1982

	Starvation	*Selenastrum minutum*	Reduced rate of respiration; Decreased photosynthetic CO_2 fixation	Theodorou et al. 1991
Iron	Limitation	*Dunaliella tertiolecta*	Decrease in cellular chlorophyll concentration	Greene et al. 1992
	Increased conc.	*Chlorella vulgaris*	High concentration of iron	Liu et al. 2008
		Haematococcus pluvialis	Increase in lipid content Increase in carotenoid formation	Kobayashi et al. 1993

(1991) investigated the polysaccharide production in *Porphyridium aerugineum* and *Spirulina platensis* respectively and both reported that there is a significant growth in polysaccharide production with an increase in light intensity. On the basis of a number of studies, Cohen (1999) suggested that light intensity also affects the lipid content mainly the proportion of polyunsaturated fatty acids (PUFA) such as eicosapentaenoic acid (EPA) of the cell. It is reported that in response to low light intensity, *Nannochloropsis* cells show high lipid content and high proportions of EPA, while some species accumulate saturated or monounsaturated (MUFA) fatty acids in response to high light intensity (Sukenik et al. 1989). Burner et al. (1989) stated that the low-light induced accumulation of PUFAs coupled with production of thylakoid membranes as they contain PUFAs as their major constituent; so in low-light conditions, as the production of total thylakoid membranes increases in the cells, is usually responsible for the enhanced production of PUFAs. It was also observed that some microalgal species showed increased PUFA levels in high light conditions; but this might be due to enhancing the desaturation of oxygen-mediated lipids in high light conditions.

Fauchot et al. (2000) observed that the presence of harmful UV-B radiation have a negative effect on the nitrogen uptake in phytoplanktonic communities of microalgae. It is also reported that if the UV-B radiation filter out, then it enhances the uptake rates of the nitrogen forms i.e., NO_3^-, NH_4^+ and urea.

Temperature

Temperature can influence the biochemical composition of microalgal cells, it has been studied that temperature has a significant effect on the composition of membrane lipids and their contents. Nishida and Murata (1996) observed that under low temperature (below an optimal level), there is an increase in the level of unsaturated fatty acids in membrane systems, leading to an enhancement of stability and fluidity of membranes, especially thylakoid membranes; resulting in the protection of the photosynthetic apparatus from photoinhibition at low temperature conditions. Murata (1989) observed that low-temperature induced the desaturation of fatty acids in *Synechocystis* sp. PCC6803; the reason for this is upregulation of the gene expression encoding the enzyme acyl–lipid desaturases. It is suggested that temperature in between a physiologically tolerant range, has more effect on lipid classes and/or their relative composition within a lipid class compared to total lipid content of the cells. Thompson et al. (1992) investigated this trend in several marine phytoplanktonic species and reported that with change of temperature, there is no such significant effect on the total lipid content in the cell. Below the optimal level, temperature also induces the enzyme production as an adaptive mechanism to maintain and regulate the photosynthesis- and respiration- rates.

Further there is an accumulation of polyols, amino acids or amino acid derivatives in response to low temperature, which could help the microalgae cell to acquire sensitivity or tolerance against chilling. Tjahjono et al. (1994) and Liu and Lee (2000) observed the incidence of total carotenoids enhancement in *Haematococcus* sp. and *Chlorococcum* sp. with an increase in temperature, this is mainly due to an astaxanthin formation in the cells. Tjahjono et al.(1994) suggested that microalgal cells might generate the active oxygen radicals in response to higher temperatures, leading to oxidative stress-induced carotenogenesis; or it could secure the temperature-dependent enzymatic reactions that participate in carotenogenesis (Liu and Lee 2000). Goldman and Mann (1980), Rhee (1982) and Harris (1988) suggested that temperature might have some influence on the carbon and nitrogen reserves of the cell and also on the cell volume. There is an incidence of minimal sized microalgal cell with the least amount of carbon and nitrogen contents at an optimal growth temperature, but deviating below or above from that optimal temperature might lead to the enhancement of the cell volume with increased carbon and nitrogen content. Darley (1982) emphasized that at a non-optimal temperature, more carbon and other essential nutrients were needed to maintain the same cell growth rate of the microalgal cell.

Salinity

Several microalgae have the capability to accumulate some osmoregulatory substances or osmoticants to cope with the increasing salinity or osmotic pressure in surroundings. Polyols are the main osmoticants that are commonly found in microalgal species. Some of the common microalgal polyols are glycerol, sorbitol, mannitol, sucrose, galactitol, trehalose and glycerol galactoside. Brown and Borowitzka (1979) observed that the glycerol level could reach up to 50% of the dry weight in *Dunaliella* sp. grown in elevated salinity conditions. It may be possible that under elevated salinity, accumulation of glycerol somehow interrelated with the breakdown of starch in the microalgal cell, but there is no clarity about the regulation of molecular mechanism that is responsible for carbon distribution between starch and glycerol (Brown and Borowitzka 1979).

Iwamoto and Sato (1986) observed that under elevated salinity conditions, *Monodus subterraneus* might induce an increase in the total lipid content, but there is a decrease in the ratio of unsaturated fatty acid EPA in total fatty acids content. Borowitzka and Borowitzka (1988) also reported an increase in total lipid content in *Dunaliella* sp. under elevated salinity conditions. Cifuentes et al. (2001) suggested that elevated salinity might have a very limited influence on the carotenogenesis of microalgae, while Borowitzka and Borowitzka (1988) reported a significant growth in carotenoid content of *Dunaliella* sp. with an increase in level of salinity.

Turbulence

Microalgal communities are significantly affected by turbulence as it brings the nutrient-rich water to the upper layers where microalgal communities reside (Ebert et al. 2001, Huisman et al. 2002), and impact the light availability in the oceanic water column (Yu et al. 2015), also facilitating the possible encounter between microalgal communities and their predators (Peters and Marrase 2000). Smayda and Reynolds (2001) suggested that small-scale turbulence could play a decisive role for different communities of microalgae, from which one community particularly dominates over the other in a dynamic environment. Sullivan et al. (2003) observed that small-scale turbulence is a frequent phenomenon in the water column, which might affect the physiology of microalgal communities.

Savidge (1981), Lazier and Mann (1989) and Karp-Boss et al. (1996) emphasized the positive outcomes of small-scale turbulence on microalgal communities, which include the changing diffusive sub-layers and regulating nutrient fluxes of cells, while Sullivan et al. (2003), Estrada and Berdalet (1997), Karp-Boss et al. (2000) and Juhl and Latz (2002) gave accounts of the negative impacts that include behavioral changes, mechanical damages and physiological impairment. Comparing many studies, Sullivan and Swift (2003) suggested that there was great diversity in the microalgal responses to turbulence; some cyanobacterial species might have some specific adaptive mechanisms to small-scale turbulence e.g., non-heterocystous *Oscillatoria* sp. showed more shear tolerance than the heterocystous *Anabaena* sp. (Reynolds et al. 1983, Steinberg and Tille-Backhaus 1990). Based on seasonal studies of mixed microalgal communities in field- and enclosure-experiments, Reynolds et al. (1983), Steinberg and Tille-Backhaus (1990), Karp-Boss et al. (1996) and Petersen et al. (1998) investigated microalgal abundance or activities can be increased or decreased by the shear effect.

Among the microalgal world, cyanobacterial growth strategies to the small-scale turbulent conditions depend up on the intensity of turbulence and their synergy with other environmental factors (Oliver and Ganf 2000, Guven and Howard 2006, Li et al. 2013, Xiao et al. 2016). In response to turbulence, cyanobacteria shows a variety of adaptive strategies such as light tolerance, buoyancy, N_2-fixation and efficient carbon fixation and phosphorous uptake (Paerl 1983, Visser et al. 1997, Sternar and Elser 2002, Burford et al. 2006, Shen and Song 2007, Klausmeier et al. 2008, Sommaruga et al. 2009), which could provide cyanobacteria an advantage over the other competitors.

Nutritional Stress

Both the deficiency and oversupply of C, N, and P could be problematic for the microalgal cell that could lead to stress (Table 3.1) and reduces the growth. If the growth rates of cell are plotted as a function of the nutrient concentrations, four states are recognized, i.e.:

(i) Deficient state: nutrients are in low concentrations, if nutrients are provided, the growth dramatically increases;

(ii) Transition state: nutrients are in their critical or optimal concentration, if further nutrients are added, the growth is affected a little;

(iii) Adequate state: no increase in the growth on addition of nutrients; luxury storage could be possible at these concentrations;
(iv) Toxic zone: reduction in growth observed on addition of nutrients.

The adequate state is generally associated with the supply of macronutrients, but not for the supply of micronutrients, because they are always limited in environments. The deficient state is associated more with the micronutrients, if their concentration expands then microalgal growth dramatically increased, but if the concentration is decreased then growth is retarded in the deficient state, which could further lead to the dominance of alien species and severe microbial infections by the bacteria, fungi and viruses; resulting in the total devastation of microalgal communities in that environment.

Nitrogen

Being an essential component, nitrogen is an integral part of all structural and functional structures of the cells and it could make up about 7–10% of cell dry weight. Microalgal cells have a limited capability to form nitrogen storage compounds in nitrogen-rich conditions, but with some exceptions like many cyanobacterial species in which nitrogen is stored as cyanophycin and phycocyanin (Simon 1971, Boussiba and Richmond 1980). In nitrogen-limited conditions, phycobilisomes undergo active and specific degradation to cope with this situation (Collier and Grossman 1992). The photosynthesis process is slowed up to the nitrogen level in the cell, if it goes below a threshold value; fixed carbon is diverted from the path of protein synthesis to speed up the synthesis and accumulation of either lipid-or carbohydrate.

Thompson (1996) reported that nitrogen-deprivation induced the accumulation of neutral lipids in the form of triacylglycerols instead of polar lipids in nitrogen-rich conditions. Borowitzka and Borowitzka (1988) observed that many microalgal species prefer carbohydrate accumulation in their cells instead of lipids under nitrogen-limited conditions, e.g., many *Dunaliella* strains accumulate large quantities of glycerol and mono-, di- and polysaccharides under nitrogen starvation. It is not clear whether nitrogen-starvation induces accumulation of neutral lipids or carbohydrates is species-specific and further if it has any physiological significance. Like in *Chlorella* genus, some strains prefer to accumulate neutral lipids, while others showed the accumulation of starch in nitrogen-starved conditions (Richmond 1986).

Besides carbohydrate and lipid accumulation, many microalgae like *Dunaliella, Haematococcus* accumulate secondary carotenoids in their cells under nitrogen-limited conditions (Ben-Amotz et al. 1982, Borowitzka et al. 1991; Zhekisheva et al. 2002); this is usually associated with reduction in the chlorophyll content of the cells. Nitrogen-starved *Dunaliella* cells prefer to accumulate β-carotene (Ben-Amotz et al. 1982). It is also observed that nitrogen-starved conditions induce the formation and accumulation of astaxanthin and its acylesters in *Haematococcus pluvialis* (Borowitzka et al. 1991). Zhekisheva et al. (2002) observed the simultaneous production of astaxanthin and fatty acids, particularly triacylglycerol in nitrogen-limited conditions; and further suggested that there is some interrelation between these processes, because of oil globules that might help in maintaining the high level of astaxanthin esters in the cells.

Phosphorus

Phosphorous is another essential component that plays a significant role in many metabolic processes that are necessary for the growth and development of the microalgal cell. In nutrient- sufficient environments, the microalgal cell can contain the phosphorous level up to 1% of dry weight (Goldman 1980). The incorporation of phosphorous into organic components of the cell occurs in the form of orthophosphate, the process is known as phosphorylation. Phosphorous is also present inorganically in the cell, and is found in the form of polyphosphates in polyphosphate granules. These polyphosphate granules act as phosphorous storage structures, often visible in the cells in phosphate-sufficient conditions, but they start to disappear under phosphorous-limited conditions (Healey 1982).

Like in nitrogen-deprived conditions, microalgal cells show similar responses and metabolic changes on facing phosphorous-limited conditions. In phosphorous depletion, a little degradation of phycobilisome occurs rather than total degradation of phycobilisome in nitrogen depletion, it might be due to loss of phycobilisome in cell division and cessation of synthesis of new phycobilisome (Collier and Grossman 1992). Healey (1982) reported that both cyanobacterial and microalgal cells start to accumulate the carbohydrate content, while there is a decrease in chlorophyll a content. Ben-Amotz et al. (1982) reported the accumulation of β-carotene in *Dunaliella* cells on facing phosphorous depletion, while there is accumulation of astaxanthinin *Haematococcus* cells (Boussiba et al. 1992), but the level of accumulation is as much as observed in nitrogen deficiency.

Iron

Iron is considered an essential trace element which acts as an integral part of metabolic processes of the cell. Due to its redox properties, it plays an important role in fundamental processes such as photosynthesis, nitrogen fixation, respiration and DNA synthesis. Iron-limited conditions induce various biochemical changes, leading to either the degradation of c-phycocyanin and chlorophyll a content (Hardie et al. 1983) or the accumulation of iron-stress-induced protein (isiA) that is primarily associated with the PS I (Bibby et al. 2001). Bibby et al. (2001) further suggested that on facing iron starvation, these isiA proteins are responsible for the formation of isiA proteins complex in some cyanobacteria. These isiA proteins complex are responsible for a significant increase in the size of the light-harvesting system of PS Ias which consists of a ring of 18 isiA molecules around a PS I trimer, that may enhance the flexibility of light-harvesting systems in cyanobacteria. This might be a cyanobacterial strategy to compensate the reduction of phycobilisome and lowering PS I levels on facing iron deficiency.

In low Fe availability, many cyanobacteria and microalgae show a decrease in the ferredoxin level, an iron-containing component of the electron transport chain, but there is a subsequent increase in flavodoxin level, a non-iron-containing electron carrier (Bottin and Lagoutte 1992, McKay et al. 1999). Further high Fe availability may also induce an oxidative stress that could lead to various physiological changes. There could be a reaction of Fe^{2+} with H_2O_2 (produced by the microalgal cells themselves), which take place inside the cell to produce hydroxyl radicals (OH·) through the Fenton reaction:

$$Fe(III) + O_2^- \rightarrow Fe(II) + O_2^-$$

$$H_2O_2 + Fe(II) \rightarrow OH^- + OH^- + Fe(III)$$

It is reported that the accumulation of iron (Fe^{2+}) enhances acetate-induced antioxidant astaxanthin formation in *Haematococcus pluvialis* (Kobayashi et al. 1993, Kobayashi 2000). Kobayashi et al. (1993) further suggested that Fe^{2+} might act as an OH^- generator through the Fenton reaction, and these OH^- or other active oxygen species ($^1O_2$, O_2^-, H_2O_2, or AO_2^-) could be responsible for enhancing the carotenogenesis in the microalgae.

Synergistic Effects of Combinations of Physical and Nutritional Stress

Microalgae show specific strategies in response to a particular stress, and further the synergistic effects of combinations of physical and nutritional stress on the composition of the microalgal cell is also required. In relation to this, there are many studies, but their effectiveness is not satisfactory. For example, Cifuentes et al. (2001) studied the effect of the increasing concentration of salinity (1–30% (w/v) NaCl) on the growth and pigment content of strains of *Dunaliella* sp.; and reported that there is no increase in the carotenoids (orange or red) pigmentation. The ratio between total carotenoid and chlorophyll is found to be always lower or equal to 1.0. Borowitzka and Borowitzka (1988) studied the combined effect of increasing salinity with high light intensity and high temperature in *Dunaliella* sp., and found the high β-carotene (carotenoid) production in the microalgal cell.

Steinbrenner and Linden (2001) presented a clear account on the synergistic effects of a multiple combination of physical, chemical and nutritional stresses on the microalgal cell pigmentation. Steinbrenner and Linden (2001) studied the production of stress-induced astaxanthin in the microalgae *Haematococcuspluvialis*, and found moderate synthesis and accumulation of astaxanthin in response to a single stress factor (either high light, salt or iron); but in facing a combination of two stress factors (high light plus salt, high light plus iron or salt plus iron), there is dramatic and sustained increase in pigment synthesis. When *Haematococcus pluvialis* culture is subjected to the combination of all three stress factors (high light plus salt plus iron), it leads to the highest production of astaxanthin.

Microalgae Mediated Wastewater Remediation

Microalgae show a great ability to survive in most polluted environments that have large amounts of nutrients, a strong color and odor, persistent nature compounds and toxic substances. There are many merits in using microalgae over other organisms such as bacteria: autotrophic nature, oxygen evolution, cell size, biomolecules excretion and a good relationship with heterotrophic bacteria that are more efficient in degrading persistent nature compounds (Fig. 3.1).

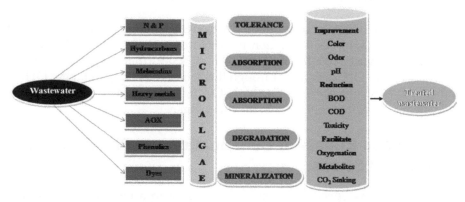

Figure 3.1: Microalgae mediated wastewater remediation.

Nutrient Sequestration

Microalgae have proved to very efficient in sequestration and assimilation of nutrients nitrogen and phosphorous from wastewater (Larsdotter 2006). The microalgal biomass consists of 6.6% nitrogen and 1.3% phosphorous in dry weight with a molar N: P ratio of 11.2 (Chisti 2013). Based on the N and P content, it is assumed that 0.3–1.15 g L⁻¹ of dry weight microalgal biomass can be generated in wastewater. Following this presumption, it is calculated that with 1 kg microalgal biomass it could be possible to sequestrate approximately 66 g of dissolved nitrogen and 13 g of phosphorous from wastewater, resulting in the generation of 0.7–3.0 m³ of cleaned wastewater. Further facing low N:P ratios, microalgae have the ability to sequestrate phosphorous in excess (luxury uptake); and stored in the cells as polyphosphate (Larsdotter 2006, Powell et al. 2009). This stored polyphosphate form of phosphorous can be used in times of phosphorous limitation and starvation conditions, which further increases the nutrient removal capacity of microalgae to provide clean wastewater. For the sequestration of nutrients, microalgae use some mechanisms that are helpful in significant reduction of nutrient concentration in wastewater:

(a) Direct removal, which involves sequestration of these nutrients in the cell through diffusion, which depend on the thickness of the surrounding wastewater column; and turbulence mixing (either natural or by the paddle wheel) could improve the mass transfer rate of the nutrients (Borowitzka 1988). In this route nitrogen forms nitrate (NO_3^-), nitrite (NO_2^-) or urea are assimilated into the cell, while some cyanobacteria have the heterocysts to perform atmospheric nitrogen fixation (Bhaya et al. 2002).

(b) Indirect removal: involves nutrient removal by microalgae mediated pH level elevation to precipitate- and strip- out the nutrients. Phosphorous is the primary nutrient which is precipitated out through the elevated pH level (Brown and Shilton 2014). In the absence of CO_2, microalgae also utilizes the bicarbonate ions (HCO_3^-) that further converts into CO_2 by the carbonic anhydrase enzyme (Borowitzka 1988, Oswald 1988), leading to an increase in pH. This elevated pH could change water chemistry and help in precipitation of phosphorous. The most preferable nitrogen form i.e., ammonia (NH_4^+) also strips through the result of the elevated pH.

O_2 release and CO_2 fixation

Through photosynthetic oxygen evolution, microalgae can provide a sustainable and eco-friendly aeration method, which can be helpful to reduce the need of mechanical aeration. It is estimated that mechanical aeration enhances the cost of the treatment, which could be nearly 50% of the total treatment cost (Tchobanoglous et al. 2003). Oswald (2003) estimated that the removal of 1000 mg BOD requires about 1 kWh of electricity, while in case of microalgal photosynthetic oxygenation needs no energy input. Further it could produce the microalgal biomass which can be used to produce biogas that is enough to generate one kWh of electricity (Oswald 2003).

Besides aeration, microalgal oxygen could provide the required oxygen to the aerobic heterotroph bacteria which is needed in the degradation of complex organic compounds and also helps in nitrogen removal through nitrification and denitrification processes. Microalgae are also able to sequestrate the CO_2 that is released through bacterial respiration; resulting in the reduction of emission of CO_2 into to the atmosphere. The ability of microalgae to fix CO_2 is about 10% more than for terrestrial plants. Hu et al. (2008) estimated that microalgae need 1.7 kg of CO_2 for obtaining 1 kg dry weight biomass. Due to this, microalgae-mediated remediation could be very useful in reducing the overall CO_2 emissions from wastewater treatment facilities.

COD reduction

Some microalgae species such as *Chlorella* sp. are able to switch the nutrition mode in response to environmental conditions. They utilize CO_2 as a main carbon source under phototrophic conditions, while they also survive in heterotrophic conditions by changing their nutrition preferences, using sugars, acetate and organic acids as an alternative carbon source (Borowitzka 1998). For the heterotrophic growth, microalgae could use the following two strategies:

(a) In dark conditions and low CO_2 concentration, microalgae are able to uptake dissolved organic carbon as their carbon source as well energy source, known as chemohetertrophic mode of nutrition; leading to reduction in COD level.

(b) In light conditions and low CO_2 concentration, microalgae use dissolved organic carbon as a carbon source but for the energy source they prefer light as an energy source, known as photohetertrophic mode of nutrition; again COD level decreases.

Absorption and degradation of toxic compounds

Industrial wastewater often contains heavy metal and a diverse range of toxic compounds such as pesticides, phenols, azo dyes and chlorinated compounds. Microalgae are found to be very efficient in the removal of heavy metal, which later either assimilates or is excreted in a non-toxic form in the environment. In relation to toxic compounds, microalgae adapt several strategies such as adsorption, accumulation and metabolization of these compounds in to less toxic compounds. Several persistent toxic compounds like melanoidins, lignin and PAHs are found quite difficult for the microalgal species to degrade, but together with heterotrophic bacteria, they can play a direct or indirect role in degradation of such compounds.

Indicator bacteria removal

Wastewater treatment faces the presence of pathogenic microscopic organisms including viruses, bacteria, parasites and protozoa; these require disinfection in the last stage of the wastewater treatment procedure. The performance of disinfection is usually calculated on the basis of the removal rate of indicator organisms such as total coliforms. It is suggested that reduction in the number of coliform bacteria could be dependent on different physical-biochemical reactions, microalgae concentration, organic matter content and solar irradiation. There are many hypotheses available in relation to the role of microalgae in bacterial reduction, this could include: excretion of antibacterial substances by the microalgae, high pH conditions in microalgae-mediated wastewater remediation, production of toxic extracellular compounds by certain species of microalgae, depletion of nutrients and organic matter in wastewater, and high oxidation reduction potential in microalgal-bacterial cultures.

Conclusion

Microalgae show extensive responses and strategies against various stresses such as light intensity, temperature, turbulence, salinity and nutrient limitation or starvation. In response to light intensity, microalgae tend to accumulate secondary carotenoids or polyols, while temperature induces the increase in the lipid content of the cell. But small scale turbulence have no such negative effect on the microalgal growth, it favors seasonal excess growth by providing essential nutrients via upwelling the nutrients in the upper layer. In response to nutrient stress, such as nitrogen and phosphorous, microalgae tend to increase their carbohydrate or lipid content, while some species accumulate secondary carotenoids or polyols in their cell.

A numbers of studies have reported the well-adapted strategy of microalgae that involve changing their cell composition and accumulation of lipids, polyols and polysaccharides to overcome stress. But there are few studies that are directly related to the evolutionary adaptation of microalgae to overcome stress tolerance, further none of these studies specify the stress-induced changes in the genomes of these adapted microalgal species. Use of next gen sequencing and directed experimental evolution approaches with such studies would be helpful in improving and understanding of the microalgal response and adaptation to various stresses.

This ability of microalgae to sustain and survive in a wide range of pH, temperature, light intensity, different carbon sources and nutrient limited or starved conditions, makes them a potential organism in the remediation of various municipal and industrial wastewater. They could solve major problems of conventional wastewater treatments like high pH, excess inorganic ions, metals and the presence of indicator organisms. Besides the essential nutrients, wastewater also contains compounds such as heavy metals and xenobiotics that are toxic for microalgal growth; microalgae are also found in absorption and degradation of such toxic compounds. Further their spontaneous auto-flocculation in the late stationary phase of their growth cycle could be easier for separation of their biomass for successive wastewater treatment.

References

Ben-Amotz, A., A. Katz and M. Avron. 1982. Accumulation of b-carotene in halotolerant algae: purification and characterization of b-carotene-rich globules from *Dunaliella bardawil* (Chlorophyceae). J. Phycol. 18: 529–537.

Bhaya, D., R. Schwarz and A.R. Grossman. 2002. Molecular responses to environmental stress. pp. 397–442. *In*: B.A. Whitton and M. Potts (eds.). The Ecology of Cyanobacteria. Kluwer Academic Publishers, Netherlands.

Bibby, T.S., J. Nield and J. Barber. 2001. Iron deficiency induces the formation of an antenna ring around trimeric photosystem I in cyanobacteria. Nature 412(6848): 743–745.

Borowitzka, M.A. and L.J. Borowitzka. 1988. Dunaliella. pp. 27–58. *In*: M.A. Borowitzka and L.J. Borowitzka (eds.). Microalgal Biotechnology. Cambridge University Press, Cambridge, UK.

Borowitzka, M.A., J.M. Huisman and A. Osborn. 1991. Cultures of the astaxanthin producing green alga *Haematococcus* pluvialis. I. Effect of nutrient on growth and cell type. J. Appl. Phycol. 3: 295–304.

Bottin, H. and B. Lagoutte. 1992. Ferredoxin and flavodoxin from the cyanobacterium *Synechocystis* sp. PCC 6803. Biochim. Biophys. Acta 1101: 48–56.

Boussiba, S. and A. Richmond. 1980. C-phycocyanion as a storage protein in the bluegreen alga *Spirulina platensis*. Arch. Microbiol. 125: 143–147.

Boussiba, S., F. Lu and A. Vonshak. 1992. Enhancement and determination of astaxanthin accumulation in the green alga *Haematococcus pluvialis*. Methods Enzymol. 213: 386–391.

Brown, A.D. and L.J. Borowitzka. 1979. Halotolerance of *Dunaliella*. pp. 139–90. *In*: M. Levandowsky and S.H. Hunter (eds.). Biochemistry and Physiology of Protozoa, Vol. 1, 2nd edition. Academic Press, New York.

Brown, N. and A. Shilton. 2014. Luxury uptake of phosphorus by microalgae in waste stabilisation ponds: current understanding and future direction. Rev. Environ. Sci. Bio/Technol. 13: 321–328.

Burford, M.A., K.L. Mcneale and F.J. Mckenzie-Smith. 2006. The role of nitrogen in promoting the toxic cyanophyte *Cylindrospermopsis raciborskii* in a subtropical water reservoir. Freshwater Biol. 51: 2143–2153.

Burner, T., Z. Dubinsky, K. Wyman and P.G. Falkowski. 1989. Photoadaptation and the package effect in *Dunaliella tertiolecta* (Chlorophyceae) J. Phycol. 25: 70–78.

Chisti, Y. 2013. Constraints to commercialization of algal fuels. J. Biotechnol. 167: 201–214.

Cifuentes, A.S., M.A. González, I. Inostroza and A. Aguilera. 2001. Reappraisal of physiological attributes of nine strains of *Dunaliella* (Chlorophyceae): growth and pigment content across a salinity gradient. J. Phycol. 37: 334–344.

Cohen, Z. 1999. *Porphyridium cruentum*. pp. 1–24. *In*: Z. Cohen (ed.). Chemicals from Microalgae. Taylor & Francis Ltd, London, UK.

Coleman, J.R. and B. Colman. 1981. Inorganic carbon accumulation and photosynthesis in a blue-green alga as a function of external pH. Plant Physiol. 67: 917–921.

Collier, J.L. and A.R. Grossman. 1992. Chlorosis induced by nutrient deprivation in *Synechococcus* sp. strain PCC 7942: not all bleaching is the same. J. Bacteriol. 174: 4718–4726.

Converti, A., A.A. Casazza, E.Y. Ortiz, P. Perego and M. delBorghi. 2009. Effect of temperature and nitrogen concentration on the growth and lipid content of *Nannochloropsis oculata* and *Chlorella vulgaris* for biodiesel production. Chem. Eng. Process. 48: 1146–1151.

Cunningham, F.X. Jr., R.J. Dennenberg, P.A. Jursinic and E. Gantt. 1990. Growth under red light enhances photosystem II relative to photosystem I and phycobilisomes in the red alga *Porphyridium cruentum*. Plant Physiol. 93: 888–895.

Darley, W.M. 1982. Algal Biology: A Physiological Approach. Blackwell Science Inc., London.

De la Noüe, J., G. Laliberté and D. Proulx. 1992. Algae and waste water. J. Appl. Phycol. 4: 247–254.

Dubinsky, Z., R. Matsukawa and I. Karube. 1995. Photobiological aspects of algal mass culture. J. Mar. Biotechnol. 2: 61–65.

Ebert, U., M. Arrayas, N. Temme, B. Sommeijer and J. Huisman. 2001. Critical conditions for phytoplankton blooms. B. Math. Biol. 63: 1095–1124.

Estrada, M. and E. Berdalet. 1997. Phytoplankton in a turbulent world. Sci. Mar. 61: 125–140.

Fauchot, J., M. Gosselin, M. Levasseur, B. Mostajir, C. Belzile, S. Demers, et al. 2000. Influence of UV-B radiation on nitrogen utilization by a natural assemblage of phytoplankton. J. Phycol. 36: 484–496.

Friedman, O., Z. Dubinsky and S. Arad (Malis). 1991. Effect of light intensity on growth and polysaccharide production in red and blue-green *Rhodophyta unicells*. Bioresour. Technol. 38: 105–110.

Gimpel, J.A., V. Henriquez and S.P. Mayfield. 2015. In metabolic engineering of eukaryotic microalgae: potential and challenges come with great diversity. Front Microbiol. 6: 1376.

Goldman, J.C. 1980. Physiological aspects in algal mass cultures. pp. 343–353. *In*: G. Shelef and C.J. Soeder (eds.). Algal Biomass. Biomedical Press. Elsevier/North-Holland.

Goldman, J.C. and R. Mann, R. 1980. Temperature influenced variations in speciation and the chemical composition of marine phytoplankton in outdoor mass cultures. J. Exp. Mar. Biol. Ecol. 46: 29–40.

Gordillo, F.J.L., C. Jiménez, F.L. Figueroa and F.X. Niell. 1998. Effects of increased atmospheric CO_2 and N supply on photosynthesis, growth and cell composition of the cyanobacterium *Spirulina platensis* (Arthrospira). J. Appl. Phycol. 10: 461–469.

Greene, R.M., R.J. Geider, Z. Kolber and P.G. Falkowski. 1992. Iron-induced changes in light harvesting and photochemical energy conversion processes in eukaryotic marine algae. Plant Physiol. 100: 565–575.

Guven, B. and A. Howard. 2006. Modelling the growth and movement of cyanobacteria in river systems. Sci. Total Environ. 368: 898–908.

Hardie, L.P., D.I. Balkwill and S.E. Stevens Jr. 1983. Effects of iron starvation on the physiology of the cyanobacterium. *Agmenellum-quadruplicatum*. Appl. Env. Microbiol. 3: 999–1006.

Harris, G.P. 1988. Phytoplankton Ecology. Chapman & Hall, New York.

Healey, F.P. 1982. Phosphate. pp. 105–124. *In*: N.G. Carr and B.A. Whitton (eds.). The Biology of Cyanobacteria. Blackwell Scientific Publications, Oxford, UK.

Healey, F.P. and L.L. Hendzel. 1979. Indicators of phosphorus and nitrogen deficiency in five algae in culture. J. Fish. Board Can. 36: 1364–1369.

Hu, Q., M. Sommerfeld, E. Jarvis, M. Ghirardi, M. Posewitz, M. Seibert, et al. 2008. Microalgaltriacylglycerols as feedstocks for biofuel production: perspectives and advances. Plant J. 54: 621–639.

Huisman, J., M. Arrayas, U. Ebert and B. Sommeijer. 2002. How do sinking phytoplankton species manage to persist? Am. Nat. 159: 245–254.

Iwamoto, H. and S. Sato. 1986. EPA production by freshwater algae. J. Am. Oil Chem. Soc. 63: 434.

Juhl, A.R. and M.I. Latz. 2002. Mechanisms of fluid shear-inducted inhibition of population growth in a red-tide dinoflagellate. J. Phycol. 38: 683–694.

Kalacheva, G., N. Zhila, T. Volova and M. Gladyshev. 2002. The effect of temperature on the lipid composition of the green alga *Botryococcus*. Microbiol. 71: 286–293.

Karp-Boss, L., E. Boss and P.A. Jumars. 1996. Nutrients fluxes to planktonic osmotrophs in the presence of fluid motion. Oceanogr. Mar. Biol. 34: 71–107.

Karp-Boss, L., E. Boss and P.A. Jumars. 2000. Motion of dinoflagellates in a simple shear flow. Limnol. Oceanogr. 45: 1594–1602.

Kilham, S., D. Kreeger, C. Goulden and S. Lynn. 1997. Effects of nutrient limitation on biochemical constituents of *Ankistrodesmus falcatus*. Freshw. Biol. 38: 591–596.

Klausmeier, C.A., E. Litchman, T. Daufresne and S.A. Levin. 2008. Phytoplankton stoichiometry. Ecol. Res. 23: 479–485.

Kobayashi, M. 2000. In vivo antioxidant role of astaxanthin under oxidative stress in the green alga *Haematococcus pluvialis*. Appl. Microbiol. Biotechnol. 54: 550–555.

Kobayashi, M., T. Kakizono and S. Nagai. 1993. Enhanced carotenoid biosynthesis by oxidative stress in acetate-induced cyst cells of a green unicellular alga *Haematococcus pluvialis*. Appl. Env. Microbiol. 59: 867–873.

Larsdotter, K. 2006. Wastewater treatment with microalgae: A literature review. Vatten 62: 31–38.

Lazier, J.R.N. and K.H. Mann. 1989. Turbulence and the diffusive layers around small organisms. Deep-Sea Res. 36: 1721–1733.

Li, F.P., H.P. Zhang, Y.P. Zhu, Y.H. Xiao and L. Chen. 2013. Effect of flow velocity on phytoplankton biomass and composition in a freshwater lake. Sci. Total Environ. 447: 64–71.

Li, Y., M. Horsman, B. Wang, N. Wu and C.Q. Lan. 2008. Effects of nitrogen sources on cell growth and lipid accumulation of green alga *Neochlorisoleo abundans*. Appl. Microbiol. Biotechnol. 81: 629–636.

Liu, B.H. and Y.K. Lee. 2000. Secondary carotenoids formation by the green alga *Chlorococcum* sp. J. Appl. Phycol. 12: 301–307.

Liu, Z.Y., G.C. Wang and B.C. Zhou. 2008. Effect of iron on growth and lipid accumulation in *Chlorella vulgaris*. Bioresour. Technol. 99: 4717–4722.

McKay, R.M.L., J. La Roche, A.F. Yakunin, D.G. Durnford and R.J. Geider. 1999. Accumulation of ferredoxin and flavodoxin in a marine diatom in response to Fe. J. Phycol. 35: 510–519.

Miyachi, S. and A. Kamiya. 1978. Wavelength effects on photosynthetic carbon metabolism in *Chlorella*. Plant Cell Physiol. 19: 277–288.

Molina-Grima, E., F. Garcia-Camacho and F.G. Acien-Fernandez. 1999. Production of EPA from *Phaeodactylum tricornutum*. pp. 57–92. *In*: Z. Cohen (ed.). Chemicals from Microalgae. Taylor & Francis Ltd, London, UK.

Moroney, J.V. and N.E. Tolbert. 1985. Inorganic carbon uptake by *Chlamydomonas reinhardtii*. Plant Physiol. 77: 253–258.

Morris, I., H. Glover and C. Yentsch. 1974. Products of photosynthesis by marine phytoplankton: the effect of environmental factors on the relative rates of protein synthesis. Mar. Biol. 27: 1–9.

Muradyan, E., G. Klyachko-Gurvich, L. Tsoglin, T. Sergeyenko and N. Pronina. 2004. Changes in lipid metabolism during adaptation of the *Dunaliella salina* photosynthetic apparatus to high CO_2 concentration. Rus. J. Plant Physiol. 51: 53–62.

Murata, N. 1989. Low-temperature effects on cyanobacterial membranes. J. Bioenerg. Biomembr. 21: 61–75.

Nakamura, Y. and M. Imamura. 1983. Change in properties of starch when photosynthesized at different temperatures in *Chlorella vulgaris*. Plant Sci. Lett. 31: 123–131.

Nakamura, Y. and S. Miyachi. 1982. Effect of temperature on starch degradation in *Chlorella vulgaris* 11 h cells. Plant Cell Physiol. 23: 333–341.

Nishida, I. and N. Murata. 1996. Chilling sensitivity in plants and cyanobacteria: the crucial contribution of membrane lipids. Annu. Rev. Plant Physiol. Plant Mol. Biol. 47: 541–568.

Oliver, R.L. and G.G. Ganf. 2000. Freshwater blooms. pp. 105–194. *In*: B.A. Whitton and M. Potts (eds.). The Ecology of Cyanobacteria: Their Diversity in Time and Space. Kluwer Academic, Dordrecht, Netherlands.

Oswald, W. 1988. Micro-algae and wastewater treatment. pp. 305–328. *In*: M.A. Borowitzka and L.J. Borowitzka (eds.). Micro-algal Biotechnology. Cambridge Univ. Press, Cambridge, UK.

Oswald, W.J. 2003. My sixty years in applied algology. J. Appl. Phycol. 15: 99–106.

Paerl, H.W. 1983. Partitioning of CO_2 fixation in the colonial cyanobacterium *Microcystis aeruginosa*: mechanism promoting formation of surface scums. Appl. Environ. Microb. 46: 252–259.

Peters, F. and C. Marrase. 2000. Effects of turbulence on plankton: an overview of experimental evidence and some theoretical considerations. Mar. Ecol. Prog. Ser. 205: 291–306.

Petersen, J.E., L.P. Sanford and W.M. Kemp. 1998. Coastal plankton responses to turbulent mixing in experimental ecosystems. Mar. Ecol. Prog. Ser. 171: 23–41.

Pittman, J.K., A.P. Dean and O. Osundeko. 2011. The potential of sustainable algal biofuel production using wastewater resources. Bioresour. Technol. 102: 17–25.

Powell, N., A. Shilton, Y. Chisti and S. Pratt. 2009. Towards a luxury uptake process via microalgae—defining the polyphosphate dynamics. Water Res. 43: 4207–4213.

Raven, J.A. and R.J. Geider. 1988. Temperature and algal growth. New Phytol. 110: 441–461.

Reynolds, C.S., S.W. Wiseman, B.M. Godfrey and C. Butterwick. 1983. Some effects of artificial mixing on the dynamics of phytoplankton populations in large limnetic enclosures. J. Plankton Res. 5: 203–234.

Rhee, G.Y. 1982. Effects of environmental factors and their interactions on phytoplankton growth. Adv. Microb. Ecol. 6: 33–74.

Richmond, A. 1986. Microalgae of economic potential. pp. 199–243. *In*: A. Richmond (ed.). Handbook of Microalgal Mass Culture. CRC Press, Inc. Boca Raton, Florida.

Sato, N., M. Hagio, H. Wada and M. Tsuzuki. 2000. Environmental effects on acidic lipids of thylakoid membranes. Biochem. Soc. Trans. 28: 912–914.

Savidge, G. 1981. Studies of the effect of small-scale turbulence on phytoplankton. J. Mar. Biol. Assoc. U.K. 61: 477–488.

Shen, H. and L.R. Song. 2007. Comparative studies on physiological responses to phosphorus in two phenotypes of bloom-forming *Microcystis*. Hydrobiologia 592: 475–486.

Simon, R.D. 1971. Cyanophycin granules from the blue-green alga *Anabaena cylindrica*: a reserve material consisting of copolymers of aspartic and arginine. Proc. Natl. Acad. Sci. U.S.A. 68: 265–267.

Singh, U.B. and A.S. Ahluwalia. 2013. Microalgae: a promising tool for carbon sequestration. Mitig. Adapt. Strateg. Glob. Change 18(1): 73–95.

Smayda, T.J. and C.S. Reynolds. 2001. Community assembly in marine phytoplankton: application of recent models to harmful dinoflagellate blooms. J. Plankton Res. 23: 447–461.

Sommaruga, R., Y.W. Chen and Z.W. Liu. 2009. Multiple strategies of bloom-forming *Microcystis* to minimize damage by solar ultraviolet radiation in surface waters. Microb. Ecol. 57: 667–674.

Steinberg, C.W. and R. Tille-Backhaus. 1990. Re-occurrence of filamentous planktonic cyanobacteria during permanent artificial destratification. J. Plankton Res. 12: 661–664.

Steinbrenner, J. and H. Linden. 2001. Regulation of two carotenoid biosynthesis genes coding for phytoene synthase and carotenoid hydroxylase during stress-induced astaxanthin formation in the green alga *Haematococcus pluvialis*. Plant Physiol. 125: 810–817.

Sterner, R.W. and J.J. Elser. 2002. Ecological Stoichiometry: The Biology of Elements from Molecules to the Biosphere. Princeton University Press, Princeton, New Jersey, US.

Sukenik, A., Y. Carmeli and T. Berner. 1989. Regulation of fatty acid composition by irradiance level in the eustigmatophyte *Nannochlor opsis* sp. J. Phycol. 25: 686–692.

Sullivan, J.M. and E. Swift. 2003. Effects of small-scale turbulence on net growth rate and size of ten species of marine dinoflagellates. J. Phycol. 39: 83–94.

Sullivan, J.M., E. Swift, P.L. Donaghay and J.E.B. Rines. 2003 Small-scale turbulence affects the division rate and morphology of two red-tide dinoflagellates. Harmful Algae 2: 183–199.

Tchobanoglous, G., F. Burton and H. Stensel. 2003. Wastewater Engineering: Treatment and Reuse. McGraw-Hill, New York.

Theodorou, M.E., I.R. Elrifi, D.H. Turpin and W.C. Plaxton. 1991. Effects of phosphorus limitation on respiratory metabolism in the green alga *Selenastrum minutum*. Plant Physiol. 95: 1089–1095.

Thompson, Jr. G.A. 1996. Lipids and membrane function in green algae. Biochem. Biophys. Acta 1302: 17–45.

Thompson, P.A., M. Guo and P.J. Harrison. 1992. Effects of temperature. I. On the biochemical composition of eight species of marine phytoplankton. J. Phycol. 28: 481–488.

Tjahjono, A.E., Y. Hayama, T. Kakizono, Y. Terada, N. Nishio and S. Nagai. 1994. Hyper-accumulation of astaxanthin in a green alga *Haematococcus pluvialis* at elevated-temperatures. Biotech. Lett. 16: 133–138.

Tredici, M.R., P. Carlozzi, G.C. Zittelli and R. Materassi. 1991. A vertical alveolar panel (VAP) for outdoor mass cultivation of microalgae and cyanobacteria. Bioresour. Technol. 38: 153–159.

Vechtel, B., W. Eichenberger and H.G. Ruppel. 1992. Lipid bodies in *Eremosphaera viridis* De Bary (Chlorophyceae). Plant Cell Physiol. 33: 41–48.

Visser, P.M., J. Passarge and L.R. Mur. 1997. Modelling vertical migration of the cyanobacterium *Microcystis*. Hydrobiologia 349: 99–109.

Visviki, I. and J. Palladino. 2001. Growth and cytology of *Chlamydomonas acidophila* under acidic stress. Bull. Environ. Contam. Toxicol. 66: 623–630.

Wan, C., B.-L. Chen, X.-Q. Zhao, and F.-W. Bai. 2019. Stress response of microalgae and its manipulation for development of robust strains. pp. 95-113. *In*: M.A. Alam and Z. Wang (eds.). Microalgae Biotechnology for Development of Biofuel and Wastewater Treatment. Springer Nature Singapore Pvt. Ltd. Singapore.

Xiao, Y., Z. Li, C. Li, Z. Zhang and J. Guo. 2016. Effect of small-scale turbulence on the physiology and morphology of two bloom-forming cyanobacteria. PLoS ONE 11(12): e0168925.

Yu, Q., Y.C. Chen, Z.W. Liu, N. van de Giesen and D.J. Zhu. 2015. The influence of a eutrophic lake to the river downstream: spatiotemporal algal composition changes and the driving factors. Water 7: 2184–2201.

Zhekisheva, M., S. Boussiba, I. Khozin-Goldberg, A. Zarka and Z. Cohen. 2002. Accumulation of oleic acid in *Haematococcus pluvialis* (Chlorophyceae) under nitrogen starvation or high light is correlated with that of astaxanthin esters. J. Phycol. 38: 325–331.

4

Municipal Wastewater

Introduction

Increased municipal wastewater generation is a global reality. As more and more of the population prefer urban centers for better livelihood and standard of living, the problem of municipal wastewater has also increased. It is estimated that 2.1 billion more people will be added to the existing cities by 2030 (UN 2012). AQUASTAT, 2014 and Sato et al. (2013) suggested that more than 330 km^3 year^{-1} of (mostly) municipal wastewater generated globally and only 60% of the produced municipal wastewater is treated. Although these figures could be contradictory for various reasons such as: (a) underperformed wastewater treatment plants in middle and low income countries, (b) non-availability of data of some urban population in countries like Nigeria (AQUASTAT 2014), (c) the definition of treated water as some countries consider only secondary and tertiary treated wastewater as treated wastewater, while in some countries this also includes primary treated wastewater. Further the economic conditions of countries directly affect the capability of wastewater treatment, it was surveyed that on an average 28% of municipal wastewater treated in lower-middle-income countries and it was only 8% in low-income countries, while high income countries treated their wastewater up to 70% (Sato et al. 2013).

This untreated or partially treated municipal wastewater creates many problems in the environment. The wastewater comprises of high loads of inorganic nutrients, leading to excess growth of algae and disrupts the whole aquatic ecosystem by depleting dissolved oxygen and affecting light penetration. Municipal wastewaters also have large populations of pathogenic microbes which cause cholera, diarrhea and act as a breeding ground for vectors such as malaria and dengue, and are collectively responsible for millions of deaths each year.

Conventional wastewater treatment involves preliminary, primary, secondary and tertiary processes. Bar screens are usually applied to get rid of large floating materials that could interfere with the subsequent treatment processes. Then primary clarifiers are used for removal of settable and floating materials (including grease and oils). During this process primary sludge is produced which is sent for a different treatment. After this homogeneous wastewater is transferred to secondary clarifiers, which involves biological degradation of organic matter and secondary sludge is also produced that mixes with primary sludge for further treatment. Finally a tertiary treatment follows which involves sand filtration or membrane filtration to remove residual suspended matter and fine particulates like metals. In the end, the disinfection (use of chlorine, ozone or ultraviolet light) process is carried out to get rid of odor and pathogenic microorganisms from the municipal wastewater. It has been suggested that wastewater treatment facilities usually do not involve the tertiary and disinfection process in poor and middle income countries.

Microalgae mediated-wastewater remediation could provide a new dimension for sequestrating excess nutrients that exist even after the biological process and further it could help in management and handling sludge generated during primary and secondary treatment processes. It has been studied that microalgae could sequestrate or remove N and P through two approaches; (a) directly by uptake and assimilation for their growth, (b) indirectly by facilitating volatilization or precipitation through alkalization of the medium. For better results, microalgae consortia or microalgae-bacteria consortia could be used for effective nutrient removal and less sludge production, microalgae and bacteria tends to support each other by exchanging their respiration byproducts O_2 and CO_2; resulting in less carbon emissions from wastewater treatment.

Further microalgal biofilms could provide many advantages in comparison to suspended microalgal or microalgae-bacterial consortia: (a) these systems are able to maintain the biomass, even while operating on a short hydraulic retention time; (b) much easier harvesting as it needs little or no operation to separate the microalgae from wastewater before discharging the wastewater to the environment (Schumacher et al. 2003, Roeselers et al. 2008); (c) further there is no need of stirring the wastewater medium, which requires less electrical energy as compared to paddle wheel in suspended microalgal systems. However, microalgal biofilm systems have some limitations like photoinhibition and diffusion of nutrients or carbon dioxide (CO_2) in lower layers of biofilms (Liehr et al. 1988).

This chapter highlights the situation and the influence of municipal wastewater on water sources and public health. It also briefly presents an account of treatment methods and their inefficiency in terms of removal efficiency, cost and sustainability. To address these issues, microalgae-mediated wastewater remediation provides us an efficient, sustainable and cost effective solution to solve this situation of municipal wastewater in a manner that could be adapted or implemented worldwide especially in poor and lower income countries.

Municipal Wastewater: Characteristics and Specific Pollutants

Municipal wastewater also referred to as domestic wastewater mostly comprising of wastewater generated from kitchens, baths, laundry and

Table 4.1: Average water quality parameters of municipal wastewater

Water quality parameter	Fito and Alemu 2019	Foladori et al. 2018	Nordlander et al. 2017	Cho et al. 2011	Li et al. 2011	Jiang et al. 2011	De Bashan et al. 2004	Yadav et al. 2002
pH	7.2	7.8	-	-	-	7.5	6.3-7.9	7.4
Conductivity (ms cm⁻¹)	0.633	-	-	-	-	-	1.633	1.74
TS (mg L⁻¹)	-	-	-	50	-	-	0.98-80	900
TDS (mg L⁻¹)	489	-	-	-	-	-	-	900
TSS (mg L⁻¹)	166	143	-	-	0.070	-	-	-
BOD (mg L⁻¹)	341.5	-	166-500	6.9	2304	-	53.5-113	169
COD (mg L⁻¹)	530	292	-	11.2	-	-	-	382
TN (mg L⁻¹)	62	66	26.4-54	18.9	116.1	110.2	55	-
NH₄⁺N (mg L⁻¹)	-	55	-	10	82.5	92	0.1-4.26	3.5
NO₃⁻N (mg L⁻¹)	-	7.1	-	6.6	-	3.9	4-5.18	28
TP (mg L⁻¹)	7.2	2.7	3.5-6.2	1.7	212	5.3	5	26
TOC (mg L⁻¹)	-	-	-	-	-	59.7	-	-

lavatories from residential, institutions or commercial areas; sometimes it also has wastewater from small scale industrial entities operating within residential or commercial areas. Further based on the source, municipal wastewater is categorically divided into two groups i.e. grey water which originates from kitchen sinks, baths, laundry etc., brown; and black water which comes from toilets which is further divided in yellow (urine with or without flush water) and brown water (toilet wastewater without urine) (NWP 2006, Elmitwalli and Otterpohl 2007, Eriksson et al. 2002, Gross et al. 2005, Hernández Leal et al. 2007, Vinnerås 2002).

Municipal wastewater comprises of mainly 99.9% water and 0.1% solid matter. These solids contain about two-third organic compounds and rest are inorganic compounds; that also have some amount of different gases. The organic compounds primarily include portions of carbohydrates, and proteins and fats in a smaller amount; while inorganic compounds are composed of inorganic nitrogen and phosphorus, sulfur, chlorides, heavy metals and other toxic compounds. Both organic and inorganic compounds could be either in a suspended or dissolved form. Most of the organic compounds such as proteins, lipids, lignocellulose, cellulose and some inorganic particulate matter exist in suspended forms; while most of the inorganic compounds such as nitrogen and phosphorous, ions, heavy metals and some organic compounds i.e., carbohydrates, fatty acids, amino acids, alcohols are in a dissolved form. Some gases such as, H_2S, methane, ammonia, O_2, CO_2 and nitrogen also disperse; in which the first three gases

Table 4.2: Specific pollutants exist in municipal wastewater
(Henze and Ledin 2001)

Specific pollutants	Components	Impacts
Biodegradable organic compounds	Oxygen depletion in rivers and lakes	Fish death, odours
Other organic compounds	Detergents, pesticides, fat, oil and grease, colouring, solvents, phenols, cyanide	Toxic effect, aesthetic inconveniences, bioaccumulation in the food chain
Nutrients	Nitrogen, phosphorus, ammonium	Eutrophication, oxygen depletion, toxic effect
Metals	Hg, Pb, Cd, Cr, Cu, Ni	Toxic effect, bioaccumulation
Micro-flora	Pathogenic bacteria, virus and worms eggs	Risk when bathing and eating shellfish
Other inorganic materials	Acids, for example hydrogen sulphide, base	Corrosion, toxic effect
Odour (and taste)	Hydrogen sulphide	Aesthetic inconveniences, toxic effect

produced during decomposition of organic matter is present in municipal wastewater.

Municipal wastewater is also characterized by the presence of a high population of microorganisms such as *Salmonella, Shigella, Vibriocholera* and mainly coliform bacteria, that could be fecal i.e., *E. coli* or non-fecal *Enterobacter, Klebsiella, Citrobacter, Salmonella, Shigella, Vibrio cholera*, which are highly pathogenic in nature. Some pathogenic protozoa (*Entamoeba histolytica*) and helminths (*Ascaris Lumbricoides, Ancylostoma duodenale, Schistosoma mansoni, Taenia saginata*) are also found in raw municipal wastewater and thrive for longer periods even after conventional primary and secondary treatment.

Impact on the Environment and Public Health

It has been well investigated that municipal wastewater has high amounts of nutrients and pathogenic microbes specifically fecal coliforms. This excess nitrogen, phosphorus and other nutrients are discharged into water bodies like lakes, reservoirs and rivers, which induce the robust growth of algae and other planktonic organisms. This process is known as eutrophication, which is responsible for the formation of algae blooms in water bodies. The process eutrophication is described below:

$$106CO_2 + 16NO_3^- + HPO_4^{2-} + 122H_2O + 18H^+ + Energy + Microelement$$

$$\rightarrow C_{106} H_{263} O_{110} N_{16} P(\text{Bioplasm of algae}) + 138O_2$$

Due to eutrophication, the turbidity of water bodies increases and affects sunlight penetration to the deeper layer. On facing eutrophication for longer periods, the population of other biota declines, resulting in an increase in the sedimentation rate; further it develops anoxic conditions and alters the species diversity of aquatic ecosystem (Morrison et al. 2001, Igbinosa and Okoh 2009, Edokpayi et al. 2015, Edokpayi 2016).

The presence of pathogenic microorganisms especially coliforms in municipal wastewater poses a great danger to public health. As this wastewater is normally used in agriculture it furthers contaminates the surface water and in some way the ground water, which is the sole source of drinking water in many developing countries. This enhances the chances of sanitation and wastewater-related diseases in these countries. A study in 2012 suggested that 842,000 deaths (in which 361,000 deaths were reported among children below 5 years old) were estimated in middle- and low-income countries, due to contaminated drinking water and inadequate sanitation facilities (Prüss-Üstün et al. 2014).

Current Treatment Technologies and Their Challenges

Conventional municipal wastewater treatment involves mechanical, biological and chemical processes. In biological treatment, the activated sludge process generally involves sequestration of nutrients. The activated sludge process affects the conditioning of different microorganisms for the oxidation of biodegradable matter. To enhance microbial activity, air is bubbled through the medium to fulfill the oxygen demand and a carbon source methanol is also added to improve the sequestration of nitrogen. Phosphorous is sequestrated through the use of metal salts to facilitate chemical precipitation. There are many apprehensions about this treatment: (a) it needs more electricity, chemicals for precipitation; (b) generation of excess sludge and CO_2 emission associated with this; (c) residual inorganic phosphorus and nitrogen in the form of ammonium and/or nitrate.

Further a post-treatment process is also needed to sequestrate this residual nitrogen and phosphorus from the municipal wastewater. It requires an external carbon source such as methanol for residual-nutrient removal as it has only small amounts of organic compounds (Yamashita and Yamamoto-Ikemoto 2014). Although the use of methanol could reach a high degree of denitrification, but it also poses a risk of further discharge of treated effluent having excessive organic carbon and its flammable nature (Bill et al. 2009).

Municipal wastewater or sewage treatment facilities generate a large amount of sludge or sewage sludge, a semisolid byproduct from treatment processes; containing a variety of organic and inorganic compounds (Bharathiraja et al. 2014). It has two basic forms of primary sludge (from primary treatment) and secondary sludge or activated sludge (from secondary treatment). The sludge contains many harmful substances such as dioxins and furans, PCBs, PAHs polychlorinated biphenyls, polycyclic aromatic hydrocarbons, phenols and their derivatives, traces of heavy metals such as chromium, lead, nickel, cadmium, and copper (Aznar et al. 2009, Szymański et al. 2011, Li et al. 2012, Xu et al. 2013).

This sludge undergoes different physical, chemical and biological treatment processes such as sludge degritting, dewatering, drying, filtration, stabilization, blending, thickening, anaerobic digestion, activated sludge and composting before being suitable for disposal or land application (Demirbas 2016). According to Panepinto et al. (2016), the aeration in the activated sludge process alone accounts for 50% of electrical consumption at wastewater treatment facilities.

There are also emissions of GHGs i.e., carbon dioxide (CO_2), methane (CH_4), and nitrous oxide (N_2O) during the conventional wastewater treatment of municipal wastewater. The main source of CO_2 emission is the biological treatment (the secondary process), where organic carbon present in municipal wastewater is either assimilated into microbial biomass or oxidized to CO_2. Further residual organic carbon in the sludge treatment is also converted into CO_2 and CH_4 during anaerobic digestion, but CH_4 could oxidize to CO_2 through biogas combustion. Another potential green house gas N_2O is also emitted during the nitrification and denitrification of nitrogenous compounds; it mainly takes place in the activated sludge units, grit and sludge storage tanks.

Microalgal Remediation

Due to their minimal requirements, robust growth and the ability to survive in different environments, microalgae are found to be very effective in the treatment of municipal wastewater. A large number of microalgal species were investigated and successfully implemented in the treatment of municipal wastewater (Table 4.3). They not only sequestrated excess nutrients, but also supported the growth of heterotrophic bacteria by providing oxygen for respiration; leading to an increase in the overall nutrient removal.

Nitrogen sequestration

Secondary treated municipal wastewater often comprises the nitrogenous load in the form of nitrate (NO_3^--N), nitrite (NO_2^--N), ammonium (NH_4^+-N) and Dissolved Organic Nitrogen (DON), in which NH_4^+-N and NO_3^--N are the leading nitrogen forms that could be primarily assimilated by microalgae. Liu et al. (2012) reported that some amount of dissolved organic nitrogen (DON) was also sequestrated by microalgae for their growth. As the assimilation of NO_3^--N involves intracellular conversion to NO_2^--N and then NH_4^+-N, requiring more energy than NH_4^+-N uptake and assimilation (Cai et al. 2013), and the studies of Gao et al. (2016), Ruiz-Marin et al. (2010) observed that microalgae prefers the NH_4^+-N form and their sequestration is faster than NO_3^--N sequestration. Further these assimilated nitrogen forms are used by the microalgal cells in the synthesis of amino acids, peptides, proteins, enzymes nucleic acids and energy transfer molecules (Fig. 4.1) (Ruiz et al. 2013).

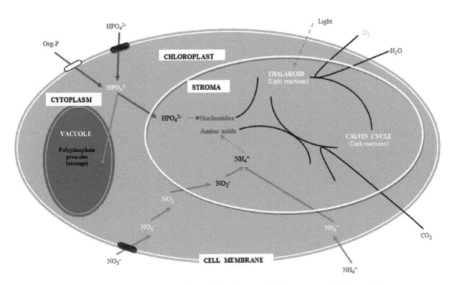

Figure 4.1: Strategies and mechanism of nitrogen and phosphorous sequestration by microalgae.

Table 4.3: Microalgal sequestration of nitrogen and phosphorous from municipal wastewater

Microalgae	Total nitrogen removal (%)	Nitrogen-ammonia removal (%)	Nitrogen-nitrate removal (%)	Total phosphorous removal (%)	References
Synechococcus sp. PCC7942			46.6-100		Hu et al. 2000
Phormidium laminosum					Garbisu et al. 1992
P. bohneri					Chevalier et al. 2000
P. tenue					Chevalier et al. 2000, Dumas et al. 1998
Planktothrix isothrix		70-85.8		83	Silva-Benavides and Torzill, 2012
Synechocystis sp.					Suzuki et al. 1995, Rai et al. 2016
Synechococcus elongatus					Aguilar-May et al. 2009
Synechocystis salina	52.3			77.8	Gonçalves et al. 2016
Microcystis aeruginosa	49.8			81.3	Gonçalves et al. 2016
Aphanothece microscopica					Queiroz et al. 2011
Spirulina platensis					Lodi et al. 2003
Botryococcus braunii			>99	>99	Sawayama et al. 1992,
			79.73	100	Sydney et al. 2011
Chlorella ellipsoidea YJ1	>90			>90	Yang et al. 2011
C. kessleri	>99			>98	Arbib et al. 2014
C. sorokiniana	80		100	0	Lizzul et al. 2014, Bohutskyi
				40	et al. 2015

Species				Reference
Chlorella sp.227	75-92		84-86	Cho et al. 2011
C. vulgaris	>99		92	Cabanelas et al. 2013,
			>99	Ji et al. 2013;
	>90		>98	Ebrahimian et al. 2014
Desmodesmus communis		100		Samori et al. 2013
Neochloris oleoabundans	>90	78-99	100	Wang et al. 2011
Ourococcus multisporus	>99		>99	Ji et al. 2013
Scenedesmus obliquus	>99		>90	Ji et al. 2013
	>90		>99	Martinez et al. 2000
			>98	Arbib et al. 2014
Scenedesmus sp. AMDD	>90		>90	Park et al. 2012
Scenedesmus sp. LX1	98.5		98	Li et al. 2010; Li et al. 2011
Chlorella sp.	100		100	Zhang et al. 2012
Phormidium sp.	>90	>90	>90	De la Noüe and Proulx 1988
S. bicellularis	100		88-100	Kaya et al. 1995
	42.1-100		19.1-99.1	
S. quadricauda	85-100			Chevalier et al 1985
Scenedesmus sp.	100		100	Zhang et al. 2008
	96		>90	He and Xue 2010

Besides the microalgal sequestration of nitrogen in secondary municipal wastewater, NH_4^+-N can also be removed by N_2 loss (through bacterial nitrification-denitrification) and NH_3 volatilization (facilitated by changes of pH and temperature). Similar to NH_4^+-N, NO_3^--N can also be removed via bacterial denitrification other than microalgal sequestration (Craggs et al. 1996, Cai et al. 2013). Although this removal of nitrogen in the form of gaseous N_2 or NH_3 helps in overall sequestration of nitrogen from municipal wastewater; it could lead to possible air pollution. To keep this N_2 or NH_3 loss at a minimum level, Wang et al. (2015) suggested that the recovery of nitrogen from harvested microalgal biomass could be more helpful.

Phosphorous sequestration

In regard to phosphorus sequestration, orthophosphate (PO_4^{3-}-P) is considered the leading form of phosphorous that is often assimilated by the microalgae for their growth (Dueñas et al. 2003). According to Cai et al. (2013), PO_4^{3-}-P is reportedly the only and preferable form that is required by the microalgae for the synthesis of nucleic acids, phospholipids, proteins, energy transfer molecules and the intermediates needed for carbohydrate metabolism (Fig. 4.1). Microalgae (including cyanobacteria) are known to be PO_4^{3-}-P accumulated as polyphosphate granules in their cells, it could be either in the case of phosphorous starvation or not; leading to more phosphorous sequestrating by microalgae through this impressive PO_4^{3-}-P uptake/removal from the municipal wastewater (Powell et al. 2009, Guzzon et al. 2008, Ruiz et al. 2013, Sukačová et al. 2015).

Besides the microalgal assimilation and impressive uptake, phosphorous is also precipitated as $Ca_3(PO_4)_2$ and $Mg_3(PO_4)_2$; this serves as another approach for phosphorous removal from municipal wastewater (Wang et al. 2014b). This is facilitated by Ca^{2+} and Mg^{2+} cations (which are very often present in secondary municipal wastewater), elevated pH (>8.5), and dissolved oxygen (produced through photosynthesis). Through this precipitation approach, precipitation of $Ca_3(PO_4)_2$ and $Mg_3(PO_4)_2$ could be responsible for phosphorous removal up to 50% of the level in secondary wastewater (He and Xue 2010, El Hamouri 2012, Gao et al. 2014, Sukačová et al. 2015).

Sludge management

The microalgae mediated activated sludge process could be a unique approach to reduce the energy demand needed for the aeration of an activated sludge process; improving overall energy efficiency. The necessary oxygen for bacterial oxidation and nitrification for biodegradable organic compounds, are provided by microalgal photosynthesis, which was earlier supplemented with mechanical aeration. Due to less demand of energy in the activated sludge process, makes this approach more energy efficient and attractive for municipal wastewater treatment. Nordlander et al. (2017) presented a microalgal-based biological treatment of municipal wastewater, formally known as microalgal based activated sludge process (MAAS). In this study, an open pond system with sludge recirculation was investigated; which used natural sunlight for the microalgal photosynthesis. Microalgal-bacterial consortia systems were also applied in the treatment of municipal wastewater sludge, and they were well feasible in thermophilic and mesophilic anaerobic digestion. There is a problem of methane emissions in sludge treatment, which could be minimized by the use of microalgal-bacterial systems to digest primary sludge.

Microalgae/Microalgae–Bacteria Consortia

Microalgae/Microalgae–bacteria consortia systems could be new dimensions in the remediation of municipal wastewater; as they provide higher nutrient removal efficiency and better endurance during nutrient scarcity or unfavorable conditions by supporting each other by sharing various metabolites in the consortium (Chinaswamy et al. 2010, Pires et al. 2013, Renuka et al. 2013, Ma et al. 2014, Jia and Yuan 2017, Alemu 2019). Mun and Guieysse (2006) suggested that in microalgae–bacteria consortia systems, the photosynthetic activity of microalgae delivers the necessary O_2 to bacteria for the oxidation of organic matter; while in turn CO_2 produced during bacterial respiration is used by the microalgae for their growth.

Further extracellular substances including proteins, lipids and nucleic acid produced during microalgal metabolism, could be a potential substrate for more bacterial growth (Gonzalez 2000, Hernandez et al. 2009, Liang et al. 2013). Apart from this O_2/CO_2 exchange between microalgae and bacteria for metabolism enhancement, Pires et al. (2013) suggested that microalgae assimilates ammonium released during bacterial degradation to improve the oxidation of nitrogenous organic compounds. However, the efficiency of these microalgae–bacteria consortia could be affected by species selection and pollutant concentration in municipal wastewater (Jia and Yuan, 2017). Ma et al.(2014) observed that as raw municipal wastewaters have a high population of naturally existing bacteria and other microorganisms, it might have a negative effect on microalgal growth and nutrient removal. To avoid this, secondary treated or completely sterilized municipal wastewater is best suitable for the microalgae–bacteria consortium for nutrient removal (Santiago et al. 2013).

Fito and Alemu (2019) studied the performance of native microalgal consortia of *Chlamydomonas*, *Chlorella* and *Scenedesmus* in nutrient removal from raw municipal wastewater. Along with bacteria present in raw municipal wastewater, this combined microalgae-bacteria consortia successfully removed 69% TKN, 59% TP, 73% PO_4^{3-}-P, 85% BOD_5 and 84% COD; as compared to 31% TKN, 56% TP, 50% PO_4^{3-}-P, 52% BOD_5 and 44% COD in bacteria remediation of municipal wastewater. Further the estimation of chlorophyll a production in this study, established that there are no such negative impacts of naturally existing bacteria in raw wastewater on microalgal growth.

Microalgal Biofilms

Microalgal biofilms are primarily known as autotrophic biofilms, that consist of microalgae (including cyanobacteria) and heterotrophic microorganisms (fungi, bacteria and protozoa) (Sukačová et al. 2015, Choudhary et al. 2017, Mantzorou et al. 2018). They could be developed on solid substrata which provides adequate humidity, illumination and is also capable of supplying nutrients to microalgae and other the microorganisms. Microalgal and bacterial cells in the biofilms are present to acclimatize themselves in polluted and extreme conditions, that ultimately leads to better survival in different environments. It is also suggested that bacteria in biofilms, would be able to resist 100–1000 times more antimicrobial substances. There are many advantages of using microalgal biofilms over suspended microalgae or microalgae-bacteria consortia:

1. Microalgae attached to substrata are more efficient in wastewater treatment (Shen et al. 2018, Ma et al. 2018, Palma et al. 2017, Hodges et al. 2017).
2. Microalgae in biofilm systems have better light availability as compared to suspended systems (Katarzyna et al. 2015, Zhang et al. 2018).
3. Microalgal biofilms require less water than those of suspended cultures (Podola et al. 2017, Shukla et al. 2017).
4. They can be operated on short hydraulic retention times as compared to suspended systems which need long hydraulic retention times (Boelee et al. 2014).
5. The biomass production capability of microalgal biofilms is found to be much greater than those of suspended cultivation systems (Berner et al. 2015, Gao et al. 2015).

Many earlier studies suggested that microalgal biofilms proved to be more efficient in N and P sequestration from municipal wastewater. The sequestration capability of microalgae observed more than 90% for NH_4^+ and NO_3^- and more than 80% for PO_4^{3-} (Gonzalez et al. 2008). Besides this, Gonzalez et al. (2008) observed that microalgal biofilm could remove COD up to 75% from diluted swine manure wastewater.

Conclusion

Considering the current situation of municipal wastewater generation, there is a need for comprehensive planning related to maximum sewer coverage to collect domestic wastewater and further treatment of this wastewater before discharging it into water bodies or reusing for agriculture and industrial purposes. Further there is a requirement for different methods and processes for the treatment of municipal wastewater, which could not only provide an effective and sustainable treatment but also be economical for lower and middle income countries who could adapt these methods without risking valuable money needed for economic development.

Conventional wastewater treatment involves many physical, chemical and biological processes that remove the maximum BOD and COD from wastewater but inorganic N and P and trace amount of heavy metals are still present. To remove these, advanced technology like membrane filtration is required, which further increases the cost of wastewater treatment. Besides this, the production of excess sludge another energy-intensive process, requires proper handling and management to dispose of or convert into value added substances like fertilizers.

Microalgae-mediated treatment of municipal wastewater provides a better opportunity to provide effective treatment and is also helpful in handling and management of excess sludge. Although microalgae proved to be very effective in N and P sequestration, it has some limitations like the need of neutral pH and need of sunlight for growth. Further most of the studies related to microalgal-based nitrogen NO_3^--N sequestration investigated were either under sterilized conditions or with microalgae monoculture alone in municipal wastewater. In reality, pH is not often maintained at neutral, which might be enhanced to over 9.0; leading to NH_3 volatilization from municipal wastewater. This phenomenon has not been thoroughly described as yet, so there is a need for more research in this field.

Microalgae-bacterial consortia could be a better option than monoculture of microalgae, as they provide various ranges of adaptability for different pH, temperature and pollutant concentrations. But the association and interaction between different microalgae and bacterial strains should be well studied and also observed in actual conditions before being applied for municipal wastewater treatment. Although microalgal biofilms solve the limitations and constraints related to nutrient removal by the use of microalgae and microalgae-bacterial consortia in municipal wastewater treatment, but the success of microalgal biofilms could depend on many factors such as achieved effluent concentrations, species control and maintaining composition and the area requirement. However, species control within a wastewater environment is challenging as microalgae are opportunistic, and attempts to control the community have failed due to contamination from native algal species.

There are three different scenarios for the use of microalgae, microalgae-bacterial consortia and microalgal biofilms, under which they are used for municipal wastewater treatment:

(a) Microalgae can thrive on inorganic N and P and does not need an organic carbon source for growth, so microalgae-mediated remediation could be used as a post-treatment method.

(b) They could also be involved in sequestrating the N and P in or after activated sludge process, as microalgal remediation could be an alternative for bacterial nitrification and denitrification and chemical or biological P removal.

(c) Or microalgae can be applied in direct treatment of municipal wastewater, which could be helpful in facilitating symbiotic association between microalgae and heterotrophic microorganisms. This beneficial association of O_2/CO_2 exchange for their growth leads to reduction of energy requirement needed for external O_2 supply for aeration of the activated sludge process.

References

Aguilar-May, B. and M. del P. Sánchez-Saavedra. 2009. Growth and removal of nitrogen and phosphorus by free-living and chitosan-immobilized cells of the marine cyanobacterium *Synechococcus elongates*. J. Appl. Phycol. 21: 353–360.

Alemu, J.F.K. 2019. Microalgae–bacteria consortium treatment technology for municipal wastewater management. Nanotechnol. Environ. 4: 1–9.

AQUASTAT. 2014. FAO global information system on water and agriculture. AQUASTAT, Wastewater section, FAO.

Arbib, Z., J. Ruiz, P. Álvarez-Díaz, C. Garrido-Pérez and J.A. Perales. 2014. Capability of different microalgae species for phytoremediation processes: wastewater tertiary treatment, CO_2 bio-fixation and low cost biofuels production. Water Res. 49: 465–474.

Aznar, M., M.S. Anselmo, J.J. Manyà and M.B. Murillo. 2009. Experimental study examining the evolution of nitrogen compounds during the gasification of dried sewage sludge. Energy Fuels 23: 3236–3245.

Berner, F., K. Heimann and M. Sheehan. 2015. Microalgal biofilms for biomass production. J. Appl. Phycol. 27: 1793–1804.

Bharathiraja, B., D. Yogendran, R.R. Kumar, M. Chakravarthy and S. Palani. 2014. Biofuels from sewage sludge – a review. Int. J. Chem. Tech. Res. 6: 4417–4427.

Bill, K.A., C.B. Bott and S.N. Murthy. 2009. Evaluation of alternative electron donors for denitrifying moving bed biofilm reactors (MBBRs). Water Sci. Technol. 60: 2647–2657.

Boelee, N.C., M. Janssen, H. Temmink, L. Taparavičiūtė, R. Khiewwijit, A. Jánoska, et al. 2014. The effect of harvesting on biomass production and nutrient removal in phototrophic biofilm reactors for effluent polishing. J. Appl. Phycol. 26: 1439–1452.

Bohutskyi, P., S. Chow, B. Ketter, M.J. Betenbaugh and E.J. Bouwer. 2015. Prospects for methane production and nutrient recycling from lipid extracted residues and whole *Nannochloropsis salina* using anaerobic digestion. Appl. Energy: 718–731.

Cabanelas, I.T.D., Z. Arbib, F.A. Chinalia, C.O. Souza, J.A. Perales, P.F. Almeida, J.I. Druzian and I.A. Nascimento. 2013. From waste to energy: Microalgae production in wastewater and glycerol. Appl. Energy 109: 283–290.

Cai, T., S.Y. Park and Y. Li. 2013. Nutrient recovery from wastewater streams by microalgae: status and prospects. Renew. Sust. Energ. Rev. 19: 360–369.

Chevalier, P. and J. de la Noüe. 1985. Efficiency of immobilized hyperconcentrated algae for ammonium and orthophosphate removal from wastewaters. Biotechnol. Lett. 7: 395–400.

Chevalier, P., D. Proulx, P. Lessard, W.F. Vincent and J. de la Noüe. 2000. Nitrogen and phosphorus removal by high latitude mat-forming cyanobacteria for potential use in tertiary wastewater treatment. J. Appl. Phycol. 12: 105–112.

Chinnasamy, S., A. Bhatnagar, R.W. Hunt and K.C. Das. 2010. Microalgae cultivation in a wastewater dominated by carpet mill effluents for biofuel applications. Bioresour. Technol. 101(9): 3097–3105.

Cho, S., T.T. Luong, D. Lee, Y.-K. Oh and T. Lee. 2011. Reuse of effluent water from a municipal wastewater treatment plant in microalgae cultivation for biofuel production. Bioresour. Technol. 102: 8639–8645.

Choudhary, P., A. Malik and K.K Pant. 2017. Algal biofilm systems: an answer to algal biofuel dilemma. pp. 77–96. In: S.K. Gupta, A. Malik and F. Bux (eds.). Algal Biofuels: Recent Advances and Future Prospects. Springer International Publishing, Cham, Switzerland.

Craggs, R.J., W.H. Adey, B.K. Jessup and W.J. Oswald. 1996. A controlled stream mesocosm for tertiary treatment of sewage. Ecol. Eng. 6: 149–169.

de la Noüe, J. and D. Proulx. 1988. Biological tertiary treatment of urban wastewaters with chitosan-immobilized *Phormidium*. Appl. Microbiol. Biotechnol. 29: 292–297.

De-Bashan, L.E., J.-P. Hernandez, T. Morey and Y. Bashan. 2004. Microalgae growth-promoting bacteria as "helpers" for microalgae: a novel approach for removing ammonium and phosphorus from municipal wastewater. Water Res. 38: 466–474.

Delgadillo-Mirquez, L., F. Lopes, B. Taidi and D. Pareau. 2016. Nitrogen and phosphate removal from wastewater with a mixed microalgae and bacteria culture. Biotechnol. Rep. 11: 18–26.

Demirbas, A. 2016. Sulfur removal from crude oil using supercritical water. Petrol. Sci. Technol. 34: 622–626.

Dueñas, J.F., J.R. Alonso, A.F. Reyand and A.S. Ferrer. 2003. Characterisation of phosphorous forms in wastewater treatment plants. J. Hazard. Mater. 97: 193–205.

Dumas, A., G. Laliberte, P. Lessard and J. Noüe. 1998. Biotreatment of fish farm effluents using the cyanobacterium *Phormidium bohneri*. Aquacult. Eng. 17: 57–68.

Ebrahimian, A. H.R. Kariminia and M. Vosoughi. 2014. Lipid production in mixotrophic cultivation of *Chlorella vulgaris* in a mixture of primary and secondary municipal waste water. Renew. Energy 71: 502–508.

Edokpayi, J.N. 2016. Assessment of the efficiency of wastewater treatment facilities and the impact of their effluent on surface water and sediments in Vhembe District, South Africa. Ph.D. thesis, University of Venda, South Africa.

Edokpayi, J.N., J.O. Odiyo, T.A.M. Msagati and E.O. Popoola. 2015. Removal efficiency of faecal indicator organisms, nutrients and heavy metals from a Peri-Urban wastewater treatment plant in Thohoyandou, Limpopo Province, South Africa. Int. J. Environ. Res. Public Health 12: 7300–7320.

El-Hamouri, B. 2012. Rethinking natural, extensive systems for tertiary treatment purposes: the high-rate algae pond as an example. Desalin. Water Treat. 4: 128–134.

Eriksson, E., K. Auffarth, M. Henze and A. Ledin. 2002. Characteristics of grey wastewater. Urban Water 4(1): 85-104.

Fito, J. and K. Alemu. 2019. Microalgae–bacteria consortium treatment technology for municipal wastewater management. Nanotechnol. Environ. Eng. 4(4):1–9.

Foladori, P., S. Petrini and G. Andreottola. 2018. Evolution of real municipal wastewater treatment in photobioreactors and microalgae-bacteria consortia using real-time parameters. Chem. Eng. J. 345: 507–516.

Gao, F., C. Li, Z.H. Yang, G.M. Zeng, J. Mu, M. Liu, et al. 2016. Removal of nutrients, organic matter, and metal from domestic secondary effluent through microalgae cultivation in a membrane photobioreactor. J. Chem. Technol. Biotechnol. 91: 2713–2719.

Gao, F., Z.H. Yang, C. Li, G.M. Zeng, D.H. Ma and L. Zhou. 2015. A novel algal biofilm membrane photobioreactor for attached microalgae growth and nutrients removal from secondary effluent. Bioresour. Technol. 179: 8–12.

Gao, F., Z.H. Yang, C. Li, Y.J. Wang, W.H. Jin and Y.B. Deng. 2014. Concentrated microalgae cultivation in treated sewage by membrane photobioreactor operated in batch flow mode. Bioresour. Technol. 167: 441–446.

Garbisu, C., D.O. Hall and J.L. Serra. 1992. Nitrate and nitrite uptake by free-living and immobilized N-starved cells of *Phormidium laminosum* J. Appl. Phycol. 4: 139–148.

Gonçalves, A.L., J.C.M. Pires and M. Simões. 2016. Biotechnological potential of *Synechocystis salina* co-cultures with selected microalgae and cyanobacteria: nutrients removal, biomass and lipid production. Bioresour. Technol. 200: 279–286.

Gonzalez, L.U.Z.E. 2000. Increased growth of the microalga *Chlorella vulgaris* when co-immobilized and co-cultured in alginate beads with the plant-growth-promoting bacterium *Azospirillum brasilense*. Appl. Environ. Microbiol. 66: 1527–1531.

Gonzalez, C., J. Marciniak, S. Villaverde, C. Leon, C., P.A. García. and R. Munoz. 2008. Efficient nutrient removal from swine manure in a tubular biofilm photo-bioreactor using algae-bacteria consortia. Water Sci. Technol. 58(1): 95–102.

Gross, A., N. Azulai, G. Oron, Z. Ronen, M. Arnold, A. Nejidat. 2005. Environmental impact and health risks associated with grey water irrigation: A case study. Water Sci. Technol. 52(8): 161–169.

Guzzon, A., A. Bohn, M. Diociaiuti and P. Albertano. 2008. Cultured phototrophic biofilms for phosphorus removal in wastewater treatment. Water Res. 42: 4357–4367.

GWI. 2009. Municipal water reuse markets 2010. Global Water Intelligence, Media Analytics Ltd, Oxford.

He, S. and G. Xue. 2010. Algal-based immobilization process to treat the effluent from a secondary wastewater treatment plant (WWTP). J. Hazard. Mater. 178: 895–899.

Henze, M. and A. Ledin. 2001. Types, characteristics and quantities of classic, combined domestic wastewaters. pp. 59–72. *In*: P. Lens, G. Zeeman and G. Lettinga (eds.). Decentralised Sanitation and Reuse; Concepts, Systems and Implementation. IWA Publishing Ltd. London, UK.

Hernandez, J., L.E. de-Bashan, D.J. Rodriguez, Y. Rodriguez and Y. Bashan. 2009. Growth promotion of the freshwater microalga *Chlorella vulgaris* by the nitrogen-fixing, plant growth-promoting bacterium *Bacillus pumilus* from arid zone soils. Eur. J. Soil. Biol. 45: 88–93.

Hernández Leal, L., G. Zeeman, H. Temmink and C. Buisman. 2007. Characterisation and biological treatment of grey water. Water Sci. Technol. 5(5): 193–200.

Hodges, A., Z. Fica, J. Wanlass, J. VanDarlin and R. Sims. 2017. Nutrient and suspended solids removal from petrochemical wastewater via microalgal biofilm cultivation. Chemosphere 174: 46–48.

Hu, Q., P. Westerhoff and W. Vermaas. 2000. Removal of nitrate from groundwater by cyanobacteria: quantitative assessment of factors influencing nitrate uptake. Appl. Environ. Microbiol. 66(1): 133–139.

Igbinosa, E.O. and A.I. Okoh. 2009. The impact of discharge wastewater effluents on the physiochemical qualities of a receiving watershed in a typical rural community. Int. J. Environ. Sci. Technol. 6(2): 175–182.

Jefferson, B., S. Judd and C. Diaper. 2001. Treatment methods for grey water. *In*: P. Lens, G. Zeeman and G. Lettinga (eds.). Decentralised Sanitation and Reuse, Concepts, Systems and Implementation. IWA Publishing Ltd. London, UK.

Jia, H. and Q. Yuan. 2017. Removal of nitrogen from wastewater using microalgae and microalgae-bacteria consortia. Cogent Environ. Sci. 31: 1–15.

Jiang, L., S. Luo, X. Fan, Z. Yang and R. Guo. 2011. Biomass and lipid production of marine microalgae using municipal wastewater and high concentration of CO_2. Appl. Energy 88: 3336–3341.

Katarzyna, L., G. Sai and O.A. Singh. 2015. Non-enclosure methods for non-suspended microalgae cultivation: literature review and research needs. Renew. Sust. Energ. Rev. 42: 1418–1427.

Kaya, V.M., J. de la Noüe and G. Picard. 1995. A comparative study of four systems for tertiary wastewater treatment by *Scenedesmus bicellularis*: new technology for immobilization. J. Appl. Phycol. 7: 85–95.

Li, L., Z.R. Xu, C. Zhang, J. Bao and X. Dai. 2012. Quantitative evaluation of heavy metals in solid residues from sub- and super-critical water gasification of sewage sludge. Bioresour. Technol. 121: 169–175.

Li, X., H.Y. Hu and J. Yang. 2010. Lipid accumulation and nutrient removal properties of a newly isolated freshwater microalga, *Scenedesmus* sp. LX1, growing in secondary effluent. New Biotechnol. 27: 59–63.

Li, Y., Y.-F. Chen, P. Chen, M. Min, W. Zhou, B. Martinez, et al. 2011. Characterization of a microalga *Chlorella* sp. well adapted to highly concentrated municipal wastewater for nutrient removal and biodiesel production. Bioresour. Technol. 102: 5138–5144.

Liang, Z., Y. Liu, F. Ge, Y. Hu, N. Tao, F. Peng, et al. 2013. Efficiency assessment and pH effect in removing nitrogen and phosphorus by algae-bacteria combined system of *Chlorella vulgaris* and *Bacillus licheniformis*. Chemosphere 92: 1383–1389.

Liehr, S.K., J.W. Eheart and M.T. Suidan. 1988. A modeling study of the effect of pH on carbon limited algal biofilms. Water Res. 22: 1033–1041.

Liu, H., J. Jeong, H. Gray, S. Smith and D.L. Sedlak. 2012. Algal uptake of hydrophobic and hydrophilic dissolved organic nitrogen in effluent from biological nutrient removal municipal wastewater treatment systems. Environ. Sci. Technol. 46: 713–721.

Lizzul, A.M., P. Hellier, S. Purton, F. Baganz, N. Ladommatos and L.Campos. 2014. Combined remediation and lipid production using *Chlorella sorokiniana* grown on wastewater and exhaust gases. Bioresour. Technol. 151: 12–18.

Lodi, A., L. Binaghi, C. Solisio, A. Converti and M. Del Borghi. 2003. Nitrate and phosphate removal by *Spirulina platensis*. J. Ind. Microbiol. Biotechnol. 30(11): 656–660.

Ma, L., F. Wang, Y. Yu, J. Liu and Y. Wu. 2018. Cu removal and response mechanisms of periphytic biofilms in a tubular bioreactor. Bioresour. Technol. 248: 61–67.

Ma, X., W. Zhou, Z. Fu, Y. Cheng, M. Min, Y. Liu, et al. 2014. Effect of wastewater-borne bacteria on algal growth and nutrients removal in wastewater-based algae cultivation system. Bioresour. Technol. 167: 8–13.

Mantzorou, A., E. Navakoudis, K. Paschalidis and F. Ververidis. 2018. Microalgae: a potential tool for remediating aquatic environments from toxic metals. Int. J. Environ. Sci. Technol. 15: 1815–1830.

Mantzorou, A. and F. Ververidis. 2019. Microalgal biofilms: a further step over current microalgal cultivation techniques. Sci. Total Environ. 651: 3187–3201.

Martinez, M.E., S. Sanchez, J.M. Jimenez, F.E. Yousfi and L. Muñoz. 2000. Nitrogen and phosphorus removal from urban wastewater by the microalga *Scenedesmus obliquus*. Bioresour. Technol. 73: 263–272.

Morrison, G., O.S. Fatoki, L. Persson and A. Ekberg. 2001. Assessment of the impact of point source pollution from the Keiskammahoek Sewage Treatment Plant on the Keiskamma River – pH, electrical conductivity, oxygen demanding substance (COD) and nutrients. Water SA. 27(4): 475–480.

Mun, R. and B. Guieysse. 2006. Algal-bacterial processes for the treatment of hazardous contaminants: a review. Water Res. 40: 2799–2815.

Nanninga, T.A. 2011. Helophyte filters: Sense or Non-Sense? A study on experiences with helophyte filters treating grey wastewater in the Netherlands. M.Sc dissertation thesis, Wageningen University, Netherlands.

Nordlander, E., J. Olsson, E. Thorin and E. Nehrenheim. 2017. Simulation of energy balance and carbon dioxide emission for microalgae introduction in wastewater treatment plants. Algal Res. 24: 251–260.

NWP, 2006. Smart sanitation solutions: examples of innovative, low-cost technologies for toilets, collection, transportation, treatment and use of sanitation products. Netherlands Water Partnership, Delft, The Netherlands.

Elmitwalli, T.A. and R. Otterpohl. 2007. Anaerobic biodegradability and treatment of grey water in upflow anaerobic sludge blanket (UASB) reactor. Water Res. 41(6): 1379–1387.

Olsson, J., S. Schwede, E. Nehrenheim and E. Thorin. 2018. Microalgae as biological treatment for municipal wastewater–effects on the sludge handling in a treatment plant. Water Sci. Technol. 78(3–4): 644–654.

Palma, H., E. Killoran, M. Sheehan, F. Berner and K. Heimann. 2017. Assessment of microalga biofilms for simultaneous remediation and biofuel generation in mine tailings water. Bioresour. Technol. 234: 327–335.

Panepinto, D., S. Fiore, M. Zappone, G. Genon and L. Meucci. 2016. Evaluation of the energy efficiency of a large wastewater treatment plant in Italy. Appl. Ener. 161: 404–411.

Park, K.C., C. Whitney, J.C. McNichol, K.E. Dickinson, S. MacQuarrie, B.P. Skrupski, et al. 2012. Mixotrophic and photoautotrophic cultivation of 14 microalgae isolates from Saskatchewan, Canada: potential applications for wastewater remediation for biofuel production. J. Appl. Phycol. 24: 339–348.

Pires, J.C.M., F.G. Martins and M. Simões. 2013. Wastewater treatment to enhance the economic viability of microalgae culture. Environ. Sci. Pollut. Res. 20: 5096–5105.

Podola, B., T. Li and M. Melkonian. 2017. Porous substrate bioreactors: a paradigm shift in microalgal biotechnology? Trends Biotechnol. 35(2): 121–132.

Powell, N., A. Shilton, Y. Chisti and S. Pratt. 2009. Towards a luxury uptake process via microalgae—defining the polyphosphate dynamics. Water Res. 43: 4207–4213.

Prüss-Ustün, A., J. Bartram, T. Clasen, J.M. Colford, O. Cumming, V. Curtis, et al. 2014. Burden of disease from inadequate water, sanitation and hygiene in low- and middle-income settings: a retrospective analysis of data from 145 countries. Trop. Med. Int. Health 19(8): 894–905.

Queiroz, M.I., M.O. Hornes, A.G. da Silva-Manetti and E. Jacob-Lopes. 2011. Single-cell oil production by cyanobacterium *Aphanothece microscopica* Nägeli cultivated heterotrophically in fish processing wastewater. Appl. Energy 88: 3438–3443.

Rai, J., D. Kumar, L.K. Pandey, A. Yadav and J.P. Gaur. 2016. Potential of cyanobacterial biofilms in phosphate removal and biomass production. J. Environ. Manage. 177: 138–144.

Renuka, N., A. Sood and S.K. Ratha. 2013. Evaluation of microalgal consortia for treatment of primary treated sewage effluent and biomass production. J. Appl. Phycol. 25: 1529–1537.

Roeselers, G., M. Loosdrecht and G. Muyzer. 2008. Phototrophic biofilms and their potential applications. J. Appl. Phycol. 20: 227–235.

Ruiz, J., Z. Arbib, P.D. Álvarez-Díaz, C. Garrido-Pérez, J. Barragán and J.A. Perales. 2013. Photobiotreatment model (PhBT): A kinetic model for microalgae biomass growth and nutrient removal in wastewater. Environ. Technol. 34: 979–991.

Ruiz-Marin, A., L.G. Mendoza-Espinosa and T. Stephenson. 2010. Growth and nutrient removal in free and immobilized green algae in batch and semi-continuous cultures treating real wastewater. Bioresour. Technol. 101: 58–64.

Samorì, G., C. Samorì, F. Guerrini and R. Pistocchi. 2013. Growth and nitrogen removal capacity of *Desmodesmus communis* and of a natural microalgae consortium in a batch culture system in view of urban wastewater treatment: Part I. Water Res. 47(2): 791–801.

Santiago, A.F., M.L. Calijuri and P.P. Assemany. 2013. Algal biomass production and wastewater treatment in high rate algal ponds receiving disinfected effluent. Environ. Technol. 34(13–14): 1877–1885.

Sato, T., M. Qadir, S. Yamamoto, T. Endo and A. Zahoor. 2013. Global, regional, and country level need for data on wastewater generation, treatment, and reuse. Agric. Water Manage. 130: 1–13.

Sawayama, S., T. Minowa, Y. Dote and S. Yokoyama. 1992. Growth of the hydrocarbon-rich microalga *Botryococcus braunii* in secondarily treated sewage. Appl. Microbiol. Biotechnol. 38: 135–138.

Schumacher, G., T. Blume and I. Sekoulov. 2003. Bacteria reduction and nutrient removal in small wastewater treatment plants by an algal biofilm. Water Sci. Technol. 47: 195–202.

Shen, Y., S. Wang, S.-H. Ho, Y. Xie and J. Chen. 2018. Enhancing lipid production in attached culture of a thermo-tolerant microalga *Desmodesmus* sp. F51 using light-related strategies. Biochem. Eng. J. 129: 119–128.

Shukla, S.K., J.V. Thanikal, L. Haouech, S.G. Patil and V. Kumar. 2017. Critical evaluation of algal biofuel production processes using wastewater. pp. 189–225. *In*: S.K. Gupta, A. Malik and F. Bux (eds.). Algal Biofuels: Recent Advances and Future Prospects. Springer International Publishing, Cham, Switzerland.

Silva-Benavides, A.M. and G. Torzill. 2012. Nitrogen and phosphorus removal through laboratory batch cultures of microalga *Chlorella vulgaris* and cyanobacterium *Planktothrix isothrix* grown as monoalgal and as co-cultures. J. Appl. Phycol. 24: 267–276.

Sukačová, K., M. Trtílek and T. Rataj. 2015. Phosphorus removal using a microalgal biofilm in a new biofilm photobioreactor for tertiary wastewater treatment. Water Res. 71: 55–63.

Suzuki, I., N. Horie, T. Sugiyama and T. Omata. 1995. Identification and characterization of two nitrogen-regulated genes of the cyanobacterium *Synechococcus* sp. strain PCC7942 required for maximum efficiency of nitrogen assimilation. J. Bacteriol. 177: 290–296.

Sydney, E.B., T.E. da Silva, A. Tokarski, A.C. Novak, J.C. de Carvalho, A.L. Woiciecohwski, et al. 2011. Screening of microalgae with potential for biodiesel production and nutrient removal from treated domestic sewage. Appl. Energy 88: 3291–3294.

Szymański, K., B. Janowska and P. Jastrzębski. 2011. Heavy metal compounds in wastewater and sewage sludge. Annu. Set. Environ. Prot. 13: 83–100.

UN. 2012. World population prospects: The 2012 revision. Department of Economic and Social Affairs, Population Division. United Nations, New York, US.

Vinnerås, B. 2002. Possibilities for sustainable nutrient recycling by faecal separation combined with urine diversion. PhD thesis. Swedish University of Agricultural Sciences, Uppsala, Sweden.

Wang, B. and C.Q. Lan. 2011. Biomass production and nitrogen and phosphorus removal by the green alga *Neochloris oleoabundans* in simulated wastewater and secondary municipal wastewater effluent. Bioresour. Technol. 102: 5639–5644.

Wang, J., W. Zhou, H. Yang, F. Wang and R. Ruan. 2015. Trophic mode conversion and nitrogen deprivation of microalgae for high ammonium removal from synthetic wastewater. Bioresour. Technol. 196: 668–676.

Xu, Z.R., W. Zhu, M. Li, H.W. Zhang and M. Gong. 2013. Quantitative analysis of polycyclic aromatic hydrocarbons in solid residues from supercritical water gasification of wet sewage sludge. Appl. Energ. 102: 476–483.

Yadav, R.K., B. Goyal, R.K. Sharma, S.K. Dubey and P.S. Minhas. 2002. Post-irrigation impact of domestic sewage effluent on composition of soils, crops and ground water—a case study. Environ. Int. 28: 481–486.

Yamashita, T. and R. Yamamoto-Ikemoto. 2014. Nitrogen and phosphorus removal from wastewater treatment plant effluent via bacterial sulfate reduction in an anoxic bioreactor packed with wood and Iron. Int. J. Environ. Res. Public Health 11: 9835–9853.

Yang, J., X. Li, H.Y. Hu, X. Zhang, Y. Yu and Y. Chen. 2011. Growth and lipid accumulation properties of a freshwater microalga, *Chlorella ellipsoidea* YJ1, in domestic secondary effluents. Appl. Energy 88: 3295–3299.

Zhang, E., B. Wang, Q. Wang, S. Zhang and B. Zhao. 2008. Ammonia-nitrogen and orthophosphate removal by immobilized *Scenedesmus* sp. isolated from municipal wastewater for potential use in tertiary treatment. Bioresour. Technol. 99: 3787–3793.

Zhang, E., B. Wang, S. Ning, H. Sun, B. Yang, M. Jin, et al. 2012. Ammonia-nitrogen and orthophosphate removal by immobilized *Chlorella* sp. isolated from municipal wastewater for potential use in tertiary treatment. Afr. J. Biotechnol. 11: 6529–6534.

Zhang, Q., X. Li, T. Ye, M. Xiong, L. Zhu, C. Liu, et al. 2018. Operation of a vertical algal biofilm enhanced raceway pond for nutrient removal and microalgae-based byproducts production under different wastewater loadings. Bioresour. Technol. 253: 323–332.

Petroleum Wastewater

Introduction

Petroleum refining or the petroleum industry is considered a large global industry, which utilizes a large volume of water; that is required primarily for cooling and steam generation processes. It is estimated that petroleum refineries consume water in the ranges of 0.7-1.2 m^3 of water per m^3 of crude oil processed (Diepolder 1992). The petroleum industry involves operations like refining crude oil, manufacturing fuels and petroleum intermediates (Al-Futaisi et al. 2007, Hu et al. 2013, Varjani and Upasani 2017). Petroleum wastewater is characterized with oil and grease content, high BOD and COD, high total solids, nitrates, sulfides, hydrocarbons, heavy metals and emissions of ammonia and volatile organic compounds (Honse et al. 2012, Jasmine and Mukherji 2015, Thakur et al. 2018).

There are many activities of the petroleum industry such as drilling, transportation, storage and refining of crude oil that is responsible for soil contamination. Further soil contamination also depends on the type of soil and oil; lighter oil could easily escape in to soil layers as compared to heavier oil (Fakhru'l-Razet al. 2009, Varjani and Upasani, 2017). Petroleum wastewater is also responsible for contamination of water resources, which adversely affect crop production, human health, aquatic ecosystem and other life forms (Poulopoulos et al. 2005, Veyrand et al. 2013, Zafra et al. 2015, Varjani and Upasani, 2017, Al-Hawash et al. 2018). Further accumulation of toxic chemicals in the water bodies could induce to a deleterious impact on living organisms either long term or short term which may be chronic or acute (Poulopoulos et al. 2005, Usman et al. 2012,Varjani et al. 2018).

The conventional treatment of petroleum wastewater involves different physical, chemical and biological treatment processes. The physicochemical processes such as gravity separation and skimming, dissolved air flotation, de-emulsification, coagulation and flocculation are generally used for the treatment. In these approaches, gravity separation along with skimming is proved to be effective in removing free oil from petroleum wastewater; further dissolved air flotation is used for improving efficiency of this method. Although it is observed that despite the individual process, integrating the various processes could be a better strategy for petroleum wastewater treatment.

Microalgae has proved to be a more potential, economical and eco-friendly for transformation and degradation of organic compounds in comparison of heterotrophic bacteria and fungi (Kumar and Singh 2016, Singh et al. 2016a, b, Kumar et al. 2017). Microalgal species especially cyanobacterial strains are more efficient in metabolization or transformation of hydrocarbons in petroleum wastewater (Kumar and Singh 2017, Kumar 2018, Kumar et al. 2018a, b, Singh et al. 2019, Kumar and Singh 2020a, b). But in case of petroleum wastewater, microalgae alone are not enough in hydrocarbon degradation, so use of some heterotrophic bacteria along with microalgae could provide a more advantageous position than individual microalgae or heterotrophic bacteria (Shashirekha et al. 1997).

Crude Oil: Composition and Types

The Chinese were the first to encounter crude oil through the surface seepage in pre-Christian times and after that many parts of the world learnt about crude oil, but these were mainly for medicinal purposes. The modern history of petroleum refining started in 1859, when Colonel E.A. Drake discovered oil in Pennsylvania (Alloway and Ayres 1993). Crude oil is a naturally occurring, yellowish-black liquid found in geological formations beneath the Earth's surface. The elemental composition of crude oil shows 79.5-87.3% C, 10.4-14.8% H, 0-8% S, 0-2% O, 0-0.1% N, and 0-0.05% metals (Fe, V, Ni, As, etc.). Besides elemental forms, crude oils are primarily composed of hydrocarbons of different molecular mass and type, ranging from 1 to 60 C atoms. The main hydrocarbon groups that are found in crude oils are isoalkanes, cycloalkanes and arenes; and some hybrids like cycloalkanes and arenes with side alkyl chains or cycloalkane-arene compounds with side chains.

Sulfur has existed as H_2S, dissolved free sulfur, mercaptans, thiophene, sulfoxides, sulfones, alkylsulfides, alkylsulfates and sulfonic acids. Oxygen and nitrogen are more common in naphtenic crude oils, oxygen primarily existed as organic carboxylic acids and phenols (limited extent); while nitrogen existed as alkylquinolines and pyridines, pyrroles, indoles and carbazoles. There is also the presence of organic metal complexes; having pyrrole $(CH_4)_4NH$ rings, which could be more common in heavier distillate fractions.

Based on the extent of similar hydrocarbon molecules, crude oils are designated as paraffinic-, naphthenic- (cycloparaffins) or aromatic- base crude oils. There is also mixed-base crude oil which contains all the three types of hydrocarbon in different proportions. In refineries, crude base stocks are composed of two or more different base crude oils.

Petroleum Industry: A Brief Description

Crude oil contains different salts mainly in the form of sodium chloride and chlorides of calcium and magnesium, which could be occur either as a water-in-oil emulsion or crystallized and suspended solids. These salts might have a negative effect in the forthcoming processes due to scaling, corrosion and catalyst deactivation. So desalting is carried out initially at the oil field and later at the petroleum refineries. It involves water washing of heated crude oil, and mixing valve or static mixers are applied to ensure a proper contact between the crude oil and the water. Then it conveyed in to a separating container, where proper separation takes place between the aqueous and organic phases. Due to the formation of emulsions, it could be possible that water can escape through the organic phase; to release this water chemical de-emulsifiers are used to enhance the emulsion breaking. Along with this, an electric field is also used across the settling container to merge the polar salty water droplets; leading to a reduction of water and salt content in crude oil.

Petroleum refining usually starts with the distillation or fractionation, in which crude oil separates into different fractions or straight run cuts. The main fractions or cuts are separated which have particular boiling-point ranges and can be categorized based on the decreasing volatility into gases, light distillates, middle distillates, gas oils and residues. Two types of distillation used: fractional distillation and vacuum distillation.

Before fractional distillation, crude oil is heated to about 350°C; which converts the crude oil into a vaporized form. Then the vaporized crude oil is fed into a vertical distillation column or atmospheric tower, whereas the hot vapor moves upward in the column, leads to reduction in its temperature by the circulating refluxes. The fractions are separated out corresponding to their boiling points, the fractions with the highest boiling point are condensed lower in the column; while the fractions with lower boiling points are condensed upper in the column. The major fractions include naphtha, gasoline, kerosene, diesel and uncondensed gases are drawn from fat successive at higher levels in the column. The uncondensed gases and Reduced Crude Oil (RCO) is successively taken out from the top and bottom.

Further the residuum or topped crude from the atmospheric tower requires distillation at higher temperatures, but it could cause thermal cracking. So reduced pressure conditions are needed to avoid thermal cracking, this is followed by vacuum distillation. In vacuum distillation, one or more vacuum distillation towers or columns are used. After this process, main fractions such as gas oils, lubricating-oil base stocks and heavy residual are obtained. The main purpose of using vacuum towers is the separation of catalytic cracking feed-stocks from surplus residuum.

Cracking is used to convert the distillation fractions of crude oil to produce the required products as per the demand in the market. It could be divided in to thermal and catalytic cracking; in which catalytic cracking is often used. Catalytic cracking involves higher temperatures of 850-950°F and lower pressure of 10-20 psi. The common catalysts such as zeolite, aluminum hydrosilicate, bauxite, silica-alumina, treated bentonite clay and fuller's earth are used in the form of powders, beads, pellets or shaped materials called extrudites. This process changes the molecular structure of hydrocarbon compounds, and further converts the heavier fractions into lighter fractions such as LPG, kerosene, gasoline, heating oil and petroleum feed stocks. In addition to cracking, there are other catalytic activities like dehydrogenation, hydrogenation and isomerization that are also followed.

Catalytic cracking is further categorized into (a) Fluid Catalytic Cracking (FCC), (b) Moving-bed catalytic cracking, and (c) Thermo for Catalytic Cracking (TCC). Fluid catalytic cracking is the most preferred process, in which distilled oil fractions are separated and broken in the presence of a finely divided catalyst. It involves mixing of preheated hydrocarbon fractions feed with hot, regenerated catalyst, leading to rising of catalyst aerated or fluidized state with oil vapors; resulting in the cracking at temperatures (900-1000°F) and pressure of 10-30 psi. To regenerate the catalyst, spent catalyst is conveyed to the catalyst stripper of the regenerator, where all the coke deposits are burned off and the used catalyst is removed.

The solvent extraction process follows to remove impurities such as inorganic salts, dissolved metals and organic compounds containing sulfur, nitrogen and oxygen, which are still present in oil feedstock; as distillation and cracking are only used to separates different fractions from the crude oil by their boiling-point ranges. In this process, solvents such as phenol, furfural, and NMP (N - Methyl Pyrledene) are usually applied, while other solvents like liquid sulfur dioxide, nitrobenzene and 2,2 dichloroethyl ether are also used. These solvents extract aromatics, naphthenes and impurities from the fractions that are produced after distillation and cracking. Further solvents are separated out from the fractions stream through heating, evaporation or fractionation processes.

Then the solvent dewaxing process is used to remove wax from either distillate or residual base stocks. Solvent dewaxing involves some stages: (a) feedstock mixing with a solvent; (b) the mixture is chilled to precipitate the wax; and (c) the solvent is recovered from the wax and dewaxed oil. The common solvents that are involved for solvent dewaxing are propane and Methyl Ethyl Ketone (MEK), but many other solvents like petroleum naphtha, benzene, sulfur dioxide, toulene, ethylene dichloride, methylene chloride and methyl isobutyl ketone are also used.

Petroleum Wastewater: Characteristics and Specific Pollutants

The composition of petroleum wastewater could depend on the type of crude oil, plant configuration and the operations involved in the refining (Saien and Nejati 2007). It mainly comprises of inorganic salts, oil and grease, sulfide, ammonia, phenols, hydrocarbons and Polycyclic Aromatic Hydrocarbons (PAHs) and BTEX (Benzene, Toluene, Ethylbenzene, Xylene) (Tobiszewski et al. 2012, Pérez et al. 2010, Wang 2015). It was observed that both aromatic and aliphatic compounds made up 75% of petroleum hydrocarbons present in petroleum wastewater (Perera et al. 2012, Jasmine and Mukherji 2015, Varjani et al. 2018). There is also the presence of heavy metals such as

Table 5.1: Charaterstics of petroleum wastewater by various researchers

Water quality parameter	Vendrmel et al. 2015	Aljuboury et al. 2014	Saber et al. 2014	Gasim et al. 2013	Tony et al. 2012	Hasan et al. 2012	Farajnezhed and Gharbani 2012	El-Naas et al. 2010
pH	8.3	6.5-9.5	6.7	8.48	7.6	7.0	7.5	9.5
Turbidity (NTU)					42	83		
TSS (mg L^{-1})	150		150		105	74	110	80
TDS (mg L^{-1})		1200-1500						
BOD (mg L^{-1})			174	3378		846		
COD (mg L^{-1})	1250	550-1600	450	7896	364	1343	1120	4050
TOC (mg L^{-1})	220-265			13.5				
Ammonia (mg L^{-1})								
Sulphides (mg L^{-1})								1222

Cd, Cr, Cu, Pb, Hg, Ni, Au and Zn present in petroleum wastewaters, and the concentration of these heavy metals primarily depends up on the geology of the well from where the extraction of crude oil takes place (Ahmadun et al. 2009).

Table 5.2: Specific pollutants from processes of petroleum refining
(Based on the El-Naas et al. 2014)

Processes	Pollutants
Crude desalting	Free oil, ammonia, sulfides and suspended solids
Crude oil distillation	Sulfides, ammonia, phenols, oil, chloride, mercaptans
Cracking	H_2S, oil, ammonia, phenols, sulfides, cyanide,
Polymerization	Sulfides, mercaptans, ammonia
Alkylation	Spent caustic, oil, sulfides
Isomerization	Low level of phenols
Reforming	Sulfide
Hydrotreating	Ammonia, sulfides, phenol
Solvent extraction & Dewaxing	

Oil and hydrocarbons

Oil and hydrocarbons are the main pollutants in petroleum wastewater. Hydrocarbons range from straight chained n-alkanes (paraffins) to naphthenes (cycloparaffins) to aromatics compounds (having a benzene ring). In addition to hydrocarbons, there is the presence of naphthenic acids, which are considered to have very toxic effects and are persistent in nature.

Salts

Salts like sodium or potassium chloride are found in petroleum wastewater. These could be naturally occurring in crude oil or often mixed into drilling fluid to avoid the reactions of crude oil with drilling fluid, that could lead to the formation of unnecessary compounds. This excess salt could have adverse effects on aquatic ecosystems.

Volatile organic compounds

These are some major VOCs in petroleum wastewater, which include BTEX (benzene, toluene, ethylbenzene, xylene), naphthalene, hexane, phenol, biphenyl, styrene, cresols, 2,2,4-Trimethylpentane, methyl tertiary-butyl ether and 1,3-Butadiene. The concentration of individual VOCs is calculated on the basis of benzene mass concentration ratio.

Heavy Metals

Heavy metals like arsenic, barium, cadmium, chromium, lead, mercury and barium are found in petroleum wastewater. The source of heavy metals in wastewater is mostly natural but some heavy metals are added during drilling as additives like barium (from barite weighing agents) and chromium (from chrome-ligno sulfonates deflocculants).

Impacts on the Environment and Public Health

Petroleum wastewater contain significant amounts of hydrocarbons, salts, heavy metals and VOCs, which have proved to be toxic to every life form in a particular ecosystem (Poulopoulos et al. 2005, Fakhru'l-Razi et al. 2009, Perera et al. 2012, Veyrand et al. 2013, Jasmine and Mukherji 2015, Thakur et al. 2018, Al-Hawash et al. 2018). On exposure of hydrocarbons, several behavioral changes are observed which are primarily related to motility in lower organisms; and burrowing, feeding and reproductive activities in higher organisms (Pathak and Mandalia 2012). The common victims of petroleum wastewater are birds, as their feathers are coated by oily water; leading them to lose their water-repellence and thermal insulation. These birds with oil coated feathers might sink, drown or die because of hypothermia. Similar to birds, fur-insulated mammals lose their capability to regulate their temperature thermally, because of the loss of insulating capability of oil-coated fur. Due to loss of thermal insulation, mammalian higher metabolic activity to maintain body temperature; leads to rapid exhaustion of fat and muscular energy reserves; resulting in the animal's death either by hypothermia or drowning.

There are three exposure routes i.e., ingestion, inhalation or dermal (skin) contact by which hydrocarbons affects human health; further their impacts could either be acute (short-term) or chronic (long-term).The acute or short term ingestion exposure might cause irritation of the mouth, throat and stomach, and is also responsible for digestive disorders. Further traces of ingested hydrocarbons could reach the lungs, leading to respiratory problems. The chronic ingestion exposure might cause damage to the liver, kidney or gastrointestinal tract. Pathak and Mandalia (2012) suggested that prolonged exposure to aromatics like benzene might be responsible for cancer of the skin, lungs and other areas of the body, even leukaemia.

Current Treatment and Challenges

There are sequences that are followed in conventional treatment: preliminary, primary, secondary and advanced processes; which involve many mechanical, physicochemical and biological methods with integrated activated sludge units (Diyáuddeen et al. 2011, El-Naas et al. 2014). The major constituent in petroleum wastewater i.e., oil and grease which are in a primarily suspended state, are mechanically removed through gravity in separation tanks and further degraded in physiochemical or biological processes. For advanced treatment, different methods like photodegradation, adsorption, coagulation-flocculation, electrocoagulation, photocatalytic oxidation, wet oxidation, catalytic vacuum distillation, fenton oxidation and ultrasonic degradation (Yan et al. 2014, El-Naas et al. 2014).

Diyáuddeen et al. (2011) and Rasalingam et al. (2014) emphasized the importance of different advanced oxidation processes, particularly photocatalytic degradation for the treatment of petroleum wastewater. These advanced oxidation processes are gaining attention due to their capability to degrade or mineralize a wide variety of organic substances. Diya'uddeen et al. (2011) demonstrated that heterogeneous photocatalysis could be a promising, efficient and cost-effective technique for wastewater treatment; but their application in petroleum wastewater treatment has proved to be limited due to insufficient information in literature related to this technique. Remya and Lin (2011) further underlined the current status and the possibility of application of microwave in petroleum wastewater treatment.

Although these methods proved to be very efficient for petroleum wastewater treatment; and further provided many advantages such as energy efficiency, environmentally safe and compatible (Yavuz and Koparal 2006, Abdelwahab et al. 2009). However, these methods also have some notable drawbacks like high capital and operating costs. It is also observed that most of them are not able to completely metabolize the pollutants, but most likely transform them into toxic byproducts (Abdelwahab et al. 2009, El-Naas et al. 2010). Further Yan et al. (2014) observed that some electrochemical treatment methods could be responsible for the formation of chlorinated organic compounds, and to remove them Yan et al. (2014) recommended the activated carbon adsorption method as a polishing treatment. Similar to this, it was also reported that ozonation treatment of phenolic compounds could trigger formation of intermediate byproducts like catechol (García-Peña et al. 2012).

Microalgal Remediation

Microalgae showed the ability to sequestrate N and P and remove organic compounds mainly alkanes, phenols, PAHs from petroleum-refining wastewater either through direct degradation or facilitating bacterial degradation (Al-Awadhi et al. 2003, Radwan et al. 2002, Carvalho et al. 2006, Monteiro et al. 2009, Chavan and Mukherji 2008). Microalgal genera such as *Scenedesmus, Chlorella, Microcoleus, Phormidium* and *Oscillatoria* were successfully reported in degradation of organic compounds and hydrocarbons in petroleum-refining wastewater (Table 5.3). It is observed that microalgae are able to degrade 39 to 60% of crude oil and black oil (Safonova et al. 1999, Chaillian et al. 2006). El-Sheekh et al. (2013) investigated the ability of *Chlorella vulgaris* and *Scenedesmus obliquus* for the degradation of crude oil in low concentration, and suggested that they could also grow under heterotrophic conditions by assimilating crude oil as their sole carbon source. It is also stated that *Scenedesmusobliquus* GH2 could be used in crude oil degradation through the development of an artificial microalgal-bacterial consortium (Tang et al. 2010a, b).

Table 5.3: Oil and hydrocarbon degradation by the microalgae

Microalgae	Crude oil/ Hydrocarbon type	References
Scenedesmus obliquus GH2	Crude oil	Tang et al. 2010a, b
Selenastrum capricornutum	Naphthalene, phenanthrene, pyrene, fluoranthene, mixed PAHs	Gavrilescu 2010, Lei et al. 2002, Lei et al. 2007
Consortium of *Chlorella sorokiniana* and *Pseudomonas migulae*	Phenanthrene	Munoz et al. 2003
Microcoleus chthonoplastes	Oil rich environment	Al-Hasan et al. 1998
Phormidium corium	Oil rich environment	Al-Hasan et al. 1998
Oscillatoria sp.	Naphthalene	Cerniglia et al. 1979, 1980 a b, Narro et al. 1992
Agmenellum quadruplicatum	Naphthalene	Cerniglia et al. 1979, 1980
Anabaena sp.	Naphthalene	Cerniglia et al. 1980b
A. quadruplicatum strain PR-6	Phenanthrene	Narro et al. 1992
S. obliquus ES-55	Phenanthrene	Safonova et al. 2005
S. platydiscus	Pyrene	Lei et al. 2002

Earlier studies related to the degradation of PAHs were carried out by Cerniglia et al. 1980a, b, Lindquist and Warshawsky (1985a, b), Schoeny et al. (1988), Jinqi and Houtian (1992), Warshawsky et al. (1995) and Semple et al. (1999). It was also investigated if microalgae are able to degrade aliphatic hydrocarbons and aromatic compounds which originate in crude and motor oils, and it was reported that the degradation efficiency of microalgae for crude oil is 38-60% and 12-41% in case of saturated aliphatic hydrocarbons and the aromatic compounds respectively; while for motor oil, it is 10-23% and 10-26% in case of saturated aliphatic hydrocarbons and the aromatic compounds respectively. Al-Hasan et al. (1998) observed that *Microcoleus*, *Phormidium* and *Oscillatoria* are able to directly degrade and oxidize n-alkanes. Al-Hasan et al. (1998), Radwan et al. (2002), Chaillian et al. (2006) reported that microalgae and microalgae–bacteria consortia could degrade 22–98% of the single hydrocarbon in concentration ranges of 400–10,000 mg L^{-1} in laboratory experiments.

Earlier studies investigated the capability of *Chlamydomonas angulosa* in relation to naphthalene degradation. Semple et al. (1999), Juhasz and Ravendra (2000) reported that microalgae have the capability of transforming low-molecular-weight PAHs i.e., naphthalene and phenanthrene into their hydroxylated intermediates. Although there are a few studies which demonstrated the microalgal ability to degrade high molecular weight PAHs. However, Juhasz and Ravendra (2000) suggested that microalgae showed the ability to degrade high-molecular-weight compounds. There are many studies which reported microalgal strains such as *Selenastrum capricornutum, Scenedesmus platydiscus, Chlorella vulgaris, Scenedesmus quadricauda, Chlorella* sp. MM³, *Oscillatoria* sp., *Cholrella* sp. (Warshawsky et al. 1988, Chan et al. 2006, Lei et al. 2007, Yan et al. 2014, Subashchandrabose et al. 2017, Aldaby and Mawad 2019); which are able to metabolize the high-molecular-weight compounds pyrene and benzo[a]pyrene.

Mechanism of Degradation of PAHs

Microalgae have been successfully reported for the accumulation and degradation of PAHs but there are few studies related to their degradation mechanism. Further the metabolic pathway for low-molecular weight PAHs such naphthalene, phenanthrene and fluoranthene have been investigated more and understood in comparison to high molecular weight PAHs like pyrene, benzo[a]pyrene (Juhasz and Naidu 2000). It was suggested the enzyme dioxogenases catalyzed the reaction of two atoms of oxygen with two atoms of carbon from the benzene ring, which is considered the first reaction intransformation of PAHs. Due to this, cis-dihydrodiole generated through the enzyme dehydrogenase, is later converted in to dihydrogenate intermediate–pyrocatechol (Cerniglia 1984).

Naphthalene

Earlier studies showed that microalgae *Chlamydomonas angulosa* were be able to accumulate naphthalene within their cells. Cerniglia et al.(1979, 1980 a, b) Semple et al. (1999) reported that microalgae including cyanobacteria have the capability to degrade naphthalene and transform it into major metabolites: cis- and trans-naphthalene dihydrodiol, 1-naphthol and 4-hydrox-4-tetralone (Fig. 5.1). Similar results by Cerniglia et al. (1982) reported the degradation of naphthalene by diatoms and 1-naphthol was found to be the major metabolite.

Phenanthrene and Fluoranthene

Narro et al. (1992) investigated the ability of microalgae *Agmenellum quadruplicatum* PR-6 to metabolize phenanthrene, and also studied the metabolites of phenanthrene. It was found that *Agmenellum quadruplicatum* PR-6 transformed phenanthrene in to trans-9,10 dihyrodiol, 1-methoxyphenanthrene and traces of phenanthrols (Fig. 5.2). Narro et al. (1992), Warshawsky et al. (1995) further reported that *Oscillatoria* sp. adapted the same pathways as *Agmenellumquadruplicatum* to form dihydrodiol; which is similar to that pathway used by the fungal and mammalian systems. Hong

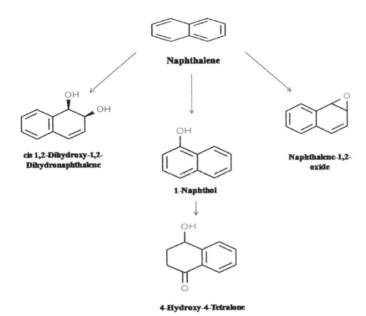

Figure 5.1: Naphthalene degradation by microalgae (Cerniglia et al. 1979, 1980a, b, 1982).

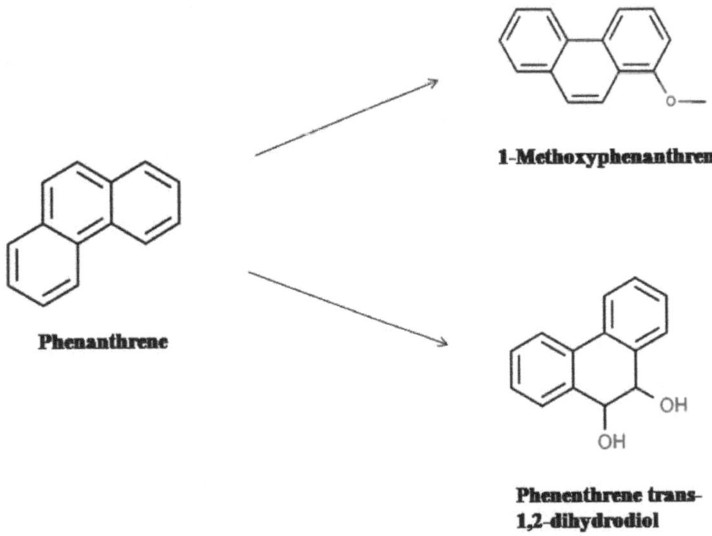

Figure 5.2: Phenanthrene degradation by microalgae (Narro et al. 1992).

et al. (2008) discovered that *Skeletonema costatum* and *Nitzschia* sp. were able to accumulate and degrade phenanthrene and fluoranthene; and found that *Nitzschia* sp. showed better accumulation and degradation abilities than *Skeletonema costatum*. After analyzing slow degradation of fluoranthene by both microalgae, Hong et al. (2008) suggested that fluoranthene proved to be more a recalcitrant PAH compound. It was also observed that the microalgal species showed better degradation efficiency for the phenanthrene-fluoranthene mixture than phenanthrene or fluoranthene alone; this could be due to stimulation of degradation of one PAH compound by the presence of other PAH.

Benzo[a]-Pyrene (BaP)

De Llasera et al. (2016) observed that *Selenastrum capricornutum* and *Scenedesmus acutus* could degrade the benzopyrène from an aqueous medium. Lindquist and Warshawsky (1985 a,b), Schoeny et al. (1988) and Warshawsky et al. (1995) investigated the degradation of benzo[a]-pyrene by microalgae *Selenastrum capricornutum*; and reported the involvement of dioxygenase enzyme system, that oxidize the benzo[a]-pyrene into cis-dihydrodiols, which further transform in to sulfate ester and glucoside conjugates. Schoeny et al. (1988) conducted laboratory experiments involving *Scenedesmus acutus, Ankistrodesmus braunii, Selenastrum capricornutum* for the degradation of benzopyrène. It was observed that white light inhibition of microalgal growth occurs and benzopyrène is transformed into mainly 3,6-quinones, while gold light supported the growth and promoted the catabolism of benzopyrene in to cis-dihydrodiols.

Role of Cyanobacteria Mats in Oil Degradation

Cyanobacteria mats inhabited primarily in sheltered, shallow coastal areas and intertidal zones; which comprise of different communities of microorganisms dispersed in many layers according to the physicochemical gradients (Sánchez et al. 2006). They are also known as microbial mats or marine mats, which play a significant role not only in natural functions and services; but also help in scavenging marine pollution especially oil degradation. Along with the dominance of cyanobacterial communities, there are different members of bacteria such as Cytophaga–Flavobacterium–Bacteroides group (CFB), proteobacteria and green non-sulfur bacteria present in cyanobacterial mats; that are able to effectively degrade both aliphatic and aromatic compounds (Abed et al. 2002, Sánchez et al. 2006).

Paerl et al. (1993), Wieland and Kühl (2006) Abed et al. (2007) and Abed (2019) emphasized the importance of cyanobacteria-heterotrophic bacteria's association in cyanobacterial mats. This relationship could be a competitive type for nutrients and other resources, or a symbiotic type supporting the growth and survival of each other (Paerl et al. 1993). These cyanobacteria and heterotrophic bacteria primarily reside in the upper most layer of sediments and cyanobacterial mats, where exchanging of active carbon and oxygen takes place (Wieland and Kühl 2006). The photosynthetic activity of cyanobacteria and microalgae in such mats is interrelated with the respiration activities of aerobic heterotrophic bacteria.

On exposure of oil pollution to a cyanobacterial mat, the heterotrophic bacteria that are considered major oil degraders, play a direct role in oil degradation, while cyanobacteria and microalgae are considered to play an indirect role by supporting the overall degradation process. There are two points of view related to the role of cyanobacteria in oil degradation, one group suggested that they play an indirect role by assisting the degradation process, while others advocated that cyanobacteria and microalgae are able to degrade oil and hydrocarbons. Radwan and Al-Hasan (2000) suggested that cyanobacteria and microalgae helps in keeping heterotrophic bacteria in their position through entrapping and immobilizing them in their mucilage. Radwan and Al-Hasan (2000), Abed et al. (2002) and Abed and Köster (2005) reported the presence of oil-degrading bacteria in the sheaths of cyanobacterial strains *Microcoleus chthonoplastes* and *Phormidium corium*. Cyanobacteria and microalgae secrete exopolymeric substances (EPS) to overcome the stress of oil pollution, which helps in aggregating the sediment particles; leading to the development of cyanobacterial mats. Further the growth of cyanobacteria might be facilitated through the nutrients and liberated CO_2 by the oil-degrading heterotrophic bacteria.

However there are a number of studies on the degradation of a wide range of hydrocarbons by different cyanobacterial and microalgal strains (as described earlier in this chapter). Vidyashankar and Ravishankar (2016) underlined the detection of catabolic genes, such as monooxygenase and dioxygenase systems (that are known for catalyzing the breakdown of hydrocarbons) in several strains of cyanobacteria and microalgae. Besides this, marine and freshwater cyanobacteria are also known to produce and accumulate hydrocarbons and branch-chain alkanes (Liu et al. 2013, Lea-Smith et al. 2015); which could help to survive and maintain the growth of oil-degrading heterotrophic bacteria. Safonova et al. (1999) reported that co-cultivation of alkane-degrading bacteria stimulates the growth of oil-tolerant microalgal strains such as *Stichococcus*, *Chlorella*, *Scenedesmus*, *Nostoc* and *Phormidium*. Similar reports suggested by Abed (2010) in which co-cultivation of *Synechocystis* PCC6803 with hexadecane degrading heterotrophic bacteria were studied and it was reported that hexadecane degrading bacteria (i.e., GM41 strain) helps in increasing the growth of Synechocystis PCC6803 and this growth is further enhanced by the addition of more hexadecane to the medium.

Conclusion

Petroleum wastewater contains a large amount of hydrocarbons which could be a great threat to the environment due to its toxic nature and persistency for a long time. There is also a significant presence of ammonia, hydrogen sulfide and traces of heavy metals in petroleum wastewater. Due to the high demand of water and easy global transport, petroleum industries prefer to be located near sea shores, leading to major pollution of marine ecosystems. It is not feasible to completely cleanup and restore marine ecosystems and further it requires more resources in terms of funds and expertise.

There are many physical, chemical, and biological processes that could be applied for the treatment of petroleum wastewater. Hydrocarbons as a major proportion lie in a suspended state, which is removed primarily by applying mechanical processes i.e., gravity separation. To degrade the hydrocarbons and other toxic compounds, many methods like adsorption photodegradation, coagulation-flocculation, electrocoagulation, photocatalytic oxidation, wet oxidation, catalytic vacuum distillation, fenton oxidation and ultrasonic degradation can be used. Although these methods were found to be effective for the removal and degradation of pollutants, but a major limitation is their high cost which makes them unsuitable for large scale use in petroleum industries.

Microorganisms like many heterotrophic bacteria proved to very effective for the degradation of hydrocarbons and other toxic compounds present in petroleum wastewater. The presence of specific catabolic genes and enzymes in microorganisms facilitates the use of hydrocarbons for their need as carbon and energy sources. Some factors like oxygen concentration, temperature, pH and nutrient availability affects the removal efficiency of microorganisms. This could be the most efficient strategy for detoxification of petroleum wastewater contaminating the environment. But there is a need of further extensive research for the multiple answers related to their mechanism of degradation to design an efficient and better system for oil bioremediation.

Microalgal-mediated remediation could provide a sustainable and effective solution for hydrocarbon degradation. It is controversial whether microalgae degrade hydrocarbon directly however they facilitate such conditions that could fasten oil degradation. There are various studies which indicate the direct involvement of microalgae in hydrocarbon degradation, but presently they are not the principal degraders. Along with heterotrophic bacteria or in symbiotic association, they could have a more sustainable effect for the treatment of petroleum wastewater. Currently major microalgal-mediated remediation involves the suspended growth treatment systems, which have several operational disadvantages related to sludge settleability and their accumulation. Such disadvantages can be removed through the use of attached growth systems or microalgal biofilm systems. They are found to more effective and have great potential as an alternative of suspended growth processes, but there is need of more elaborative research in relation of treatment of petroleum wastewater.

References

Abdelwahab, O., N.K. Amin and E.S.Z. El-Ashtoukhy. 2009. Electrochemical removal of phenol from oil refinery wastewater. J. Hazard. Mater. 163(2–3): 711–716.

Abed, R.M.M., N.M.D. Safi, J. Köster, D. de Beer, Y. El-Nahhal, J. Rullkötter, et al. 2002. Microbial diversity of a heavily polluted microbial mat and its community changes following degradation of petroleum compounds. Appl. Environ. Microbiol. 68: 1674–1683.

Abed, R,M.M. and J. Köster. 2005. The direct role of aerobic heterotrophic bacteria associated with cyanobacteria in the degradation of oil compounds. Int. Biodeterior. Biodegrad. 55: 29–37.

Abed, R.M.M., K. Kohls and D. de Beer. 2007. Effect of salinity changes on the bacterial diversity, photosynthesis and oxygen consumption of cyanobacterial mats from an intertidal flat of the Arabian Gulf. Environ. Microbiol. 9: 1384–1392.

Abed, R.M.M. 2010. Interaction between cyanobacteria and aerobic heterotrophic bacteria in the degradation of hydrocarbons. Int. Biodeterior. Biodegrad. 64: 58–64.

Abed, R.M.M. 2019. Phototroph-heterotroph oil-degrading partnerships. pp. 1–14. *In*: T.J. McGenity (ed.). Microbial Communities Utilizing Hydrocarbons and Lipids: Members, Metagenomics and Ecophysiology. Handbook of Hydrocarbon and Lipid Microbiology. Springer Nature Switzerland AG, Basel, Switzerland.

Ahmadun, F.R., A. Pendashteh, L.C. Abdullah, D.R.A. Biak, S.S. Madaeni and Z.Z. Abidin. 2009. Review of technologies for oil and gas produced water treatment. J. Hazard. Mater. 170: 530–551.

Al-Awadhi, H., R.H. Al-Hasan, N.A. Sorkhoh, S. Salamah and S.S. Radwan. 2003. Establishing oil-degrading biofilms on gravel particles and glass plates. Int. Biodeter. Biodegr. 5: 181–185.

Aldaby, E.S.E. and A.M.M. Mawad. 2019. Pyrene biodegradation capability of two different microalgal strains. Global NEST J. 21(3): 290–295.

Al-Futaisi, A., A. Jamrah, B. Yaghi and R. Taha. 2007 Assessment of alternative management techniques of tank bottom petroleum sludge in Oman. J. Hazard. Mater. 141: 557–564.

Al-Hasan, R.H., D. Al-Bader, N.A. Sorkhoh and S.S. Radwan. 1998. Evidence for n-alkane consumption and oxidation by filamentous cyanobacteria from oil contaminated coasts of the Arabian-Gulf. Marine Biol. 130: 521–527.

Al-Hawash, A.B., M.A. Dragh, S. Li, A. Alhujaily, H.A. Abbood, X. Zhang, et al. 2018. Principles of microbial degradation of petroleum hydrocarbons in the environment. Egyptian J. Aquat. Res. 44: 71–76.

Aljuboury, D.D.A., P. Palaniandy, H.B.A. Aziz and S. Feroz. 2014. Organic pollutants removal from petroleum refinery wastewater with nanotitania photo-catalyst and solar irradiation in Sohar oil refinery. J. Innov. Eng. 2(3): 1–12.

Alloway, B.J. and D.C. Ayres. 1993. Chemical Principles of Environmental Pollution, 1st edition. Chapman and Hall Publishers, India.

Carvalho, A.P., L.A. Meireles and F.X. Malcata. 2006 Microalgal reactors: a review of enclosed system designs and performances. Biotechnol. Prog. 22: 1490–1506.

Cerniglia, C.E., D.T. Gibson and C. van Baalen. 1979. Algal oxidation of aromatic hydrocarbons: formation of 1-naphthol from naphthalene by *Agmenellum quadruplicatum*, strain PR-6. Biochem. Biophys. Res. Commun. 88: 50–58.

Cerniglia, C.E., C. van Baalen and D.T. Gibson. 1980a. Metabolism of naphthalene by the cyanobacterium *Oscillatoria* sp., strain JCM. J. Gen. Microbiol. 116: 485–494.

Cerniglia, C.E., D.T. Gibson and C. van Baalen. 1980b. Oxidation of naphthalene by cyanobacteria and microalgae. J. Gen. Microbiol. 116: 495–500.

Cerniglia, C.E., D.T. Gibson and C. van Baalen. 1982. Naphthalene metabolism by diatoms from the Kachemak Bay region of Alaska. J. Gen. Microbiol. 128: 987–990.

Cerniglia, C.E. 1984. Microbial metabolism of polycyclic aromatic hydrocarbons. Adv. Appl. Microbiol. 30: 31–71.

Chaillian, F., M. Gugger, A. Saliot, A. Coute and J. Oudot. 2006. Role of cyanobacteria in the biodegradation of crude oil by a tropical cyanobacterial mat. Chemosphere 62: 1574–1582.

Chan, S.M.N., T. Luan, M.H. Wong and N.F.Y. Tam. 2006. Removal and biodegradation of polycyclic aromatic hydrocarbons by *Selenastrum capricornutum*. Environ. Toxicol. Chem. 25(7): 1772–1779.

Chavan, A. and S. Mukherji. 2008. Treatment of hydrocarbon-rich wastewater using oil degrading bacteria and phototrophic microorganisms in rotating biological contactor: effect of N : P ratio. J. Hazard. Mater. 154: 63–72.

De Llasera, M.P.G., J. de Jesús Olmos-Espejel, G. Díaz-Flores and A. Montaño-Montiel. 2016. Biodegradation of benzo(a) pyrene by two freshwater microalgae *Selenastrum capricornutum* and *Scenedes musacutus*: a comparative study useful for bioremediation. Environ. Sci. Pollut. Res. 23: 3365–3375.

Diepolder, P. 1992. Is "zero discharges" realistic? Hydrocarb. Process. 71(10): 129.

Diya'uddeen B.H., W.M.A.W. Daud and A.R. Abdul Aziz. 2011. Treatment technologies for petroleum refinery effluents: a review. Process Saf. Environ. 89(2): 95–105.

El-Bestawy, E.A., A.Z. Abd El-Salam and A.E.R.H. Mansy. 2007. Potential use of environmental cyanobacterial species in bioremediation of lindane-contaminated effluents. Int. Biodeterior. Biodegrad. 59: 180–192.

El-Naas, M.H., S. Al-Zuhair and M.A. Alhaija. 2010. Removal of phenol from petroleum refinery wastewater through adsorption on date-pit activated carbon. Chem. Eng. J. 162(3): 997–1005.

El-Naas, M.H., S. Al-Zuhair, M.A. Alhaija. 2010. Reduction of COD in refinery wastewater through adsorption on date-pit activated carbon. J. Hazard. Mater. 173: 750–757.

El-Naas, M.H., M.A. Alhaija and S. Al-Zuhair. 2014. Evaluation of a three-step process for the treatment of petroleum refinery wastewater. J. Environ. Chem. Eng. 2(1): 56–62.

El-Sheekh, M.M., R.A. Hamouda and A.A. Nizam. 2013. Biodegradation of crude oil by *Scenedesmus obliquus* and *Chlorella vulgaris* growing under heterotrophic conditions. Int. Biodeter. Biodegr. 82: 67–72.

Fakhru'l-Razi, A., A. Pendashteh, L.C. Abdullah, D.R.A. Biak, S.S. Madaeni and Z.Z. Abidin. 2009. Review of technologies for oil and gas produced water treatment. J. Hazard. Mater. 170: 530–551.

Farajnezhad, H. and P. Gharbani. 2012. Coagulation treatment of wastewater in petroleum industry using poly aluminum chloride and ferric chloride. Inter. J. Res. Review. Appl. Sci. 13: 306–310.

García-Peña, E.I., P. Zarate-Segura, P. Guerra-Blanco, T. Poznyak and I. Chairez. 2012. Enhanced phenol and chlorinated phenols removal by combining ozonation and biodegradation. Water Air Soil Poll. 223(7): 4047–4064.

Gasim, H.A., S.R.M. Kutty, M. Hasnain-Isa and L.T. Alemu. 2013. Optimization of anaerobic treatment of petroleum refinery wastewater using artificial neural networks. Res. J. Appl. Sci. Eng. Tech. 6: 2077–2082.

Gavrilescu, M. 2010. Environmental biotechnology: Achievements, opportunities and challenges. Dynamic Biochem. Process. Biotech. Mol. Biol. 4: 1–36.

Hasan, D.U.B., A.R.A. Aziz and W.M.A.V. Daud. 2012. Oxidative mineralisation of petroleum refinery effluent using fenton-like process. Chem. Eng. Res. Des. 90(2): 298–307.

Hong, Y.W., D.X. Yuan, Q.M. Lin and T.L. Yang. 2008. Accumulation and biodegradation of phenanthrene and fluoranthene by the algae enriched from a mangrove aquatic ecosystem. Mar. Pollut. Bull. 56: 1400–1405.

Honse, S.O., S.R. Ferreira, C.R.E. Mansur and E.F. Lucas. 2012. Separation and characterization of asphaltenic sub-fractions. Quim Nova 35: 1991–1994.

Hu, G., J. Li and G. Zeng. 2013. Recent development in the treatment of oily sludge from petroleum industry. J. Hazard. Mater. 261: 470–490.

Jasmine, J. and S. Mukherji. 2015. Characterization of oily sludge from a refinery and biodegradability assessment using various hydrocarbon degrading strains and reconstitute consortia. J. Environ. Manag. 149: 118–125.

Jinqi, L. and L. Houtian. 1992. Degradation of azo dyes by algae. Environ. Pollut. 75(3): 273–278.

Juhasz, A.L. and N. Ravendra. 2000. Bioremediation of high molecular weight polycyclic aromatic hydrocarbons: a review of the microbial degradation of banzo[a]pyrene. Int. Biodeterior. Biodegrad. 45: 57–88.

Kumar, A. 2018. Assessment of Cyanobacterial Diversity in Paddy Fields and Their Capability to Degrade the Pesticides. Babasahaeb Bhimrao Ambedkar University, Lucknow, India.

Kumar, A. and J.S. Singh. 2016. Microalgae and cyanobacteria biofuels: a sustainable alternate to crop-based fuels. pp. 1–20. *In*: J.S. Singh, D.P. Singh (eds.). Microbes and Environmental Management. Studium Press Pvt. Ltd. New Delhi, India.

Kumar, A. and J.S. Singh. 2017. Cyanoremediation: a green-clean tool for decontamination of synthetic pesticides from agro- and aquatic-ecosystems. pp. 59–83. *In*: J.S. Singh, G. Seneviratne (eds.). Agro-Environmental Sustainability, Vol. II: Managing Environment Pollution. Springer Int., Cham, Switzerland.

Kumar, A., S. Kaushal, S.A. Saraf and J.S Singh. 2017. Cyanobacterial biotechnology: an opportunity for sustainable industrial production. Clim. Chang. Environ. Sustain. 5(1): 97–110.

Kumar, A., S. Kaushal, S.A. Saraf and J.S Singh. 2018. Microbial bio-fuels: a solution to carbon emissions and energy crisis. Front.Biosci. (Landmark) 23: 1789–1802.

Kumar, A., S. Kaushal, S.A. Saraf and J.S Singh. 2018. Screening of Chlorpyrifos (CPF) tolerant cyanobacteria from paddy field soil of Lucknow, India. Int. J. Appl. Adv. Sci. Res. 3(1): 100-105.

Kumar, A. and J.S. Singh. 2020. Biochar coupled rehabilitation of cyanobacterial soil crusts: a sustainable approach in stabilization of arid and semiarid soils. pp. 167–191. *In*: J.S. Singh, C. Singh (eds.). Biochar Applications in Agriculture and Environment Management. Springer Int., Cham, Switzerland.

Kumar, A. and J.S. Singh. 2020. Microalgal bio-fertilizers. *In*: E. Jacob-Lopes, M.M. Maroneze, M.I. Queiroz, L.Q. Zepka (eds.). Handbook of Microalgae-based Processes and Products. Academic Press, Cambridge, US, In Press.

Lea-Smith, D.J., S.J. Biller, M.P. Davey, C.A.R. Cotton, B.M. Perez Sepulveda, A.V. Turchyn, et al. 2015. Contribution of cyanobacterial alkane production to the ocean hydrocarbon cycle. Proc. Natl. Acad. Sci. USA 112: 13591–13596.

Lei, A.P., Y.S. Wong and N.F.Y. Tam. 2002. Removal of pyrene by different microalgal species. Water Sci. Technol. 46(11–12): 195–201.

Lei, A.-P., Z.-L. Hu, Y.-S. Wong and N.F.-Y. Tam. 2007. Removal of fluoranthene and pyrene by different microalgal species. Bioresour. Technol. 98: 273–280.

Lindquist, B. and D. Warshawsky. 1985a. Identiﬁcation of the 11,12-dihydro-11,12-dihydroxybenzo(a)pyrene as a major metabolite by the green alga, *Selenastrum capricornutum*. Biochem. Biophys. Res. Commun. 130: 71–75.

Lindquist, B. and D. Warshawsky. 1985b. Stereospeciﬁcity in algal oxidation of the carcinogen benzo(a)pyrene. Experientia 41: 767–769.

Liu, A., T. Zhu, X. Lu and L. Song. 2013. Hydrocarbon profiles and phylogenetic analyses of diversified cyanobacterial species. Appl. Energ. 111: 383–393.

Monteiro, C.M., P.M.L. Castro and F.X. Malcata. 2009. Use of microalga *Scendesmus obliquus* to remove cadmium cations from aqueous solutions. World J. Microbiol. Biotechnol. 25: 1573–1578.

Muñoz, R., B.Guieysse and B. Mattiasson. 2003. Phenanthrene biodegradation by an algal-bacterial consortium in two-phase partitioning bioreactors. Appl. Microbiol. Biotechnol. 61: 261–267.

Narro, M.L., C.E. Cerniglia, C. Van Baalen and D.T. Gibson. 1992. Metabolism of phenanthrene by the marine cyanobacterium *Agmenellum quadruplicatum* PR-6. Appl. Environ. Microbiol. 58: 1351–1359.

Paerl, H.W., B.M. Bebout, S.B. Joye and D.J. Des Marais. 1993. Microscale characterization of dissolved organic-matter production and uptake in marine microbial mat communities. Limnol. Oceanogr. 38: 1150–1161.

Pathak, C. and H.C. Mandalia. 2012. Petroleum industries: environmental pollution effects, management and treatment methods. Int. J. Separation Environ. Sci. 1(1): 55–62.

Perera, F.P., D. Tang, S. Wang, J. Vishnevetsky, B. Zhang, D. Diaz, et al. 2012. Prenatal polycyclic aromatic hydrocarbon (PAH) exposure and child behavior at age 6-7 years. Environ. Health Perspect. 120: 921–926.

Pérez, G., A.R. Fernández-Alba, A.M. Urtiaga and I. Ortiz. 2010. Electro-oxidation of reverse osmosis concentrates generated in tertiary water treatment. Water Res. 44: 2763–2772.

Poulopoulos, S.G., E.C. Voutsas, H.P. Grigoropoulou and C.J. Philippopoulos. 2005. Stripping as a pretreatment process of industrial oily wastewater. J. Hazard. Mater. 117: 135–139.

Radwan, S.S. and R.H. Al-Hasan. 2000. Oil pollution and cyanobacteria. pp. 307–319. *In*: B.A. Whitton and M. Potts (eds.). The Ecology of Cyanobacteria. Kluwer Academic Publishers, Dordrecht, The Netharlands.

Radwan, S.S., R.H. Al-Hasan, S. Salamah and S. Al-Dabbous. 2002. Bioremediation of oily sea water by bacteria immobilized in biofilms coating macroalgae. Int. Biodeter. Biodegr. 50: 55–59.

Rasalingam, S., R. Peng and R. Koodali. 2014. Removal of hazardous pollutants from wastewaters applications of TiO_2-SiO_2 mixed oxide materials. J. Nanomater. 2014 (617405): 1–42.

Remya, N. and J-G. Lin. 2011. Current status of microwave application in wastewater treatment—a review. Chem. Eng. J. 166: 797–813.

Saber, A., H. Hasheminejad, A. Taebi and G. Ghaffari. 2014. Optimization of fenton-based treatment of petroleum refinery wastewater with scrap iron using response surface methodology. Appl. Water Sci. 4: 283–290.

Safonova, E., K. Kvitko, P. Kuschk, M. Möder and W. Reisser. 2005. Biodegradation of phenanthrene by the green alga *Scenedesmus obliquus* ES-55. Eng. Life Sci. 5: 234–239.

Safonova, E.T., I.A. Dmitrieva and K.V. Kvitko. 1999. The interaction of algae with alcanotrophic bacteria in black oil decomposition. Resour. Conserv. Recycl. 27: 193–201.

Saien, J. and H. Nejati. 2007. Enhanced photocatalytic degradation of pollutants in petroleum refinery wastewater under mild conditions. J. Hazard. Mater. 148(1–2): 491–495.

Sanchez, O., I. Ferrera, N. Vigues, T.G. de Oteyza, J. Grimalt and J. Mas. 2006. Role of cyanobacteria in oil biodegradation by microbial mats. Int. Biodeterior. Biodegrad. 58: 186–195.

Schoeny, R., T. Cody, D. Warshawsky and M. Radike. 1988. Metabolism of mutagenic polycyclic aromatic hydrocarbons by photosynthetic algal species. Mutat. Res. 197: 289–302.

Semple, K.T. and R.B. Cain. 1996. Biodegradation of phenols by the alga *Ochromonas danica*. Appl. Environ. Microb. 62: 1265–1273.

Semple, K.T., R.B. Cain and S. Schmidt. 1999. Biodegradation of aromatic compounds by microalgae. FEMS Microbiol. Lett. 170: 291–300.

Shashirekha, S., L. Uma and G. Subramanian. 1997. Phenol degradation by the marine cyanobacterium *Phormidium valderianum* BDU 30501. J. Ind. Microbiol. Biotechnol. 19: 130–133.

Singh, J.S., A. Kumar, A.N. Rai and D.P. Singh. 2016. Cyanobacteria: a precious bioresource in agriculture, ecosystem, and environmental sustainability. Front. Microbiol. 7: 529.

Singh, J.S., S. Koushal, A. Kumar, S.R. Vimal and V.K. Gupta. 2016. Book review: Microbial inoculants in sustainable agricultural productivity, Vol. II: Functional application. Front. Microbiol. 7: 2105.

Singh, J.S., A. Kumar and M. Singh. 2019. Cyanobacteria: a sustainable and commercial bio-resource in production of bio-fertilizer and bio-fuel from wastewaters. Environ. Sustain. Indic. 3: 100008.

Subashchandrabose, S.R., P. Logeshwaran, K. Venkateswarlu, R. Naidu and M. Megharaj. 2017. Pyrene degradation by *Chlorella* sp. MM3 in liquid medium and soil slurry: possible role of dihydrolipoamide acetyltransferase in pyrene biodegradation. Algal Res. 23: 223–232.

Takáčová, A., M. Smolinská, J. Ryba, T. Mackuľak, J. Jokrllová, P. Hronec, et al. 2014. Biodegradation of benzo[a]pyrene through the use of algae. Cent. Eur. J. Chem. 12(11): 1133–1143.

Tang, X., L.Y. He, X.Q. Tao, Z. Dang, C.L. Guo, G.N. Lu and X.Y. Yi. 2010. Construction of an artificial microalgal-bacterial consortium that efficiently degrades crude oil. J. Hazard. Mater. 181: 1158–1162.

Thakur, C., V.C. Srivastava, I.D. Mall and A.D. Hiwarkar. 2018. Mechanistic study and multi response optimization of the electrochemical treatment of petroleum refinery wastewater. Clean: Soil, Air, Water 46(3): 1700624: 1–19.

Tobiszewski, M., S. Tsakovski, V. Simeonov and J. Namieśnik. 2012. Chlorinated solvents in a petrochemical wastewater treatment plant: an assessment of their removal using self-organising maps. Chemosphere 87: 962–968.

Tony, M.A., P.J. Purcell and Y. Zhao. 2012. Oil refinery wastewater treatment using physicochemical, fenton and photo-fenton oxidation processes. J. Environ. Sci. Health. A. Tox. Hazard. Subst. Environ. Eng. 47(3): 435–440.

Usman, M., P. Faure, K. Hanna, M. Abdelmoula and C. Ruby. 2012. Application of magnetite catalyzed chemical oxidation (fenton-like and persulfate) for the remediation of oil hydrocarbon contamination. Fuel 96: 270–276.

Varjani, S.J. 2017. Microbial degradation of petroleum hydrocarbons. Bioresour. Technol. 223: 277–286.

Varjani, S.J. and V.N. Upasani. 2017. A new look on factors affecting microbial degradation of petroleum hydrocarbon pollutants. Int. Biodeterior. Biodegrad. 120: 71–83.

Varjani, S.J. and M.C. Sudha. 2018. Treatment technologies for emerging organic contaminants removal from wastewater. pp. 91-115. In: S. Bhattacharya, A.B. Gupta, A. Gupta and A. Pandey (eds) Water remediation. Springer Nature, Singapore.

Vendramel, S., J.P. Bassin, M. Dezotti and G.L. Sant'Anna Jr. 2015. Treatment of petroleum refinery wastewater containing heavily polluting substances in an aerobic submerged fixed-bed reactor. Environ. Tech. 36: 2052–2059.

Veyrand, B., V. Sirot, S. Durand, C. Pollono, P. Marchand, G. Dervilly-Pinel, et al. 2013. Human dietary exposure to polycyclic aromatic hydrocarbons: results of the second French total diet study. Environ. Int. 54: 11–17.

Vidyashankar, S. and G.A. Ravishankar. 2016. Algae-based bioremediation: bioproducts and biofuels for biobusiness. pp. 457–483. *In*: M.N.V. Prasad (ed.). Bioremediation and Bioeconomy. Elsevier, Amsterdam, Netherlands.

Wang, J., W. Zhou, H. Yang, F. Wang and R. Ruan. 2015. Trophic mode conversion and nitrogen deprivation of microalgae for high ammonium removal from synthetic wastewater. Bioresour. Technol. 196: 668–676.

Warshawsky, D., M. Radike, K. Jayasimhulu and T. Cody. 1988. Metabolism of benzo(a)pyrene by a dioxygenase enzyme system of the freshwater green alga *Selenastrum capricornutum*. Biochem. Biophys. Res. Commun. 152(2): 540–544.

Warshawsky, D., T.H. Keenan, R. Reilman, T.E. Cody and M.J. Radike. 1990. Conjugation of benzo[a]pyrene metabolites by the freshwater green alga *Selenastrum capricornutum*. Chem. Biol. Interact. 73: 93–105.

Warshawsky, D., T. Cody, M. Radike, R. Reilman, B. Schumann, K. LaDow, et al. 1995. Biotransformation of benzo[a]pyrene and other polycyclic aromatic hydrocarbons and heterocyclic analogs by several green algae and other algal species under gold and white light. Chem. Biol. Interact. 97: 131–148.

Wieland, A. and M. Kühl. 2006. Regulation of photosynthesis and oxygen consumption in a hypersaline cyanobacterial mat (Camargue, France) by irradiance, temperature and salinity. FEMS Microbiol. Ecol. 55: 195–210.

Yan, Z., H. Jiang, X. Li and Y. Shi. 2014. Accelerated removal of pyrene and benzo [a] pyrene in freshwater sediments with amendment of cyanobacteria-derived organic matter. J. Hazard. Mater 272: 66–74.

Yavuz, Y. and A.S. Koparal. 2006. Electrochemical oxidation of phenol in a parallel plate reactor using ruthenium mixed metal oxide electrode. J. Hazard. Mater. 136(2): 296–302.

Zafra, G., A. Moreno-Montano, A. Absalon and D. Cortes-Espinosa. 2015. Degradation of polycyclic aromatic hydrocarbons in soil by a tolerant strain of *Trichoderma asperellum*. Environ. Sci. Pollut. Res. 22: 1034–1042.

Distillery and Sugar Mill Wastewater

Introduction

Distilleries and sugar mills are known to generate a large amount of wastewater, in which the maximum of its raw materials are discharged as wastewater. It is estimated that about 15 liters of wastewater (spent wash) is released to produce 1 liter of alcohol (Ravikumar et al. 2007, Kharayat 2012). The severity of distillery and sugar mill wastewater depends on the composition, which varies according the plants, type of processes and the quality of the product. Further the presence of various recalcitrant compounds like melonidins in the distillery and sugar mill wastewater makes them very difficult for treatment. Melonoidins are complex organic pollutants that are generated through the Maillard reaction, and discharged in large amounts in distillery and sugar mill operations; they found to be highly resistant to the microbial decomposition (Kumar and Chandra 2006).

The conventional treatment of distillery wastewater involves the physicochemical processes that are primarily used for the removal of various pollutants. Further biological treatment is also followed for the degradation of pollutants, but the main focus is on bacteria or fungal-mediated treatment of the distillery and sugar mill wastewater. Due to the inefficiency and complex nature, conventional treatment is not enough for the efficient and appropriate treatment of the distillery and sugar mill wastewater. Although there are additional advanced processes like flocculation, coagulation, filtration and oxidation methods that have proved to be significant in treatment, but the high cost and the need of skilled operators makes them unaffordable and unfeasible for widespread use.

Microalgal remediation could be a potential and affordable approach in the treatment of distillery and sugar mill wastewater (Kumar and Singh 2016, Singh et al. 2016a, b, Kumar et al. 2017, Kumar and Singh 2017). Initially it was proposed that microalgae could be used for the treatment of domestic/municipal wastewater, later it was also found to be very effective in the different wastewaters including distillery and sugar mill wastewater (Kumar 2018, Kumar et al. 2018a, b, Singh et al. 2019, Kumar and Singh 2020a, b). Due to the minimum growth requirements and wide range of survivability, microalgae could be used for the removal, biodegradation or biotransformation of colored compounds (Olaizola 2003). Further biotechnological implication in the field of microalgal remediation, makes it an interesting and viable means of bioremediation in the current scenario (Sasso et al. 2012, Hallmann 2007). The use of immobilized or dead biomass of microalgae could provide an advantage for the treatment of hazardous contaminants (Moreno-Garrido 2008, Muñoz and Guieysse 2006, Singh and Dhar 2007).

Distillery and Sugar Mills: A Brief Description

Raw materials sugarcane and sugar beet are washed with warm water to remove impurities such as mud and dust. This cleaning of the sugarcane involves slicing big pieces into small ones; from where it is transferred in to rolling mills to press the sugarcane and extract the juices from it. To increase the juice %, pressed sugarcane is further wetted in hot water and again pressed to recover the maximum juice. The extracted juice is conveyed to juice clarifiers for purification, and bagasse is generally sent to boiler houses for fuel purposes.

The extracted juice or sugar extract is composed of 15–20% of solid, 10–15% sucrose and a purity of 70–90%. The sugar extract or solution is heated up to 70–75°C to concentrate the extract upto 15% of the original solution. Then the concentrated solution is conveyed to another clarifier and heated up to 110–120 °C with different materials such as lime, salts, silicates, invert sugar, amino acids, enzymes and some organic acids to achieve the 7.5 pH. This concentrated sugar solution is separated out into a liquid-form syrup and a suspended solid form of mud, which is coagulated by phosphate and the maximum is removed. The syrup still has a significant amount of water, which is further conveyed to vacuum boilers for evaporation of water. After the vacuum evaporation, the syrup concentration reaches up to 60% as compared to 15% in clarifiers.

The concentrated syrup is sent to a single effect vacuum pan, where evaporation of the syrup takes place until it reaches up to a super saturation and the syrup is transformed into grains (sugar). More water is added to increase the growth and size of grains, this process is continued till the vacuum pan is filled with sugar crystals and syrup (dense mass). The sugar crystals are removed from the pan and sent to a crystallizer to cool it. The dense mass is also called masticate.

The dense mass or masticate is conveyed to centrifugation, where the remaining syrup in masticate is separated through high spinning. This separated syrup which is also known as molasses is again circulated to the crystallizer to recover the maximum sugar grains, to form more crystals. The crystals free from molasses and masticates are treated as commercial sugar, which further passes to drier and standard sizes of a mesh. Molasses and masticates are sent to the distillery to make vinegar, citric acid and other products.

Primary sugar appears to be yellow-brownish in color, due to the presence of plant pigment, reaction with amino acid and reduction of sucrose. The color is generally removed through the use of bone char and carbon material. Due to high crystallization and drying, sugar crystals lose the moisture in it, which drench the sugar again. To remove extra moisture from sugar, conditioning is applied, in which air-current is passed to the sugar. The refined sugar is stocked up in silos and according to the requirements in the market, packaged in different capacity bags and send to markets from time to time.

Distilleries use different materials such as sugarcane/sugar beet juice, molasses, fruits and grains. They can be integrated with sugar production or operated in separation. The grains are the main material which is primarily used for alcohol production. It involves many steps such as, grinding, saccharification of starch, fermentation and distillation. In case of molasses and fruit juices, the first two steps are not needed, only fermentation and distillation are required.

The grains are conveyed in to a rotating brushing machine to remove the chaff from the grains. The cleaned grains are then grinded in a hammer mill to make flour. Before the saccharifiction of starch, flour is mixed with water in the mash turn. Then the mash undergoes heating to reach 100°C by the direct addition of steam, which is further cooled down to about 69°C. The saccharification process starts with the addition of malt-germinated barley, and this process operates at 61°C for about 1 hour. Then the temperature of mash comes down to 30°C before being pumped into fermentation tanks.

The cooled mash is mixed with water to reduce the sugar concentration, as more sugar concentration acts as a limiting factor for the growth of yeast, leading to insufficient conversion of the sugar into alcohol. For the fermentation of mash, baker's yeast is primarily used and is continued for about 72 hours. The advantage of using baker's yeast is that it not only transforms the sugar into alcohol, but also facilitates the synthesis of other aromatic entities.

The fermented mash is pre-heated into a condenser, before it goes to distillation columns. In the distillation process, condensed fermented mash is heated up to 79°C to evaporate the alcohol (boiling point 79°C). The steam coming from the mash carries the volatile components along with it, which is cooled down in the first condenser; then conveyed to the second condenser to bring down the mash temperature up to room temperature. After the first distillation, about 57% volume of alcohol is produced, known as phlegm and the residue remains at the bottom of the columns, which is removed as waste products and could be used as nutritious fodder for cattle. The phlegm is stored in tanks, for further processing. Further a second distillation also follows to convert the phlegm into malt wine or grain alcohol.

Distillery and Sugar Mill Wastewater: Characteristics and Specific Pollutants

Distillery wastewater is composed of high organic and inorganic contents

Table 6.1: Characteristics of distillery and sugar mill wastewater

Water quality parameter	Chowdhary et al. 2017	Jadhav et al. 2013	Padmapriya et al. 2012	Doušková et al. 2010	Simate et al. 2011	Prasad and Srivastava 2009	Travieso et al. 2008	Melamane et al. 2007
pH	3.0-4.1	6.8	5.01	3.5	3-12	4.2-4.3	5.3	3.53-5.4
Conductivity (ms cm^{-1})	0.346		3.195			1.733		
Alkalinity (mg L^{-1})							7500	3.53-62.4
TS (mg L^{-1})	51.5-100		2386	38.3	500-8750		70615	11.4-32
TDS (mg L^{-1})			2258	33.4	2020-5940	5500-5700		
TSS (mg L^{-1})		80	370	4.9	2901-3000		7690	2.4-5.0
BOD (mg L^{-1})	30	681	417		1200-3600	700-7500	38600	0.21-8.0
COD (mg L^{-1})	100-120	3549	412	54.2	2000-6000	10000-11000	76960	3.1-40
TN (mg L^{-1})				51	25-80			0.1-64
TP (mg L^{-1})					10-15			0.24-65.7
TOC (mg L^{-1})				32.9				2.5-6.0
Phenol (mg L^{-1})								0.028-0.474

Note: TS: Total Solids; TDS: Total Dissolved Solids; TSS: Total Suspended Solids; BOD: Biological Oxygen Demand; COD: Chemical Oxygen Demand; TN: Total Nitrogen; NH$_4^+$-N: Ammonium Nitrogen; NO$_3^-$-N: Nitrate Nitrogen; TP: Total Phosphorous; TOC: Total Organic Content.

which are discharged from different processes carried out during the manufacturing of sugar and alcohol. It also contains a wide variety of natural pigments and colorants such as carotenoids, betalains, riboflavins, melanins, quinines, caramels, melanoidins and alkaline degradation of hexoses (ADPH) (Dai and Mumper 2010). Although polyphenolic compounds (colorants) in the distillery and sugar mill wastewater are also known for their antioxidantal, antimicrobial and anticarcinogenic activities (Borrelli et al. 2003, Silvan et al. 2006). The quantity of these polyphenols in distillery and sugar mill wastewater depends on the sources of molasses and the sugar concentrations in the feed flow.

Table 6.2: Specific pollutants in distillery and sugar mill wastewater

Industry	Types	Process	Wastewater
Integrated sugar and ethanol production	Sugar mill	Extraction of Juice	Bagasse, press mud, molasses, melanoidins, ADPHs, lignins & tannins,
		Clarification	
		Evaporation	
		Crystallization	
		Centrifugation	
		Refining	
	Distillery	Grinding (For grains)	
		Saccharification (For grains)	
		Fermentation	
		Distillation	

Melanoidins

Melanoidins are brown nitrogenous polymers and co-polymers which are formed through polymerization reactions of highly reactive intermediates during the Maillard reaction. In the Maillard reaction, amino compounds react to the presence of monosaccharides and other carbonyl compounds, leading to the formation of these melanoidins through a wide range of reactions like rearrangements, cyclizations, isomerization, dehydrations and condensation (Martins et al. 2001, Kim and Lee 2009).

Alkaline Degradation Products of Hexoses (ADPH)

In alkaline solutions, monosaccharides present in the sugar extract undergo many reversible reactions: ionization, mutarotation, enolization and isomerization; which lead to the formation of enediol anions. These enediol species are known to be the possible intermediates in the conversion of monosaccharides to ADPHs through the ionization, mutarotation, enolization and isomerization reactions. These ADPHs, along with the melonidins contribute up to 80% of color in distillery and sugar mill wastewater.

Caramels

Caramels are color products that are formed through the thermal degradation of sugars normally monosaccharides. It occurs when concentrated sugar syrup is heated at a temperature of above 210°C, there is a generation of yellow or brown color compounds through the intramolecular rearrangements, the process is known as caramelization. They are colloidal in nature and have a tendency to remain primarily on the surface of sugar crystals, which are responsible for downgrading the quality of white sugar.

Distiller Grains and Slop

In a distillery, there is a presence of unfermented grain particles after settling and decanting of the liquid, known as wet distiller's grains. The decanted liquid from the wet distiller's grains is known as thin slop or wet distiller's soluble, which is produced in huge amounts during the alcohol production process in distilleries. They are characterized with a dark color and very high BOD.

Impacts on Environment and Public Health

The distillery wastewater are characterized with high BOD, COD, organic matter and the presence of phenols and lignin derivatives; and traces of heavy metals due to the use of several chemicals (Chaudhary and Arora 2011).When this distillery wastewater is discharged into adjoining aquatic and terrestrial habitats, it creates a serious threat to ecosystems and their organisms (Agarwal and Pandey 1994). The water bodies receiving distillery waste containing color and high organic load are affected in two ways: (a) color affects the penetration of light in aquatic ecosystems, which in turn affect the aquatic fauna and flora; (b) high organic load leads to the low pH and eutrophication which decreases the dissolved oxygen level in the contaminated aquatic bodies (Fitz Gibbon et al. 1998, Kumar and Chandra 2004, Ramakritinan et al. 2005, Bharagava and Chandra 2010, Yadav and Chandra 2012).These factors could be responsible for a large scale of fish mortality in aquatic ecosystems (Ramakritinan et al. 2005).

It also causes soil pollution and is responsible for acidification of the soil on the untreated discharge of distillery wastewater. Mohana et al. (2009) and Bharagava and Chandra (2010) observed that untreated sugar mill wastewater or Post Methylated Distillery Effluent (PMDE) from distilleries are responsible for inhibition of seed germination, reduction of soil alkalinity and cause manganese deficiency in the soil. Further colored compounds like melanoidins have antioxidant properties, which could be more toxic to all soil organisms including microorganisms, therefore, it should undergo treatment before discharge into the environment (Dahiya et al. 2001, Chandra et al. 2008).

Regarding human health, it is reported that distillery wastewater causes irritation of the eyes, headaches, skin allergies, vomiting sensation, fever, and stomach pain. Due to contamination of this wastewater, it poses a serious threat to the water quality in affected regions. Currently there is great concern about the chemical nature of melanoidins and their effects on human health. It is also reported that effects of melanoidins could be mutagenic, cytotoxic and carcinogenic (Somoza 2005, Silvan et al. 2006, Plavsic et al. 2006).

Current Treatment and Its Challenges

Physicochemical processes are conventionally used for the treatment of distillery and sugar mill wastewater. The common physical treatment methods are flotation and sedimentation along with preliminary screening and flow equalization mixing in the wastewater treatment (Thakur 2006). The main purpose of physical treatment of the wastewater is the reduction of the suspended/settable solids from wastewater, which are either removed through inexpensive sedimentation or by the use of gravity force to separate suspended solids, oil and grease from distillery and sugar mill wastewater (Jayanti and Narayanan 2004).

Coagulation and flocculation are the main physicochemical methods that are primarily used for rapid and cost-effective removal of pollutants from distillery and sugar mill wastewater (Tatsi et al. 2003, Mohana et al. 2009). Pikaev et al. (2001) and Beltrain de Heredia et al. (2005) investigated the use of a coagulant and integrated Fenton-coagulant/flocculation process, and found that they significantly removed the COD from distillery and sugar mill wastewater. Agrawal and Pandey (1994) and Chaudhari et al. (2007) observed the significant effect of coagulants with many salts such as ferrous sulfate, ferric chloride, calcium oxide and aluminum chloride, in color and COD reduction. Along with physicochemical processes, anaerobic and aerobic biological treatment processes significantly reduced the BOD and COD levels of distillery and sugar mill wastewater, but there was no major improvement in the color levels, which could be due to the presence of melanoidins like recalcitrant compounds (Satayawali and Balakrishnan 2008). These physical and biological processes have some disadvantages such as high cost, use of chemicals, a generation of a large amount of solid waste that might further lead to the formation of harmful byproducts (Kanimozhi and Vasudevan 2010).

There are many advanced processes that are also used for the treatment of distillery and sugar mill wastewater. Nataraj et al. (2006) emphasized the use of membrane based nanofiltration and reverse osmosis processes in the efficient treatment of the distillery and sugar mill wastewater. These advanced methods have proved to be very effective in the removal of pollutants from wastewater. However their high building and operating cost makes them unviable for industries.

Microalgal Bioremediation

Distillery and sugar mill wastewater have brownish black, high turbidity, high BOD and COD and the presence of significant amounts of lignin and tannin which are very persistent in nature and not easily degradable for microorganisms except for a few fungal species. Further there is the presence of toxic compounds such as melanoidins, caramels, Alkaline Degradation Products of Hexoses (ADPH) and unfermented distiller grain slop that makes them difficult to treat by physicochemical and biological (use of bacteria and fungi) methods. To overcome this problem, microalgae-mediated remediation could provide an effective, sustainable and economical approach for the treatment of distillery and sugar mill wastewater.

There are a number of cyanobacterial strains such as *Oscillatoria* sp., *Lyngbya* sp., *Synechocystis*, *Pithophora* sp., *Scenedesmus* sp, *Chlorella vulgaris* and *Spirulina platensis*, which have been successfully used for the removal of color, solids, BOD from the waste water (Table 6.3) (Patel et al. 2001, Vijayakumar et al. 2005, Dhamotharan et al. 2009, Murugesan et al. 2010, Rao et al. 2011, Ganapathiselvam et al. 2011). Although microalgae are not found to be very effective in the degradation of lignin and tannins that are present in distillery and sugar mill wastewater. However many microalgal strains like *Phormidium ambiguum*, *Chroococcus minutes*, *Oscillatoria willei*, *Phormidium valderianum* and *Anabaena azollae* exist for the degradation of lignin and are also reported for the presence of laccase enzymes that known for the degradation of lignin in the microalgae (Semple and Cain 1996, Semple et al. 1999, Kalavathi et al. 2001, Papazi and Kotzabasis 2007, Anbuselvi and Jeyanthi 2009).

Table 6.3: Microalgae in distillery wastewater treatment

Microalgae	*Removal of parameters*	*References*
Oscillatoria boryana BDV 92181	54% decolorization at 5% (v/v)	Campbell and Laudenbach 1993
Oscillatoria sp.	Color reduction of 96-98%	Patel et al. 2001
Lyngbya sp.	Color reduction of 81%	Patel et al. 2001
Synechocystis sp.	Color reduction of 26%	Patel et al. 2001
Chlorella vulgaris	COD and BOD removals of 83.2% and 88.0%, respectively	Della Greca et al. 1992 Kanimozhi and Vasudevan 2010
Spirulina maxima		Scalbert 1991
Oscillatoriaboryana BDV 92181	Decolorization of pure pigment (0.1% w/v) by about 75%	Campbell and Laudenbach 1993
Scenedesmus sp.		Padmapriya et al. 2012

Padmapriya et al. (2012) observed that on inoculation of *Scenedesmus* sp., the color of wastewater turns to greenish from brownish black, and there is a 93.83 and 3.19% reduction in the turbidity and Total Dissolved Solids (TDS) respectively. It is reported that microalgae *Pithophora* sp. is able to reduce the turbidity up to 48.68%; and *Scenedesmus* sp. is able to reduce the Total Solids (TS), total and Total Suspended Solids (TSS) up to 7.88 and 90.63% respectively in distillery wastewater (Dhamotharan et al. 2009). It was also reported that *Chlorella vulgaris* helps in reduction of the level of TS, TDS and TSS up to 46.25, 63.75 and 1.3% in different wastewater (Murugesan et al. 2010, Rao et al. 2011).

Due to excess growth of microalgae, there is an increase in the pH from 5.01 to 7.85 of distillery wastewater. Similar results were also observed by Vijayakumar et al. (2005) in which *Oscillatoria* sp. was able to increase the pH in dye-containing wastewater. Further there is a reduction in the electrical conductivity of the untreated wastewater up to 2.82%, it could be because of the use of inorganic compounds by the microalgae. It was observed that low electrical conductivity could indicate the presence of organic compounds in wastewater. It was investigated that *Scenedesmus* sp. and *Chlorella vulgaris* were be able to reduce the total hardness of the wastewater up to 8.92 and 50% (Padmapriya et al. 2012, Rao et al.2011). Ganapathiselvam et al. (2011) investigated the ability of microalgae *Nostoc* in sequestration of calcium, magnesium, whose removal proved to be quite difficult in conventional systems such as in settlement tanks and aerobic digester; but *Nostoc* sp. Successfully sequestrate the Ca, Mg from the wastewater. Padmapriya et al. (2012) reported similar results involving *Scenedesmus* sp. to sequestrate approximately 95.30% of Fe, 11.54% of Ca, 14.41% of Mg and 15% of K from distillery wastewater. However there is an increase of 12.26% in the sodium level in treated distillery wastewater. On microalgal treatment, there is a significant reduction in the BOD and COD of distillery wastewater. It is in the range of 53 to 97.13% in case of BOD and 68 to 97.87% in case of COD by using microalgae like *Scenedesmus* sp. and *Spirulina platensis* (Manju and Soumya 2010, Ganapathiselvam et al. 2011, Padmapriya et al. 2012).

Removal of Lignin and Tannins

Lignins and tannins are phenolic polymers that show similarity in structure but have some differences in characteristics. Lignins are amorphous, heterogeneous and hydrophobic polymers that make up the middle lamella of the plant cell wall, while tannins are astringent, relatively homogenous and less hydrophobic that are known to precipitate out with proteins. They primarily provide the structural support to wood in plants especially in trees. Lignins and tannins are quite stable and prove to be toxic to many members of the microbial communities, which makes them very difficult to degrade biologically. Currently biodegradation of lignin and tannins have received much attention due to their greater structural complexity and abundance in distillery wastewater (Robinson et al. 2001, Murugan and Al-Sohaibani

2010). It is reported that anaerobic conditions are found to be not effective in breaking the aromatic rings in lignin and tannins, but aerobic conditions have proved to be helpful in biodegradation. But due to partial degradation and the slow rate, it is also difficult to degrade lignin and tannins in the aerobic conditions (Heider and Fuchs 1996, Kuhad et al. 1997).

In relation to microalgae-mediated remediation, there are some reports available which describe the biodegradation of lignin and tannin in distillery wastewater. In earlier reports, Bharati et al. (1992) observed that *Phormidium ambiguum* and *Chroococcus minutus* were able to reduce the lignin content up to 73.0% in a span of 5 days in pulp and paper wastewater. For the degradation of lignin, microalgae need to produce extracellular enzymes like peroxidases and laccase. The microalgae *Anabaena azollae*, are known for their use as a biofertilizer; also have the capability of lignolysis of coir waste (Malliga et al. 1996). Anbuselvi and Jeyanthi (2009) conducted a study involving the use of microalgal strains such as *Phormidium*, *Oscillatoria*, and *Anabaenaazollae* for the degradation of coir waste, which showed that they are able to degrade lignin and hemicelluloses up to 89 and 92% respectively.

Melanoidin Degradation

Melanoidins are colored compounds that are significantly responsible for the strong color of distillery wastewater. Microalgae are known for their ability to degrade complex compounds and oxygenation of wastewater; which makes them an excellent agent in removing the color, BOD and COD from different wastewaters including distillery wastewater (Satayawali and Balakrishnan, 2008). Kalavathi et al.(2001) screened cyanobacterial genera like *Oscillatoria*, *Phormidium*, *Spirulina* and *Synechococcus* for decolorization/degradation of anaerobically treated distillery wastewater in open field conditions.

Kalavathi et al. (2001) reported that *Oscillatoria* showed the maximum degradation efficiency, which is followed by *Phormidium*, *Spirulina* and *Synechococcus* describing the lowest degradation efficiency. It is also reported that *Oscillatoria boryana* BDV 92181 is able to use melanoidins for their nitrogen and carbon needs, which leads to the overall decolorization of distillery wastewater (Kalavathi et al. 2001). The degradation efficiency of *Oscillatoria boryana* BDV 92181 goes from 75-60% in respect to their concentration range of 0.1-0.5% w/v pure melanoidin; but beyond 0.5% w/v, there is a decrease in the degradation efficiency of the microalgae, it could be inhibition of growth of the microalgae by interfering with the penetration of light (Kalavathi et al. 2001). Patel et al. (2001) observed that *Oscillatoria* sp., *Lyngbya* sp. and *Synechocystis* sp. were able to decolorize distillery wastewater and also reported that these microalgal strains achieved 96, 81 and 26% color reduction, respectively. Further using of a consortium of all the three strains showed the maximum decolorization of 98%. Color reduction of distillery wastewater occurred through the initial absorption process and then degradation of colored compounds in subsequent stages (Patel et al. 2001).

Valderrama et al. (2002) conducted laboratory experiments using *Chlorella vulgaris* and macrophyte *Lemna minuscula* for the sequential treatment of two types of wastewater: first, untreated washing water from all the processes in the sugar mill; and the second, the slops wastewater from an ethanol distillery. In this experiment, the growth of microalgae *Chlorella vulgaris* is inhibited by the second type of wastewater, due to its high COD and strong color. The use of macrophyte *Lemna minuscula* only achieves up to 10% color reduction, while pretreatment with *Chlorella* (4 day period) followed by *Lemna minuscula* (6 day period) achieves up to 52% color reduction.

Enzymes Involved in Lignin and Melonidin Degradation

Laccases (LACs) are considered as the main enzymes, predominantly reported in basidiomicetous fungi; which is involved in the decolorization of distillery wastewater. Besides this, a melanoidin-induced laccase gene expression is inducted in to the white-rot fungus *Trametes* sp. I-62 demonstrated the role of laccase in the decolorization of distillery wastewater (González et al. 2008). But there are a few studies which reported the presence of laccase (LACs) and polyphenol oxidases (PPOs) in microalgae and cyanobacteria. Shashirekha et al. (1997) reported the presence of polyphenol oxidase and laccase activity in *Phormidium valderianum* in response to increasing the phenol concentration. There is a five-fold and two-fold increase in polyphenol oxidase and laccase activity of the *Phormidiumvalderianum* when phenols concentration of increased from 25 mg^{L-1} to 50 mg^{L-1} (Shashirekha et al. 1997). Further accumulation of protein concentration in the medium increases d indicating the novo synthesis of phenol-degrading enzymes.

Palanisami et al. (2010) observed the constitutive expression of LACs and PPOs in microalgal genera *Oscillatoria* and *Phormidium* for the decolorization of the lignin model polymeric dye Poly R-478. Palanisami et al. (2010) also reported that *Phormidium valderianum* showed the maximum decolorization values (65%) among the other species of *Oscillatoria* and *Phormidium*. Saha et al. (2010) reported that *Oscillatoria willei* performed the decolorization of lignin model Poly R-478 up to 52% in 7 days under limited nitrogen conditions, and there was a detection of increased activity of ligninolytic and oxidative enzymes such as LACs, PPOs, superoxide dismutase, peroxidase, lignin peroxidase, catalase and ascorbate peroxidase. The role of other microalgal enzymes in relation to degradation of melanoidins has also been examined. Stewart et al. (1975) investigated the role of glutamine synthetase enzyme, an efficient ammonia scavenger, which is known for the assimilation of ammonia from the amino acid component of the pigment. Kalavathi et al. (2001) observed the high glutamine synthetase activity in microalgae that was growing in media containing both pigment and effluent.

Hayase et al. (1984), Patil and Kapadnis (1995), Campbell and Laudenbach (1995), Asada (1999), Leonowicz et al.(2001) and Kalavathi et al.

(2001) emphasized the importance of Reactive Oxygen Species (ROS) such as superoxide anion (O_2^-), hydroxyl radical (OH^-), and hydrogen peroxide (H_2O_2) in melonidin degradation. These ROS are produced in all photosynthetic organisms in response to oxidative stress caused by pollutants (Asada 1999). Patil and Kapadnis (1995) suggested that H_2O_2 production in microalgae could also lead to melanoidin degradation. There are observations of H_2O_2 induced melonidin degradation, when H_2O_2 is *in vitro* added to melanoidin containing wastewater; showing 97% color reduction (Patil and Kapadnis 1995).

Kalavathi et al. (2001) reported that *Oscillatoria* is found to produce strong oxidizing agent H_2O_2 at the rate of 0.538 µmoles µg chlorophyll-Lh-1. In addition to this, there is a possibility of hydroxyl ions produced by *Oscillatoria*, which might help in degradation of melanoidin pigment into nitrogen and a carbon source for their growth. Hayase et al. (1984) and Campbell and Laudenbach (1995) suggested that hydroxyl ions production in the reaction mixture influenced the decolorization and degradation of synthetic melanoidin. These hydroxyl ions are generated through the involvement of the superoxide dismutase enzyme, leading to a shift of pH to alkaline values. Superoxide dismutase catalyzes the dismutation of the superoxide anion radical in H_2O_2 and O_2, which helps the microalgal cell to protect from toxic effects of melanoidins degradation.

Leonowicz et al. (2001) stated that superoxide radical ions are toxic, highly reactive and often generated during the redox cycling of quinine in the LACs producing organisms. So this enzyme not only scavenges the harmful superoxide radicals to protect the cell, but also could also provide the main cofactor of peroxidase enzymes i.e., H_2O_2 (Leonowicz et al. 2001). Kalavathi et al. (2001) suggested that photosynthesis induced production of hydrogen peroxide entities that could react with hydroxyl anions to produce perhydroxyl anion (HOO-); which are known as strong nucleophilies. These perhydroxyl anions (HOO-) can also facilitate the decolorization of meanoidin. It was also observed that melanoidin degradation only takes place in light conditions, which may possibly be due to the requirement of active oxygen species, that are generated through the photolysis of water by microalgae (Kalavathi et al. 2001).

Conclusion

Distilleries and sugar mills discharge 91-93% of their raw materials as wastewater, thus the selection of the treatment method should depend on two objectives: (a) effectiveness in the reduction of pollution load; (b) affordability in building and operation. Although physicochemical and biological methods significantly reduce the BOD and COD of wastewater, but are not very successful in the removal of color and dissolved nutrients. The strong color of distillery and sugar mill wastewater is due to the presence of melanoidins, which is found to be biologically degradable especially with pure microbial cultures; but there is little information about the mechanism of degradation/transformation of melanoidins during anaerobic treatment.

In spite of being environment-friendly and cost-competitive, the major problem with microbial decolorization is that distillery and sugar mill wastewater is not found to be the best suited for organisms such as fungi; which have been proved to be successful in degrading melanoidins. It is because fungi are not easily adapted in aquatic habitats and further distillery and sugar mill wastewater usually deplete the oxygen in wastewater. Several advanced processes like membrane filtration and the oxidation process proved to be effective but their high cost and need of skilled operators limited their usability. Therefore, there is need to improve the constraints in the existing treatment methods and the development of integrated processes to facilitate a complete solution for the effective treatment of distillery and sugar mill wastewater.

Microalgae are found to be the most survivable microscopic communities in different wastewaters, including distillery and sugar mill wastewater. Due to their natural resistance and selective strategy towards environmental pollutants make them potential candidates, while there is little information about the mechanism that is involved in the degradation of melanoidins, lignin and tannins and other coloring substances. To increase the effectivity of microalgal strains, there is need for more research on native microalgal strains that are more adaptable to local climatic conditions; which ultimately enhances the elimination of pollutants from wastewater. Considering the benefits of microalgae/microalgae-bacteria consortia, it could be a good strategy for treatment, but their cumulative effort in degradation of pollutants of distillery and sugar mill wastewater needs to be further examined.

References

Agarwal, C.S. and C.S. Pandey. 1994. Soil pollution by spent wash discharge: depletion of manganese (II) and impairment of its oxidation. J. Environ. Biol. 15: 49–53.

Anbuselvi, S. and R. Jeyanthi. 2009. A comparative study on the biodegradation of coir waste by three different species of marine cyanobacteria. J. Appl. Sci. Res. 5: 2369–2374.

Asada, K. 1999. The water-water cycle in chloroplasts: scavenging of active oxygen and dissipation of excess photons. Ann. Rev. Plant Physiol. Plant Mol. Biol. 50: 601–639.

Beltrain de Heredia, J., J.R. Domingues and E. Partido. 2005. Physico-chemical treatment for depuration of wine distillery wastewater (vinasses). Water Sci. Technol. 51(1): 159–166.

Bharagava, R.N. and R. Chandra. 2010. Biodegradation of the major color containing compounds in distillery wastewater by an aerobic bacterial culture and characterization of their metabolites. Biodegrad. 21: 703–711.

Bharati, S.G., A.S. Salanki, T.C. Taranath and M.V.R.N. Acharyulu. 1992. Role of cyanobacteria in the removal of lignin from the paper mill wastewaters. Bull. Environ. Contam. Toxicol. 49: 738–742.

Borrelli, R.C., C. Mennella, F. Barba, M. Russo, G.L. Russo, K. Krome, et al 2003. Characterization of coloured compounds obtained by enzymatic extraction of bakery products. Food Chem. Toxicol. 41: 1367–1374.

Campbell, W.S. and D.E. Laudenbach. 1993. Characterization of superoxide dismutase genes from the cyanobacterium *Plectone boryanum* UTEX 485. In: The use of cyanobacteria to explore basic biological processes. The cyanobacterial work shop, USA.

Campbell, W.S. and D.E. Laudenbach. 1995. Characterization of superoxide dismutase genes from the cyanobacterium. J. Bacteriol. 177(4): 964–972.

Chandra, R., R.N. Bharagava and V. Rai. 2008. Melanoidins as major colourant in sugarcane molasses based distillery effluent and its degradation. Bioresour. Technol. 99: 4648–4660.

Chaudhari, P.K., I.M. Mishra and S. Chandb. 2007. Decolorization and removal of chemical oxygen demand (COD) with energy recovery: treatment of biodigester effluent of a molasses-based alcohol distillery using inorganic coagulants. Colloid Surface A. 296: 238–247.

Chaudhary, R. and M. Arora. 2011. Study on distillery effluent: chemical analysis and impact on environment. Int. Adv. Eng. Technol. 2(2): 352–356.

Chowdhary, P., A. Yadav, G. Kaithwas and R.N. Bharagava. 2017. Distillery wastewater: a major source of environmental pollution and its biological treatment for environmental safety. pp. 409–435. In: R. Singh and S. Kumar (eds.). Green Technologies and Environmental Sustainability. Springer International Publishing AG, Cham, Switzerland.

Dahiya, J., D. Singh and P. Nigam. 2001. Decolorization of synthetic and spent wash melanoidins using the white-rot fungus *Phanerochaete chrysosporium* JAG-40. Bioresour. Technol. 78: 95–98.

Dai, J. and R.J. Mumper. 2010. Plant phenolics: extraction, analysis and their antioxidant and anticancer properties. Molecules 15: 7313–7352.

Della Greca, M., P. Monaco, A. Pollio and L. Previtera. 1992. Structure activity relationships of phenylpropanoids as growth inhibitors of the green alga *Selenastrum capricornutum*. Phytochemistry 31: 4119–4123.

Dhamotharan, R., S. Murugesan, M.C. Sridharan and M. Yoganandam. 2009. Biological decolourization and removal of metal from dye industry effluent by microalgae. Biosci. Biotech. Res. Asia 6(1): 111–120.

Douskova, I., F. Kaštánek, Y. Maléterová, P. Kaštánek, J. Doucha and V. Zachleder. 2010. Utilization of distillery stillage for energy generation and concurrent production of valuable microalgal biomass in the sequence: biogas-cogeneration-microalgae-products. Energy Convers. Manag. 51: 606–611.

Fitz-Gibbon, F., D. Singh and G. McMullan. 1998. The effect of phenolics acids and molasses spent wash concentration on distillery wastewater remediation by fungi. Process Biochem. 33: 799–803.

Ganapathiselvam, V., R. Baskaran and P.M. Mohan. 2011. Microbial diversity and bioremediation of distilleries effluent. J. Res. Biol. 3: 153–162.

González, T., M.C. Terrón, S. Yagüe, H. Junca, J.M. Carbajo, E.J. Zapico, et al. 2008. Melanoidin-containing wastewaters induce selective laccase gene expression in the white-rot fungus *Trametes* sp. I-62. Res. Microbiol. 159: 103–109.

Hallmann, A. 2007. Algal transgenics and biotechnology. Transgenic Plant J. 1: 81–98.

Hayase, F., S.B. Kim and H. Kato. 1984. Decolorization and degradation products of the melanoidins by hydrogen peroxide. Agric. Biol. Chem. 48: 2711–2717.

Heider, J. and G. Fuchs. 1996. Anaerobic metabolism of aromatic compounds. Eur. J. Biochem. 243: 577–596.

Jadhav, P.G., N.G. Vaidya and S.B. Dethe. 2013. Characterization and comparative study of cane sugar industry waste water. Int. J. Chem. Phys. Sci. 2: 19–25.

Jayanti, S. and S. Narayanan. 2004. Computational study of particle-eddy interaction in sedimentation tank. J. Environ. Eng. 130(1): 37–49.

Kalavathi, D.F., L. Uma and G. Subramanian. 2001. Degradation and metabolization of the pigment-melanoidin in distillery effluent by the marine cyanobacterium *Oscillatoria boryana* BDU 92181. Enz. Microb. Technol. 29: 246–250.

Kanimozhi, R. and N. Vasudevan. 2010. An overview of wastewater treatment in distillery industry. Int. J. Environ. Eng. 2: 159–184.

Kharayat, Y. 2012. Distillery wastewater: bioremediation approaches. J. Integr. Environ. Sci. 9(2): 69–91.

Kim, J.S. and Y.S. Lee. 2009. Enolization and racemization reactions of glucose and fructose on heating with amino-acid enantiomers and the formation of melanoidins as a result of the Maillard reaction. Amino Acids 36: 465–474.

Kuhad, R.C., A. Singh and K.E.L. Eriksson. 1997. Microorganisms and enzymes involved in the degradation of plant fibre cell walls. Adv. Biochem. Eng. Biotechnol. 57: 45–125.

Kumar, A. 2018. Assessment of Cyanobacterial Diversity in Paddy Fields and Their Capability to Degrade the Pesticides. Babasahaeb Bhimrao Ambedkar University, Lucknow, India.

Kumar, A. and J.S. Singh. 2016. Microalgae and cyanobacteria biofuels: a sustainable alternate to crop-based fuels. pp. 1–20. *In*: J.S. Singh, D.P. Singh (eds.). Microbes and Environmental Management. Studium Press Pvt. Ltd. New Delhi, India.

Kumar, A. and J.S. Singh. 2017. Cyanoremediation: a green-clean tool for decontamination of synthetic pesticides from agro- and aquatic-ecosystems. pp. 59–83. *In*: J.S. Singh, G. Seneviratne (eds.). Agro-Environmental Sustainability, Vol. II: Managing Environment Pollution. Springer Int., Cham, Switzerland.

Kumar, A. and J.S. Singh. 2020. Biochar coupled rehabilitation of cyanobacterial soil crusts: a sustainable approach in stabilization of arid and semiarid soils. pp. 167–191. *In*: J.S. Singh, C. Singh (eds.). Biochar Applications in Agriculture and Environment Management. Springer Int., Cham, Switzerland.

Kumar, A. and J.S. Singh. 2020. Microalgal bio-fertilizers. *In*: E. Jacob-Lopes, M.M. Maroneze, M.I. Queiroz, L.Q. Zepka (eds.). Handbook of Microalgae-based Processes and Products. Academic Press, Cambridge, US, In press.

Kumar, A., S. Kaushal, S.A. Saraf and J.S Singh. 2017. Cyanobacterial biotechnology: an opportunity for sustainable industrial production. Clim. Chang. Environ. Sustain. 5(1): 97–110.

Kumar, A., S. Kaushal, S.A. Saraf and J.S. Singh. 2018. Microbial bio-fuels: a solution to carbon emissions and energy crisis. Front. Biosci. (Landmark) 23: 1789–1802.

Kumar, A., S. Kaushal, S.A. Saraf and J.S. Singh. 2018. Screening of Chlorpyrifos (CPF) tolerant cyanobacteria from paddy field soil of Lucknow, India. Int. J. Appl. Adv. Sci. Res. 3(1): 100–105.

Kumar, P. and R. Chandra. 2004. Detoxification of distillery effluent through *Bacillus thuringiensis* (MTCC 4714) enhanced phytoremediation potential of *Spirodela polyrrhiza* (L.) Schliden. Bull. Environ. Contam. Toxicol. 73: 903–910.

Kumar, P. and R. Chandra. 2006. Decolorization and detoxification of synthetic molasses melanoidins by individual and mixed cultures of *Bacillus* spp. Bioresour. Technol. 97: 2096–2102.

Leonowicz, A., N. Cho, J. Luterek, A. Wilkolazka, M. Wotjas, A. Matus, et al. 2001. Fungal laccase: properties and activity on lignin. J. Basic. Microbiol. 41: 185–227.

Malliga, P., L. Uma and G. Subramanian. 1996. Lignolytic activity of the cyanobacterium *Anabaena azollae* ML2 and the value of coir waste as a carrier for BGA biofertilizer. Microbios 86: 175–183.

Manju, M.R. and M.N. Soumya. 2010. Biodegradation of industrial effluents using *Spirulina platensis*. Pollut. Res. 29(1): 149–152.

Martins, S., W.M.F. Jongen and M. van Boekel. 2001. A review of maillard reaction in food and implications to kinetic modeling. Trends Food Sci. Technol. 11: 364–373.

Melamane, X.L., P.J. Strong and J.E. Burgess. 2007. Treatment of wine distillery wastewater: A review with emphasis on anaerobic membrane reactors. S. Afr. J. Enol. Vitic. 28(1): 25–36.

Mohana, S., B.K. Acharya and D. Madamwar. 2009. Distillery spent wash: treatment technologies and potential applications. J. Hazard. Mater. 163(1): 12–25.

Moreno-Garrido, I. 2008. Microalgae immobilization: current techniques and uses. Bioresour. Technol. 99: 3949–3964.

Muñoz, R. and B. Guieysse. 2006. Algal-bacterial processes for the treatment of hazardous contaminants: a review. Water Res. 40: 2799–2815.

Murugan, K. and S.A. Al-Sohaibani. 2010. Biocompatible removal of tannin and associated color from tannery effluent using the biomass and tannin Acylhydrolase (E.C. 31120) enzyme of mango industry solid waste isolate *Aspergillus candidus* MTTC 9628. Res. J. Microbiol. 5: 262–271.

Murugesan, S. and R. Dhamotharan. 2009. Bioremediation of thermal wastewater by *Pithophora* sp. Curr. World Environ. 4 (1): 137–142.

Murugesan, S., P. Venkatesh and R. Dhamotharan. 2010. Phycoremediation of poultry wastewater by micro alga. Biosci. Biotech. Res. Comm. 3(2): 142–147.

Nataraj, S.K., K.M. Hosamani and T.M. Aminabhavi. 2006. Distillery wastewater treatment by the membrane-based nano-filtration and reverse osmosis processes. Water Res. 40(12): 2349–2356.

Olaizola, M. 2003. Commercial development of microalgal biotechnology: from the test tube to the marketplace. Biomol. Eng. 20: 459–466.

Padmapriya, C., S. Murugesan and R. Dhamotharan. 2012. Phycoremediation of distillery waste by using the green microalga *Scenedesmus* sp. Int. J. Appl. Environ. Sci. 7(1): 25–29.

Palanisami, S., S.K. Saha and L. Uma. 2010. Laccase and polyphenol oxidase activities of marine cyanobacteria: a study with Poly R-478 decolorization. World J. Microbiol. Biotechnol. 26: 63–69.

Papazi, A. and K. Kotzabasis. 2007. Bioenergetic strategy of microalgae for the biodegradation of phenolic compounds-exogenously supplied energy and carbon sources adjust the level of biodegradation. J. Biotechnol. 129: 706–716.

Papazi, A. and K. Kotzabasis. 2008. Inductive and resonance effects of substituents adjust the microalgal biodegradation of toxic phenolic compounds. J. Biotechnol. 135: 366–373.

Patel, A., P. Pawar, S. Mishra and A. Tewari. 2001. Exploitation of marine cyanobacteria for removal of color from distillery effluent. Ind. J. Environ. Prot. 21: 1118–1121.

Patil, N.B. and B.P. Kapadnis. 1995. Decolorization of melanoidin pigment from distillery spent wash. Ind. J. Environ. Health. 37: 84–87.

Pikaev, A.K., A.V. Ponimarev, A.V. Bludenko, V.N. Minin and L.M. Elizarvar. 2001. Combined electron beam and coagulation purification of molasses distillery slopes. Radiat. Phys. Chem. 61(1): 81–87.

Plavsic, M., B. Cosovic and C. Lee. 2006. Copper complexing properties of melanoidins and marine humic material. Sci. Total Environ. 366: 310–319.

Prasad, R.K. and S.N. Srivastava. 2009. Sorption of distillery spent wash onto fly ash: kinetics and mass transfer studies. Chem. Eng. J. 146: 90–97.

Ramakritinan, C.M., A.K. Kumaraguru and M.P. Balasubramanian. 2005. Impact of distillery effluent on carbohydrate metabolism of freshwater fish, *Cyprinus carpio*. Ecotoxicol. 14: 693–707.

Rao, P.H., R.R. Kumar, B.G. Raghavan, V.V. Subramanian and V. Subramanian. 2011. Application of Phycoremediation technology in the treatment from a leather processing chemical manufacturing facility. Water SA, 37(1): 7–14.

Ravikumar, R., R. Saravanan, N.S. Vasanthi, J. Swetha, N. Akshaya, M. Rajthilak, et al. 2007. Biodegradation and decolorization of biomethanated distillery spent wash. Ind. J. Sci. Technol. 1(2): 1–6.

Robinson, T., G. Mc-Mullan, R. Marchant and P. Nigan. 2001. Remediation of dyes in textile effluent: a critical review on current treatment technologies with a proposed alternative. Bioresour. Technol. 77: 247–255.

Saha, S.K., P. Swaminathan, C. Raghavan, L. Uma and G. Subramanian. 2010. Ligninolytic and antioxidative enzymes of a marine cyanobacterium *Oscillatoria willei* BDU 130511 during Poly R-478 decolorization. Bioresour. Technol. 101: 3076–3084.

Sasso, S., G. Pohnert, M. Lohr, M. Mittag and C. Hertweck. 2012. Microalgae in the postgenomic era: a blooming reservoir for new natural products. FEMS Microbiol. Rev. 36: 761–785.

Satayawali, Y. and M. Balakrishnan. 2008. Wastewater treatment in molasses-based distillery for COD and color removal: a review. J. Environ. Manage. 86: 481–497.

Scalbert, A. 1991. Antimicrobial properties of tannins. Phytochemistry 30: 875–883.

Semple, K.T. and R.B. Cain. 1996. Biodegradation of phenolic by *Ochromonasdanica*. Appl. Environ. Microbiol. 62: 1265–1273.

Semple, K.T., R.B. Cain and S. Schmidt. 1999. Biodegradation of aromatic compounds by microalgae. FEMS Microbiol. Lett. 170: 291–300.

Shashirekha, S., L. Uma and G. Subramanian. 1997. Phenol degradation by the marine cyanobacterium *Phormidium valderianum* BDU 30501. J. Ind. Microbiol. Biotechnol. 19: 130–133.

Silvan, J.M., J.V.D. Lagemaat and A. Olano. 2006. Analysis and biological properties of amino acid derivates formed by maillard reaction in foods. J. Pharm. Biomed. Anal. 41: 1543–1551.

Simate, G.S., J. Cluett, S.E. Iyuke, E.T. Musapatika, S. Ndlovu, L.F. Walubita, et al. 2011. The treatment of brewery wastewater for reuse: state of the art. Desalination 273: 235–247.

Singh, J.S., A. Kumar, A.N. Rai and D.P. Singh. 2016. Cyanobacteria: A precious bio-resource in agriculture, ecosystem, and environmental sustainability. Front. Microbiol.7: 529.

Singh, J.S., S. Koushal, A. Kumar, S.R. Vimal and V.K. Gupta. 2016. Book review: Microbial inoculants in sustainable agricultural productivity, Vol. II: Functional Application. Front. Microbiol. 7: 2105.

Singh, J.S., A. Kumar and M. Singh. 2019. Cyanobacteria: a sustainable and commercial bio-resource in production of bio-fertilizer and bio-fuel from wastewaters. Environ. Sustain. Indic. 3: 100008.

Singh, N.K. and D.W. Dhar. 2007. Nitrogen and phosphorous scavenging potential in microalgae. Ind. J. Biotechnol. 6: 52–56.

Somoza, V. 2005. Five years of research on health risks and benefits of maillard reaction products: an update. Mol. Nutr. Food Res. 49: 663–672.

Stewart, W.D.P., A. Haystead and M.W.N. Dharmawardene. 1975. Nitrogen assimilation and metabolism in blue-green algae. Cambridge University Press, Cambridge, UK.

Tatsi, A.A., A.L. Zouboulis, K.A. Matis and P. Samaras. 2003. Coagulation, flocculation pretreatment of sanitary landfill leachates. Chemosphere 53(7): 737–744.

Thakur, I.S. 2006. Industrial Biotechnology: Problems and Remedies. I.K. International Pvt. Ltd. New Delhi, India.

Travieso, L., F. Benítez, E. Sánchez, R. Borja, M. León, F. Raposo, et al. 2008. Assessment of a microalgae pond for post-treatment of the effluent from an anaerobic fixed bed reactor treating distillery wastewater. Environ. Technol. 29(9): 985–992.

Valderrama, L.T., C.M. Del Campo, C.M. Rodriguez, L.E. de-Bashan and Y. Bashan. 2002. Treatment of recalcitrant wastewater from ethanol and citric acid production using the microalga *Chlorella vulgaris* and the macrophyte *Lemna minuscula*. Water Res. 36: 4185–4192.

Vijayakumar, S., N. Thajuddin and C. Manoharan. 2005. Role of cyanobacteria in the treatment of dye industry effluent. Poll. Res. 24(1): 69–74.

Yadav, S. and R. Chandra. 2012. Biodegradation of organic compounds of molasses melanoidin (MM) from biomethanated distillery spent wash (BMDS) during the decolourisation by a potential bacterial consortium. Biodegradation 23(4): 609–620.

Tannery Wastewater

Introduction

The tannery industry involves high water intake operations such as pre-tanning and tanning processes, which are responsible for the use of a large quantity of freshwater; it is estimated that for tanning of 100 kg grams of skin or leather requires 32,000 liters of freshwater (Swartz et al. 2017). From the past several centuries, vegetable tannins have been extensively used for the tanning process, but in 1858 Friedrich Knapp and Hylten Cavalin invented the chrome tanning process; which later became the most common tanning process in tanneries. Although both the vegetable and chrome tanning process involves a large amount of water, but chrome tanning uses of lot of chemicals mainly chromium (III) oxide salts.

Tannery wastewater is characterized with dark brown, organic load, high COD, large amounts of dissolved and suspended solids, sulfates, sulfides, chlorides and chromium salts, which makes the wastewater more toxic and harmful for the environment (Kongjao et al. 2008). When untreated tannery wastewater is discharged into water bodies, it decreases the dissolved oxygen and causes toxicity to the flora and fauna of the aquatic ecosystem (Almomani and Örmeci 2016). This wastewater could reach the soil through unlined ponds, leakage in pipes and drains; and further seeps to the lower soil layers, and could cause groundwater contamination. The chromium ion i.e., chromium (Cr(VI)) is more toxic than Cr(III) and their increasing concentration could have deleterious effects on human health like nasal irritation, chronic bronchitis, problems with the liver and kidney functions, internal hemorrhage, cancers of skin, lung and stomach and damage in DNA by interface with DNA polymerase (Chhikara et al. 2010).

Conventional treatment includes physical, chemical and biological processes that are carried out to remove pollutants from tannery wastewater (Di-Iaconi et al. 2001, Schrank et al. 2003, Metcalf and Eddy 2003, Farabegoli, et al. 2004, Song et al. 2004, Dogruel et al. 2004). To remove metal especially chromium, some methods such as chemical precipitation, adsorption, lime coagulation, reverse osmosis, solvent extraction, electro dialysis and ion exchange; are conventionally applied (Ahluwalia and Goyal 2007). The major problem with conventional treatment is that it requires excess energy and chemicals for the treatment, proving it to very expensive and ineffective for high metal concentration (Stern et al. 2003, Kumar et al. 2015); and is further responsible for the generation of excessive sludge (Chu 2001). Along with Conventional and physicochemical, currently there are many works related to the use of bacteria, fungi and microalgae for the removal of metals from tannery wastewater (Fernández et al. 2018).

Microalgal bioremediation has proved to be more efficient and the best alternative to conventional methods that were previously applied for the removal of metals (Kumar and Singh 2016, Singh et al. 2016a, b, Kumar et al. 2017, Kumar and Singh 2017, Kumar 2018, Kumar et al. 2018a, b, Singh et al. 2019, Kumar and Singh 2020a, b). They only require light energy, water and inorganic nutrients for growth and metabolism; in comparison to bacteria or fungi, which need organic carbon sources and electron acceptor. Gupta et al. (2001), Singh et al. (2001), Donmez and Aksu (2002) suggested that like in other microbes, microalgae sequester the chromium and other metal ions through adsorption and absorption processes. This mechanism of binding the metal ions depends on the metal ion and their ionic charges, microalgal species and further the chemical composition and characteristics like pH of the wastewater. Thus microalgae-mediated metal removal could provide a potential and low cost alternative to the conventional wastewater treatment (Donmez and Aksu 2002).

Tannery Industry: A Brief Description

Skins and hides reach dry or wet salted conditions in tanneries. To reinstate their original condition, skins and hides are soaked in water to rehydrate and remove the salt. Soaking is primarily carried out in either in pits, in which skins and hides are flattened and loose water to eliminate the salt and other adhering materials. Further there is a need of changing the water, and first change of water is considered as dirt soaking. After two or three changes of water, raw skins and hides return back to their original condition. Despite this, skins and hides get rid of the adhering blood, dung and other soluble protein such as albumin and globulin.

Rehydrated hides and skins undergo a liming process, in which a mixture of lime and sodium sulfide performs two steps: the first step involves removing the hair and flesh, known as the unhairing process; in second step liming is carried out to improve the fiber structures of the skins and hides by suitable plumping and swelling. For the liming process, pits or paddles are used for the hides, but in the case of sheep and goat-skins a paint liming method is applied; also known as the hair shaving process. After the two steps of liming or reliming, hides and skins are well fleshed and scudded; and the remaining dirt and short hair are also removed.

To remove lime from the surface, pelt washing primarily with ammonium salts viz., chlorides or sulfates to achieve the pH to 8–8.5. This process is quite different for soft and hard types of leathers, soft leathers undergo complete deliming, while in the case of heavy and firm leather like soles of leather, moderate deliming is carried out to leave a streak of lime in the middle. For this process, drums filled with about 100% water are used; and subsequently a further bating process is followed in the same bath.

After the deliming process, bating is carried out to remove the unwanted components including short hair, proteinous products, epidermis and scud from the pelts. For this, proteolytic enzymes mostly trypsin is applied, which especially removes the interfibrillar proteins; providing proper grain texture to the leather and is helpful in achieving softness and flexibility to the finished leather. Further pelts are thoroughly washed, to remove the salts formed in the deliming and bating process.

To bring down the pH to 2.5-3.0, a pickling process is followed; pelt is treated with a mixture of formic and sulfuric acids, and further 10% sodium chloride solution is also applied to suppress the swelling which occurs due to a drop in pH. For this, water saturated pelt (containing water 80% of pelt weight) is used in the pickling process. This process is used to achieve rapid penetration and uniform distribution during chrome tanning process. If vegetable tanning is applied, then partial pickling is carried out to attain the pH 4-4.5.

Some part of the fat and natural grease in hides and skins is already removed in the liming process. But still some skins like wool and sheep skins have a lot of fats, so a degreasing process is necessary to remove this fat. Sometimes a degreasing process can be followed after deliming, but it has the disadvantage of increased pH during deliming and to overcome, this repickling would be necessary after degreasing to reduce the pH to 2.8 as it does to chrome tanning. It is more suitable to apply degreasing after the pickling process.

Tanning is considered an essential and the most significant process in the processing of leather; as it provides bacterial resistance to the leather and improves its durability. Further it gives the specific characteristics such as hydrothermal stability, fullness and charge character to the leather. Due to the charge character, it absorbs and retains more post-tanning chemicals and auxiliaries. For the tanning process, many chemicals including materials for chrome or vegetable tanning, salts of aluminium and zirconium, aldehydes like formaldehyde and gluteraldehyde and oils mainly fish oil etc., are used. The choice of chrome or vegetable tanning depends on the needs of manufacturing different types of finished leathers.

During tanning, there is a problem of the presence of excess acid existing in wet blue/semi-chrome leathers, which could hamper the following processes like dyeing, fat liquoring and retanning. To counteract this acid a neutralization process is required, in which the following chemicals such as sodium formate, sodium sulfite, sodium bicarbonate, ammonium bicarbonate, neutralizing syntan are usually applied. Further the degree of neutralization depends on the desired properties of the finished leathers.

Then a retanning procedure is carried out for following purposes; (a) filling tanning material to the loose portions of leather; (b) providing the tightness of the grain texture; (c) retaining the fullness, round feel and body to the leather. In this process, retanning materials such as vegetable tanning agents, protein based tanning material, acrylic and other resin tanning substances, whitening syntans, phenolic syntans, and polyurethane syntans are used.

To provide different colors to the tanned leather, a dyeing process is performed. Dyes are different types which are described in Chapter 10. In dyeing of leather, acid, direct and metal complex dyes are commonly used. To get darker shades like blacks and some browns, basic dyes are also applied. Reactive dyes are rarely applied in leather processing.

Then a fat liquoring process is carried out to provide softness, flexibility, feel, drape, run, to the leather. Fat liquors are prepared through emulsification of oils and fats in water by sulfation or the sulfochlorination process. This is also helpful in enhancing the strength property of the leather. All the three processes of retanning, dyeing and fat liquoring are mostly performed together in the same bath, dyes and fat liquors are fixed through the addition of formic acid. Further leathers are dried before the finishing process.

In the end, a finishing process is followed, in which the finalized formulations are applied to grain texture leather to enhance the aesthetic and sale value of the leather. Finishing formulations contain coloring agents (pigments and dyes), binders (casein or acrylates or polyurethane based), wax emulsions, fillers and nitro cellulose or cellulose acetate butyrate or other hard resins. If the leather has any grain defects, they are roofed up by a protective coat; leading to enhance the cutting value of leather. The finishing process also facilitates surface protection against rubbing, abrasion and staining.

Tannery Wastewater: Characteristics and Specific Pollutants

Start Tannery wastewater is characterized by a basic, dark brown color and contains a large amount of organic compounds (Kabdasli et al. 2002, Kongjao et al. 2008). Important pollutants that occur in tannery wastewater include tannins, chromium, chlorides, sulfate and sulfides. In addition to these pollutants, some organic and synthetic chemicals such as pesticides, dyes and finishing agents are also found in wastewater. These pollutants, in the discharged effluent, pose serious threats to the environment.

Nutrients

Textile wastewater are found to be abundant in nitrogen, especially in organic nitrogen i.e., proteinaceous materials, but the phosphorus proportion is very low. The source of nitrogen in the tannery wastewater exists from soaking, liming and deliming processes. During the soaking and liming processes, proteinaceous parts of the hides and skins are released in the wastewater stream; while the deliming process is responsible for discharge of ammonium salts.

Table 7.1: Characteristics of tannery wastewater. (Source: Kongjao et al. 2008)

Water quality parameter	Saranya and Shanthakumar 2019	Da Fontoura et al. 2017	Das et al. 2017	Adam et al. 2015	Ajayan et al. 2015	Jahan et al. 2014	Shasirekha et al. 2008	Meric et al. 2005	Tadesse et al. 2004
pH	7.2	7.5	7.45	3.09	5.6	8.3		3.99-5.02	
TS (mg L^{-1})			5000		4528		17740		
TDS (mg L^{-1})	27370		4333		2775	21300	13985		
TSS (mg L^{-1})			500	2398	1753	1250	3755	1400-1800	2517-5071
BOD (mg L^{-1})	1500	1400	1350	2287	326	4464			2810-3580
COD (mg L^{-1})	4800	4000	4000	7421	872	12840	1296	6475-7085	7937-11157
TN/NH$_4^+$-N (mg L^{-1})	760		2734.16					56.06-71.53	136-156
PO$_4^{3-}$/Phosphates (mg L^{-1})	1.7	6.6	6.01			17.1			9-16
TOC (mg L^{-1})		1692							
Sulfate/Sulfides (mg L^{-1})			178.69	50			1824	1616-1849	840-1279
Chlorides (mg L^{-1})	9967	7683				13.8	1660	2550-3050	
Chromium (ppm)	20.9			0.91	12.8	10.3			

Note: TS: Total Solids; TDS: Total Dissolved Solids; TSS: Total Suspended Solids; BOD: Biological Oxygen Demand; COD: Chemical Oxygen Demand; TN: Total Nitrogen; NH$_4^+$-N: Ammonium Nitrogen; NO$_3^-$-N: Nitrate Nitrogen; TP: Total Phosphorous; TOC: Total Organic Content.

Table 7.2: Specific pollutants from various processes in the textile industry

Process	Wastewater
Soaking	Salt, dirt, organic substances
Liming	Lime, H_2S
Deliming	Ammonia, salts
Bating	Keratin, proteoglycans, organics
Picking	Acid, salt
Tanning	Cr, salts, low pH
Neutralization	Salts, Cr
Retanning	Salt, phenol, aldehydes
Dyeing	Dyes
Fat liquoring	Oil & fats

Suspended solids

Suspended solids are generated in all stages of leather processing mainly from the liming process and existing in large amounts in tannery wastewater. These solids are categorized into gross solids and fine solids; gross solids comprises of large pieces of leather cuttings, trimmings and gross shavings, fleshing residues, solid hair debris; while fine solids include protein residues, fine leather particles, residues from various chemical discharges and reagents from different waste liquors.

Neutral salts

Sulfates and chlorides are the primary neutral salts that are found in tannery wastewater. Sulfates salts come from the use of sulfuric acids or products with a high (sodium) sulfate content, mainly from the chrome tanning materials and retanning agents. Further sulfate content including sodium sulfates are generated during the removal of sulfide content by the oxidation process. Chlorides salts are generated from the use of sodium chloride, which are applied in large quantities of common salt in preservation of hide and skin or the pickling process.

Oils and grease

Hides and skins have natural fats and oils, which are released in the wastewater stream during the soaking, liming and degreasing processes. Some textile processes also use a significant amount of oil and fats during fat liquoring and finishing processes. Further some fatty substances by reactions of with other chemicals are also present in wastewater. This floating grease and fatty substances could combine to each other to form mats and also binds other pollutants to create a messy structure, which could affect the wastewater treatment.

Chromium and other metals

Chromium is the main pollutant species which makes the wastewater highly toxic. The main forms of chromium i.e., trivalent chromium (Cr^{3+} or Cr III) and hexavalent chromium (Cr^{6+} or Cr VI), that are largely used in chrome tanning and retanning processes. Further trivalent chromium are mixed with other processes and react with other components such as proteins; leading to the formation of protein-chrome precipitates being formed, resulting in retaining the sludge. Other metals like aluminium and zirconium are also present above the permissible limits in tannery wastewater.

Solvents

Solvents like white spirit or trichloroethylene are discharged from leather degreasing and finishing processes. These solvents in tannery wastewaters when discharged in water bodies and streams could form a microfilm on the water surface, leading to disruption in penetration of sunlight and mixing oxygen in the aquatic ecosystem. Further solvents are also degraded into different components; some of which might inhibit the bacterial activity and also persist in the aquatic ecosystem for a longer duration of time.

Impacts on Environment and Public Health

Tannery wastewater have high COD, high salinity, organic nitrogen, chromium, NH_4^+ and sulfide content that makes them very complex, toxic with the presence of persistent or many harmful substances (EC 2003, INETI 2000, Lefebvre and Moletta 2006). Due to these constituents it contaminates water and soil, leading to affect crop yield, toxicity to aquatic and terrestrial organisms and also to humans (Velma et al. 2009, Siyanbola et al. 2011). A number of toxicity studies have underlined the harmful effect of tannery wastewater that involves the use of *Vibrio fisheri*, (Jochimsen and Jekel 1997), *Daphnia magna*, (Tisler et al. 2004), sea urchin (*Paracentrotus lividus*), (Tu̇nay et al. 1999) and marine microalga *Dunaliella tertiolecta* (Marttinen et al. 2002, De Nicola et al. 2004). Zayed et al. (1998) observed that tannery wastewater could influence some metabolic processes, due to its ability to coordinate with different organic compounds, which leads to the inhibition of metalloenzyme systems.

López-Luna et al. (2012) described that Cr (III) have a tendency to bind to cell walls of plants, and are unlikely concentrated in roots; while Cr (VI) are a readily available entity in the soil because of the unstable nature under normal soil conditions. Further availability of chromium depends on the soil and pH texture. It is well known that Cr (III) is an essential component of a balanced human and animal diet for preventing adverse effects in the metabolism of glucose and lipids (Hou et al. 2007). Abdul et al. (2012) suggested that chromium could be either beneficial or toxic to plants, animals and humans; it depends on its oxidation state and concentration (Zayed and Terry 2003).

Current Treatment and Challenges

Conventional treatment of tannery wastewater includes various physical, chemical and biological processes i.e., coagulation, flocculation, reverse osmosis, ion exchange and activated carbon adsorption; other oxidation processes i.e., ozonation, photocatalysis, fenton process, wet-air oxidation follow (Di-Iaconi et al. 2001, Szpyrkowicz et al. 2005, Oller et al. 2011, Mandal et al. 2010, Srinivasan et al. 2012, Lofrano et al. 2013). In physicochemical methods, coagulation/flocculation is considered the most common and reliable technique, in which aluminium and ferric salts are added to the wastewater for removal of particles of dyes, salts and metals either as precipitates or suspended particles (Song et al. 2004, Riaz et al. 2012). But some researchers have also pointed out several drawbacks of using aluminium and ferric salts such as residual metal in treated waters. Igwe et al. (2005) observed that sawdust and coconut fiber meal could be used as an adsorbent for the removal of heavy metal ions from tannery wastewater. Further Igwe and Abia (2003) also reported the ability of maize cob to adsorb the metal ions and found that the mode of adsorption could be particle diffusion or film diffusion depending on the metal ion species and the adsorbent material.

Microalgal Remediation

Microalgae are successfully applied for bioremediation of metal contaminated sites or wastewater especially tannery wastewater (Swamy 2000, Shashirekha 2005, 2008, Pandi et al. 2009, Wang et al. 2010, Min et al. 2011, Renuka et al. 2013), because of their tolerance to the heavy metals, a larger cell surface area for interaction with heavy metal and the presence of various functional groups on the cell wall that have an affinity to metal ions in tannery wastewater (Malik 2004). Biosorption of heavy metals, using microalgae provides a potential, low cost and eco-friendly alternative to conventional processing methods (Gupta and Rastogi 2008). They also facilitate an oxygen-rich environment by evolving oxygen, leading to enhance bacterial oxidation, resulting in more pollutant removal from tannery wastewater (Swamy 2000).

Chellam and Sampathkumar (2012), Ajayan and Selvaraju 2012, Jaysudha and Sampathkumar (2014), Balaji et al. (2015) and Ajayan et al. (2015) investigated the ability of microalgal genera *Oscillatoria, Phormidium, Chlorella, Scenedesmus, Ulothrix, Chlamydomonas* in growth of tannery effluent and accumulated chromium (Table 7.3). Despite metal tolerance and biosorption, *Chlorella vulgaris* successfully reported the removal of nutrients and dyes from tannery wastewater.

Shi et al. (2002), Shashirekha et al. (2005) and Dwivedi et al. (2010) studied the potential of microalgae to remove chromium and other toxic metal from tannery wastewaters. Travieso et al. (1999) reported that Cr removal efficiency of microalgae *Scenedesmus* sp. increased rapidly in lower concentrations and then remained nearly constant throughout the experiment. It was also reported that *Scenedesmus* sp. could remove 52% of Cr in tannery wastewater

Table 7.3: Microalgae in heavy metal sequestration

Heavy metals	Source	Cyanobacterial sp.	References
Cd	Sewage water aqueous solution	*Nostoc linckia, Nostoc Rivularis, Tolypothrix tenuis*	El-Enany and Issa (2000)
Co	Sewage and industrial wastewater	*Nostoc muscorum, Anabaena subcylindrica*	El-Sheekh et al. (2005)
Cr	Metal contaminated soil	*Nostoc calcicola, Chroococus* sp.	Anjana et al. (2007)
Cu	Sewage and industrial wastewater	*Nostoc muscorum, Anabaena subcylindrica*	El-Sheekh et al. (2005)
Hg	Wet biomass	*Spirulina platensis, Aphanothece flocculosa*	Cain et al. (2008)
Mn	Sewage and industrial wastewater	*Nostoc muscorum, Anabaena subcylindrica*	El-Sheekh et al. (2005)
Pb	Sewage and industrial wastewater Pb Residue of Cyanobacteria cell after polysaccharide extraction	*Nostoc muscorum, Anabaena subcylindrica, Gloeocapsa* sp.	El-Sheekh et al. (2005) Raungsomboon et al. (2006)
Zn	Sewage water	*Nostoc linckia, Nostoc rivularis*	El-Enany and Issa (2000)

in a duration of 12 days. Further removal efficiency also depends on the form of chromium metal, and Cr (III) removal proved to be more preferable than Cr (VI). Brady et al. (1994) observed that *Scenedesmus* and *Selenastrum* are able to survive at even 100 ppm levels, but it is not necessary that they are also able to survive at 100 ppm level of Cr (VI). A number of studies also reported the ability of *Nostoc* sp., *Spirulina* sp. and *Synechococcus* sp. for bioremediation of tannery wastewater containing significant amount of chromium ions (Chen and Yang 2005, Deng et al. 2007, Gupta and Rastogi 2008, Chhikara et al. 2010). Zhang et al. (2008) observed that *Scenedesmus* sp. also proved to be very efficient in the removal of inorganic nutrients and metals from artificial and real domestic and industrial secondary treated wastewater.

Raize et al. (2004), Chojnacka et al. (2005) observed that unlike living cells, dead microalgal cells are also able to perform extracellular metal-binding through adsorption, complexation, chemisorption, chelation and reduction. Aksu and Kutsal (1990), Aksu (1998) suggested that using dead microalgal biomass for metal removal could provide many advantages: (a) there is no effect of increasing toxic metal concentration, (b) easy to handle, (c) there is a no need of any nutrients for dead microalgal biomass and can be used for repeated cycles.

Metal Biosorption Mechanism

Microalgal biosorption of metals is considered is a comprehensive mechanism and involves a set of several independent processes that lead to the uptake of the metal (Brown et al. 2000; Lombardi et al. 2002). Javanbakht et al. (2014) suggested that the biosorption mechanism comprises of several processes like surface adsorption, chemisorption, complexation, ion exchange, precipitation, or metal hydroxide condensation. Javanbakht et al. (2014), Talaro and Chess (2015) reported that the biosorption mechanism primarily starts with physico-chemical interactions between metal ions and the functional groups on the cell surface, which leads to extracellular or intracellular metal binding with microalgae (Hörcsik et al. 2006, Mwinyihija 2010, Pacheco et al. 2015, Kumar et al. 2015). The functional groups responsible for metal binding include carboxyl, amino, phosphate, sulfate, imidazole, sulfydryl, phenol, thioether, carbonyl, amide and hydroxyl moieties; which are associated with polysaccharides, lipids and proteins in microalgal biomass (Fig. 7.1). Volesky and Holan (1995) observed that there are some functional groups that might not be able to participate in metal binding, because of their

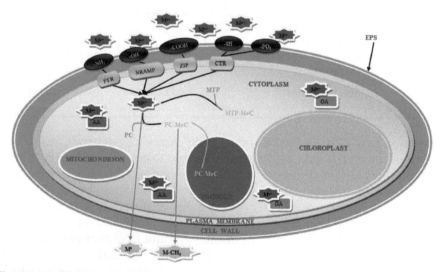

Figure 7.1: Various strategies of microalgae in heavy metal sequestration by the microalgae (Mn^+ – metal ions; functional groups: $-NH_2$ – amino group, $-OH$ – hydroxyl group, $-COOH$: carboxylic acid group, $-SOH$ – sulfahydryl group, PO_3 – phosphate group, that helps in attaching the metal ions; metal-ion transporters: FTR – Fe Transporter, NRAMP – Natural Resistance Associated Macrophage Protein, ZIP-Zrt, Irt-like transporter; CTR – Cu Transporter; PC – Phytochelatin, MTP – Metallothionein Protein, AA – Amino Acids, OA – Organic Acids, PC-MeC – Phytochelatin Metal-ion Complex; MTP-MeC – Metallothionein Protein Metal-ion Complex; non-toxic forms M^0-neutral metal, M-CH$_3$ – methylated metal that might be extruded outside; EPS – Extracellular Polymeric Substances). (*Source:* Modified from Perales-Vela et al. 2006, Toress et al. 2008, Monteiro et al. 2012, Blaby-Haas and Merchant 2012, Kumar et al. 2015).

steric, conformational or other barriers. It has also been well researched that many microalgae are known to produce Extracellular Polymeric Substances (EPS) comprising mainly of polysaccharides, which act as a sheath for the cell surface or are further released into the surroundings. These EPS also play a significant role in extracellular sorption of metal ions from tannery wastewater (Pereira et al. 2011).

When microalgal cells face exposure of metal ions beyond their cellular needs, it could interfere with their regular metabolism. To resist this adverse condition, microalgal cells facilitate the extracellular or intracellular metal binding (Monteiro and Castro 2012). Perez-Rama et al. (2001) emphasized that intracellular metal binding facilitated by intracellular compounds such as class III metallothioneins or phytochelatins, also help in transportation of metal ions to cell organelles mainly vacuoles or polyphosphate bodies (Pawlik-Skowronska 2003). Costa and Franca (2003), Jjemba (2004), Worms et al. (2006) and Levy et al. (2008) suggested that microalgal cells could also exudate this metal efflux back into the wastewater using active transport.

Role of pH in Metal Biosorption

The cell wall of microalgal comprises of some functional groups like amine, amide and carboxylicgroup, and depending up on the pH of wastewater, these groups could be protonated or deprotonated (López et al. 2000, Igwe and Abia 2003, Igwe et al. 2005, García-Rosales et al. 2012). It is observed that if pH of the solution increases, the negative charge on the surface of the microalgal cells also increases; leading to protonation of functional groups. And this process continues until all relevant functional groups are deprotonated, which helps in better electrochemical attraction and leads to adsorption of cations or metal ions. Bedell and Darnall (1990) emphasized the reason behind greater metal uptake with pH increase, it could be escalated as an efficient competition between metal ions and protons H^+ that takes place to bind more sites on the microalgae cell wall. In reverse of this, if pH decreases (increase in concentration of protons H^+), it promotes a positive charge on the microalgal cell wall, which leads to protonation of functional groups; resulting in more attraction of anions present in the solution.

Role of Immobilized Microalgae

The immobilized microalgal cells have some advantages like the possibility to reuse and act as a separating barrier between microalgal cells and the liquid wastewater (Bayramoğlu and Arica 2009, Arica and Bayramoglu 2005). This is achieved through entrapment of the microalgal cells into a matrix of polymers like alginate, chitin, chitosan and cellulose derivatives (Bai and Abraham 2003, Bajpai et al. 2004).

Adhiya et al. (2002) and Khattar et al. (2007) investigated the ability of immobilized microalgal species like *Chlamydomonas reinhardtii, Anabaena doliolum, Chlorella vulgaris, Scenedesmus obliquus, Anacytis nidulans* to sequestrate chromium ion from tannery wastewaters. It also studied the potential of alginate-immobilized *Anabaena doliolum* and *Chlorella vulgaris* to remove chromium in regard to their cell density and pH of the growth medium. It was found that increasing cell density also enhances chromium absorption; although intermediate cell density is considered the most suitable for metal removal. For maximum chromium absorption, neutral pH proved to be most favorable, and if there is an increase or decrease in pH, the effectiveness of microalgal cell also decreases. Khattar et al. (1999) studied the agar-immobilized *Anacytis nidulans* in chromium rich wastewater and reported an increase of metal absorption in comparison to free living microalgal cells. Similar results were reported by Pellon et al. (2003) which showed 95% increase of chromium removal by immobilized *Scenedesmus obliquus*.

Conclusion

In compliance to achieving strict environmental regulations, there is need for effective treatment of tannery wastewater. The biggest concern with tannery wastewater is the significant concentration of chromium metal that has proved to be very toxic to the environment and public health. A number of treatment technologies mainly physicochemical and biological processes are available and investigated for the removal of color and chromium, but they are inefficient and further responsible for release of several salts that are used in metal precipitation. There are advanced processes like filtration and reverse osmosis that are found to be efficient in metal removal, their high installation and operating cost makes them unviable and unaffordable for industries and treatment facilities.

Due to the minimal requirement and high absorptive capacity, microalgal remediation could be excellent approach for color, chromium and nutrients. Before selecting a microalgal strain for the treatment, some factors such as wastewater volume, metal concentration and the need of living/ dead microalgae cells for metal sequestration are required. Both living and dead biomass have some of their own advantages, living microalgal cells are environment-friendly, user-friendly, have recyclability, faster growth rate and applicable for wastewater with relatively low heavy metal concentration; while dead microalgal cells are of low cost, have more ease of handling, reusability, outstanding uptake capacity and applicable for very high or low pH and temperature.

Apart from the excellent metal uptake capability, there is a need of more research related to the mechanism of microalgal metal sequestration for developing efficient biosorption. These are the some advantages of using microalgae for the remediation of heavy metals containing tannery wastewater:

(i) Low capital and operating costs in comparison to use of physical and chemical processes;

(ii) Sequestration of inorganic nitrogen, phosphorous, sulfur and other inorganic salts;

(iii) Degradation of a variety of persistent and toxic organic compounds into non-toxic or less toxic metabolites

(iv) It gives operational flexibility in order to handle and manage a wide range of wastewater flows and characteristics

(v) Reduces overall toxicity in the aquatic systems by removing heavy metals and nutrients; and also improves the dissolved oxygen.

References

Abdul, G., S. Mansour, A. Okab and R. Ahmed. 2012. Exploiting of the phenoxazine as first-ever use ligand in rapid spectrophotometric methods for the determination of chromium (VI) in environmental samples. Analele Universitatii Ovidius Constanta—Seria Chimie 23(2): 180–186.

Adam, S., P.S. Kumar, P. Santhanam, S.D. Kumar and P. Prabhavathi. 2015. Bioremediation of tannery wastewater using immobilized marine microalga *Tetraselmis* sp.: experimental studies and pseudo-second order kinetics. J. Mar. Biol. Oceanogr. 4(1): 1–11.

Adhiya, J., X. Cai, R.T. Sayre and S.J. Traina. 2002. Binding of aqueous cadmium by the lyophilized biomass of *Chlamydomonas reinhardtii*. Colloids Surf., A Physicochem. Eng. Asp. 210: 1–11.

Ahluwalia, S.S. and D. Goyal. 2007. Microbial and plant derived biomass for removal of heavy metals from wastewater. Bioresour. Technol. 98: 2243–2257.

Ajayan, K.V. and M. Selvaraju. 2012. Heavy metal induced antioxidant defense system of green microalgae and its effective role in phycoremediation of tannery effluent. Pak. J. Biol. Sci. 15: 1056–1062.

Ajayan, K.V., M. Selvaraju, P. Unnikannan and P. Sruthi. 2015. Phycoremediation of tannery waste water using microalgae *Scenedesmus* species. Int. J. Phytoremediation 17(10): 907–916.

Aksu, S. 1998. Biosorption of heavy metal by microalgae in batch and continuous system. pp. 37–50. *In*: Y.S. Wong and N.F.Y. Tam (eds.). Wastewater Treatment with Algae. Springer-Verlag and Landes Bioscience, Berlin.

Aksu, Z. and T. Kutsal. 1990. A comparative study for biosorption characteristics of heavy metal ions with *C. vulgaris*. Environ. Technol. 11: 979–987.

Almomani, F.A. and B. Örmeci. 2016. Performance of *Chlorella vulgaris*, *Neochloris oleoabundans*, and mixed indigenous microalgae for treatment of primary effluent, secondary effluent and centrate. Ecol. Eng. 95: 280–289.

Anjana, K., A. Kaushik, B. Kiran and R. Nisha. 2007. Biosorption of Cr(VI) by immobilized biomass of two indigenous strains of cyanobacteria isolated from metal contaminated soil. J. Hazard. Mater. 148: 383–386.

Arica, M.Y. and G. Bayramoglu. 2005. Cr(IV) biosorption from aqueous solution using free and immobilized biomass of *Lentinus sajorcaju*: preparation and kinetic characterization. Colloids Surf., A Physicochem. Eng. Asp. 253: 203–211.

Bai, R.S. and T.E. Abraham. 2003. Studies on Cr(VI) adsorption-desorption using immobilized fungal biomass. Bioresour. Technol. 87: 17–26.

Bajpai, J., R. Shrivastava and A.K. Bajpai. 2004. Dynamic and equilibrium studies on adsorption of Cr(VI) ions onto binary biopolymeric beads of cross-linked alginate and gelatin. Colloids Surf., A Physicochem. Eng. Asp. 236: 83–92.

Balaji, S., T. Kalaivani, B. Sushma, C.V. Pillai, C.M. Shalini and C. Rajasekaran. 2015. Characterization of sorption sites and differential stress response of microalgae isolates against tannery effluents from Ranipet industrial area – an application towards phycoremediation. Int. J. Phytoremediat. 18(8): 747–753.

Bayramoğlu, G. and M.Y. Arica. 2009. Construction a hybrid biosorbent using *Scenedesmus quadricauda* and Ca-alginate for biosorption of Cu(II), Zn(II) and Ni(II): kinetics and equilibrium studies. Bioresour. Technol. 100: 186–193.

Bedell, G.W. and D.W. Darnall. 1990. Immobilization of nonviable, biosorbent, algal biomass for the recovery of metal ions. pp. 313–326. *In*: B. Volesky (ed.). Biosorption of Heavy Metals. CRC Press, Boca Raton, Florida.

Blaby-Haas, C.E. and S.S. Merchant. 2012. The ins and outs of algal metal transport. Biochim. Biophys. Acta 1823: 1531–1552.

Brady, D., B. Letebele, J.R. Duncan and P.D. Rose. 1994. Bioaccumulation of metals by *Scenedesmus, Selenastrum* and *Chlorella* algae. Water SA 20: 213–218.

Brown, P.A., S.A. Gill and S.J. Allen. 2000. Metal removal from wastewater using peat. Water Res. 34(16): 3907–3916.

Cain, A., R. Vannela and L.K. Woo. 2008. Cyanobacteria as a biosorbent for mercuricion. Bioresour. Technol. 99: 6578–6586.

Chellam, C.T. and P. Sampathkumar. 2012. Bio-removal of nutrients in tannery effluent water using marine micro algae, *Chlorella marina*. Proceedings of International Forestry and Environment Symposium, University of Sri Jayewardenepura, Vol. 17, Sri Lanka.

Chen, J.P. and L. Yang. 2005. Chemical modification of *Sargassum* sp. for prevention of organic leaching and enhancement of uptake during metal biosorption. Ind. Eng. Chem. Res. 44: 9931–9942.

Chhikara, S., A. Hooda, L. Ran and R. Dhankar. 2010. Chromium (VI) Biosorption by immobilized *Aspergillus niger* in continuous flow system with special reference to FTIR analysis. J. Environ. Biol. 31: 561–566.

Chojnacka, K., A. Chojnacki and H. Gorecka. 2005. Biosorption of Cr^{3+}, Cd^{2+}, and Cu^{2+} ions by blue-green algae *Spirulina* sp.: kinetics, equilibrium and the mechanism of the process. Chemosphere 59(1): 75–84.

Chu, W. 2001. Dye removal from textile dye wastewaters using recycled alum sludge. Water Res. 35(13): 3147–3152.

Costa, A.C.A. and F.P. Franca. 2003. Cadmium interaction with micro algal cells, cyanobacteria cells, and seaweeds; toxicology and biotechnological potential for wastewater treatment. Mar. Biotechnol. 5(2): 149–156.

Da Fontoura, J.T., G.S. Rolim, M. Farenzena and M. Gutterres. 2017. Influence of light intensity and tannery wastewater concentration on biomass production and nutrient removal by microalgae *Scenedesmus* sp. Process Saf. Environ. 111: 355–362.

Das, C., K. Naseera, A. Ram, R.M. Meena and N. Ramaiah. 2017. Bioremediation of tannery wastewater by a salt-tolerant strain of *Chlorella vulgaris*. J. Appl. Phycol. 29: 235–243.

De Nicola, E., M. Gallo, M. Iaccarino, S. Meric, R. Oral, T. Russo, et al. 2004. Hormetic vs. toxic effects of vegetable tannin in a multi-test study. Arch. Environ. Contam. Toxicol. 46: 336–344.

Deng, L., X. Zhu, X. Wang, Y. Su and H. Su. 2007. Biosorption of copper (II) from aqueous solutions by green alga *Cladophora fascicularis*. Biodegradation 18: 393–402.

Di-Iaconi, C., A. Lopez, G. Ricco and R. Ramadori. 2001. Treatment options for tannery wastewater alkalinization with or without post-ozonation. Ann. Chim. 91: 587–594.

Dogruel, S., G.E. Ates, B.F. Germirli and D. Orhon. 2004. Ozonation of nonbiodegradable organics in tannery wastewater. J. Environ. Sci. Health Part A. 39: 1705–1715.

Donmez, G. and Z. Aksu. 2002 Removal of chromium(VI) from saline wastewaters by *Dunaliella* species. Proc. Biochem. 38: 751–762.

Dwivedi, D., S. Srivastava, S. Mishra, A. Kumar, R.D. Tripathi, U.N. Rai, et al. 2010. Characterization of native microalgal strains for their chromium bioaccumulation potential: phytoplankton response in polluted habitats. J. Hazard. Mater. 173: 95–101.

El-Enany, A.E. and A.A. Issa. 2000. Cyanobacteria as a biosorbent of heavy metalsin sewage water. Environ. Toxicol. Pharmacol. 8: 95–101.

El-Sheekh, M.M., W.A. El-Shouny, M.E.H. Osman and E.W.E. El-Gammal. 2005. Growth and heavy metals removal efficiency of *Nostoc muscorum* and *Anabaena subcylindrica* in sewage and industrial wastewater effluents. Environ. Toxicol. Pharmacol. 19: 357–365.

European Commission (EC). 2003. Integrated Pollution Prevention and Control (IPPC) -Reference document on best available techniques for the tanning of hides and skins. European IPPC Bureau. Seville, Spain.

Farabegoli, G., A. Carucci, M. Majone and E. Rolle. 2004. Biological treatment of tannery wastewater in the presence of chromium. J. Environ. Manage. 71(4): 345–349.

Fernández, P.M., S.C. Viñarta, A.R. Bernal, E.L. Cruz and L.I.C. Figueroa. 2018. Bioremediation strategies for chromium removal: current research, scale-up approach and future perspectives. Chemosphere 208: 139–148.

García-Rosales, G., M.T. Olguin, A. Colín-Cruz and E.T. Romero-Guzmán. 2012. Effect of the pH and temperature on the biosorption of lead(II) and cadmium(II) by sodium-modified stalk sponge of *Zea mays*. Environ. Sci. Pollut. Res. Int. 19(1): 177–185.

Gupta, V.K. and A. Rastogi. 2008. Equilibrium and kinetic modelling of cadmium(II) biosorption by nonliving algal biomass *Oedogonium* sp. from aqueous phase. J. Hazard. Mater. 153: 759–766.

Gupta, V.K., A.K. Shrivastava and N. Jain. 2001. Biosorption of chromium(VI) from aqueous solution by green algae *Spirogyra* species. Water Res. 35: 4079–4085.

Hörcsik, Z., V. Oláh, A. Balogh, I. Mészáros, L. Simon and G. Lakatos. 2006. Effect of chromium(VI) on growth, element and photosynthetic pigment composition of *Chlorella pyrenoidosa*. Acta Biol. Szeged. 50: 19–23.

Hou, W., X. Chen, G. Song, Q. Wang and C.C. Chang. 2007. Effects of copper and cadmium on heavy metal polluted waterbody restoration by duckweed (*Lemna minor*) Plant Physiol. Biochem. 45(1): 62–69.

Igwe, J.C. and A.A. Abia. 2003. Maize cob and husk as adsorbents for removal of cadmium, lead and zinc ions from wastewater. The Physical Sci. 2: 210–215.

Igwe, J.C., E.C. Nwokennaya and A.A. Abia. 2005. The role of pH in heavy metal detoxification by biosorption from aqueous solutions containing chelating agents. Afr. J. Biotechnol. 4(10): 1109–1112.

INETI. 2000. Guia Tecnico do Sector dos Curtumes. Instituto Nacional de Engenharia e Tecnologia ´ Industrial, Lisboa, Portugal.

Jahan, M.A.A., N. Akhtar, N.M.S. Khan, C.K. Roy, R. Islam and Nurunnabi. 2014. Characterization of tannery wastewater and its treatment by aquatic macrophytes and algae. Bangladesh J. Sci. Ind. Res. 49(4): 233–242.

Javanbakht, V., S.A. Alavi and H. Zilouei. 2014. Mechanisms of heavy metal removal using microorganisms as biosorbent. Water Sci. Technol. 69(9): 1775–1787.

Jaysudha, S. and P. Sampathkumar. 2014. Nutrient removal from tannery effluent by free and immobilized cells of marine microalgae *Chlorella salina*. Int. J. Environ. Biol. 4: 21–26.

Jjemba, P.K. 2004. Interaction of metals and metalloids with microorganisms in the environment. pp. 257–270. *In*: P.K. Jjemba (ed.). Environmental Microbiology – Principles and Applications. Science Publishers, New Hampshire, USA.

Jochimsen, J.C. and M.R. Jekel. 1997. Partial oxidation effects during the combined oxidative and biological treatment of separated streams of tannery wastewater. Water Sci. Technol. 35: 337–345.

Kabdasli, I., O. Tunay, M.S. Cetin and T. Olmez. 2002. Assessment of magnesium ammonium phosphate precipitation for the treatment of leather tanning industry wastewaters. Water Sci. Technol. 46: 231–239.

Khattar, J.I.S., T.A. Sarma and D.P. Singh. 1999. Removal of chromium ions by agar immobilized cells of the cyanobacterium *Anacystis nidulans* in a continuous flow bioreactor. Enzyme Microb. Technol. 25(7): 564–568.

Khattar, J.I.S., T.A. Sarma and A. Sharma. 2007. Optimization of chromium removal by the chromium resistant mutant of the cyanobacterium *Anacystis nidulans* in a continuous flow bioreactor. J. Chem. Technol. Biotechnol. 82: 652–657.

Kongjao, S., S. Damronglerd and M. Hunsom. 2008. Simultaneous removal of organic and inorganic pollutants in tannery wastewater using electrocoagulation technique. Korean J. Chem. Eng. 25: 703–709.

Kumar, A. 2018. Assessment of Cyanobacterial Diversity in Paddy Fields and Their Capability to Degrade the Pesticides. Babasahaeb Bhimrao Ambedkar University, Lucknow, India.

Kumar, A. and J.S. Singh. 2016. Microalgae and cyanobacteria biofuels: a sustainable alternate to crop-based fuels. pp. 1-20. *In*: J.S. Singh, D.P. Singh (eds.). Microbes and Environmental Management. Studium Press Pvt. Ltd. New Delhi, India.

Kumar, A. and J.S. Singh. 2017. Cyanoremediation: a green-clean tool for decontamination of synthetic pesticides from agro- and aquatic-ecosystems. pp. 59–83. *In*: J.S. Singh, G. Seneviratne (eds.). Agro-Environmental Sustainability, Vol. II: Managing Environment Pollution. Springer Int., Cham, Switzerland.

Kumar, A. and J.S. Singh. 2020. Biochar coupled rehabilitation of cyanobacterial soil crusts: a sustainable approach in stabilization of arid and semiarid soils. pp. 167–191. *In*: J.S. Singh, C. Singh (eds.). Biochar Applications in Agriculture and Environment Management. Springer Int., Cham, Switzerland.

Kumar, A. and J.S. Singh. 2020. Microalgal bio-fertilizers. *In*: E. Jacob-Lopes, M.M. Maroneze, M.I. Queiroz, L.Q. Zepka (eds.). Handbook of Microalgae-based Processes and Products. Academic Press, Cambridge, US, In Press.

Kumar, A., S. Kaushal, S.A. Saraf and J.S. Singh. 2017. Cyanobacterial biotechnology: an opportunity for sustainable industrial production. Clim. Chang. Environ. Sustain. 5(1): 97-110.

Kumar, A., S. Kaushal, S.A. Saraf and J.S. Singh. 2018. Microbial bio-fuels: a solution to carbon emissions and energy crisis. Front.Biosci. (Landmark) 23: 1789–1802.

Kumar, A., S. Kaushal, S.A. Saraf and J.S. Singh. 2018. Screening of Chlorpyrifos (CPF) tolerant cyanobacteria from paddy field soil of Lucknow, India. Int. J. Appl. Adv. Sci. Res. 3(1): 100–105.

Kumar, K.S., H.-U. Dahms, E.-J. Won, J.-S. Lee and K.-H. Shin. 2015. Microalgae – a promising tool for heavy metal remediation. Ecotoxicol. Environ. Saf. 113: 329–352.

Lefebvre, O. and R. Moletta. 2006. Treatment of organic pollution in industrial saline wastewater: a literature review. Water Res. 40: 3671–3682.

Levy, J.L., B.M. Angel, J.L. Stauber, W.L. Poon, S.L. Simpson, S.H. Cheng et al. 2008. Uptake and internalisation of copper by three marine microalgae: comparison of copper-sensitive and copper-tolerant species. Aquat. Toxicol. 89(2): 82–93.

López, A., N. Lázaro, J.M. Priego and A.M. Marqués. 2000. Effect of pH on the biosorption of nickel and other heavy metals by *Pseudomonas fluorescens* 4F.J. Ind. Microbiol. Biotechnol. 24: 146–151.

Lofrano, G., S. Meriç, G.E. Zengin and D. Orhon. 2013. Chemical and biological treatment technologies for leather tannery chemicals and wastewaters: a review. Sci. Total Environ. 461: 265–281.

Lombardi, A.T., A.V.H. Vieira and L.A. Sartori. 2002. Mucilaginous capsule adsorption and intracellular uptake of copper by *Kirchneriella aperta* (Chlorococcales). J. Phycol. 38: 332–337.

López-Luna, J., M.C. González-Chávez, F.J. Esparza-Garca and R. Rodríguez-Vázquez. 2012. Fractionation and availability of heavy metals in tannery sludge-amended soil and toxicity assessment on the fully-grown *Phaseolus vulgaris* cultivars. J. Environ. Sci. Health Part A 47(3): 405–419.

Malik, A. 2004. Metal bioremediation through growing cells. Environ. Inter. 30: 261–278.

Mandal, T., D. Dasgupta, S. Mandal and S. Datta. 2010. Treatment of leather industry wastewater by aerobic biological and fenton oxidation process. J. Hazard Mater. 180: 204–211.

Marttinen, S.K., R.H. Kettunen, K.M. Sormunen, R.M. Soimasuo and J.A. Rintala. 2002. Screening of physical-chemical methods for removal of organic material, nitrogen and toxicity from low strength landfill leachates. Chemosphere 46: 851–858.

Meric, S., E. De Nicola, M. Iaccarino, M. Gallo, A. Di Gennaro, G. Morrone, et al. 2005. Toxicity of leather tanning wastewater effluents in sea urchin early development and in marine microalgae. Chemosphere 61: 208–217.

Metcalf and Eddy. 2003. Wastewater Engineering, McGraw Hill Publications. 2. Third Edition, Tata McGraw-Hill Publishing Company Limited, New Delhi.

Min, M., L. Wang, Y. Li, M.J. Mohr, B. Hu, W. Zhou, et al. 2011. Cultivating *Chlorella* sp. in a pilot-scale photobioreactor using centrate wastewater for microalgae biomass production and wastewater nutrient removal. Appl. Biochem. Biotechnol. 165: 123–137.

Monteiro, C.M., P.M.L. Castro and F.X. Malcata. 2012. Metal uptake by microalgae: underlying mechanisms and practical applications. Biotechnol. Prog. 28(2): 299–311.

Mwinyihija, M. 2010. Ecotoxicological Diagnosis in the Tanning Industry. Springer-Verlag, New York, US.

Oller, I., S. Malato and J. Sánchez-Pérez. 2011. Combination of advanced oxidation processes and biological treatments for wastewater decontamination – a review. Sci. Total Environ. 409: 4141–4166.

Pacheco, M.M., M. Hoeltz, M.S. Moraes and R.C. Schneider. 2015. Microalgae: cultivation techniques and wastewater phycoremediation. J. Environ. Sci. Health A 50: 585–601.

Pandi, M., V. Shashirekha and S. Mahadeswara. 2009. Bioabsorption of chromium from retan chrome liquor by cyanobacteria. Microbiol. Res. 164: 420–428.

Pawlik-Skowronska, B. 2003. Resistance, accumulation and allocation of zinc in two ecotypes of the green alga *Stigeoclonium tenue* Kutz. coming from habitats of different heavy metal concentrations. Aquat. Bot. 75(3): 189–198.

Pellon, A., F. Benitez, J. Frades, L. Garcia, A. Cerpa and F.J. Alguacil. 2003. Using microalgae *Scenedesmus obliquus* in the removal of chromium present in plating wastewaters. Revista de Metalurgia (Spain) 39: 9–16.

Perales-Vela, H.V., J.M. Peña-Castro and R.O. Cañizares-Villanueva. 2006. Heavy metal detoxification in eukaryotic microalgae. Chemosphere 64: 1–10.

Pereira, S., E. Micheletti, A. Zille, A. Santos, P. Moradas-Ferreira, P. Tamagnini, et al. 2011. Using extracellular polymeric substances (EPS) producing cyanobacteria for the bioremediation of heavy metals: do cations compete for the EPS functional groups and also accumulate inside the cell? Microbiol. 157(2): 451–458.

Perez-Rama, M., C.H. Lopez, J.A. Alonso and E.T. Vaamonde. 2001. Class III metalothioneins in response to cadmium toxicity in the marine microalga *Tetraselmis suecica* (Kylin) Butch. Environ. Toxicol. Chem. 20(9): 2061–2066.

Raize, O., Y. Argaman and S. Yannai. 2004. Mechanisms of biosorption of different heavy metals by brown marine macroalgae. Biotechnol. Bioeng. 87(4): 451–458.

Raungsomboona, S., A. Chidthaisonga., B. Bunnagb., D. Inthornc and N.W. Harveya. 2006. Production, composition and Pb^{2+} adsorption characteristics of capsular polysaccharides extracted from a cyanobacterium *Gloeocapsa gelatinosa*. Water Res. 40: 3759–3766.

Renuka, N., A. Sood, S.K. Ratha, R. Prasanna and A.S. Ahluwalia. 2013. Nutrient sequestration, biomass production by microalgae and phytoremediation of sewage water. Int. J. Phytoremediat. 15: 789–800.

Riaz, M.S., M.A. Hanif, S. Noureen, M.A. Khan, T.M. Ansari, H.N. Bhatti, et al. 2012. Coagulation/Flocculation of tannery wastewater using immobilized natural coagulants. J. Environ. Prot. Ecol. 13(3A): 1948–1957.

Saranya, D. and S. Shanthakumar. 2019. Green microalgae for combined sewage and tannery effluent treatment: Performance and lipid accumulation potential. J. Environ. Manage. 241: 167–178.

Schrank, S.G., H.J.M. Jos, R.F.P. Moreira and S.F. Schroder. 2003. Fentons oxidation of various based tanning materials. Desalination 50: 411–423.

Shashirekha, V., M. Pandi and S. Mahadeswara. 2005. Bioremediation of tannery effluent sand chromium containing wastes using cyanobacterial species. J. Am. Leather Chem. As. 11: 419–426.

Shashirekha, V., M.R. Sridharan and S. Mahadeswara. 2008. Biosorption of trivalent chromium by free and immobilized blue green algae: kinetics and equilibrium studies. J. Environ. Sci. Health A Toxic/Hazard Subst. Environ. Eng. 43: 390–401.

Shi, W., J. Becker, M. Bischoff, R.F. Turco and A.E. Konopka. 2002. Association of microbial community composition and activity with Pb, Cr and hydrocarbon contamination. Appl. Environ. Microbiol. 68: 3859–3866.

Singh, J.S., A. Kumar, A.N. Rai and D.P. Singh. 2016. Cyanobacteria: a precious bio-resource in agriculture, ecosystem, and environmental sustainability. Front. Microbiol. 7: 529.

Singh, J.S., S. Koushal, A. Kumar, S.R. Vimal and V.K. Gupta. 2016. Book review:

Microbial inoculants in sustainable agricultural productivity, Vol. II: Functional Application. Front. Microbiol. 7: 2105.

Singh, J.S., A. Kumar and M. Singh. 2019. Cyanobacteria: a sustainable and commercial bio-resource in production of bio-fertilizer and bio-fuel from wastewaters. Environ. Sustain. Indic. 3: 100008.

Singh, S., B.N. Rai and L.C. Rai. 2001. Ni(II) and Cr(VI) sorption kinetics by *Microcystis* in single and multimetatlic system. Proc. Biochem. 36: 1205–1213.

Siyanbola, T.O., K.O. Ajanaku, O.O. James, J.A.O. Olugbuyiro and J.O. Adekoya. 2011. Physico-chemical characteristics of industrial effluents in Lagos state, Nigeria. Global J. Pure Appl. Sci. 1: 49–54.

Song, Z., C.J. Williams and R.G.J. Edyvean. 2004. Treatment of tannery wastewater by chemical coagulation. Desalination 164: 249–259.

Srinivasan, S.V., G.P.S. Mary, C. Kalyanaraman, P.S. Sureshkumar, K.S. Balakameswari, R. Suthanthararajan, et al. 2012. Combined advanced oxidation and biological treatment of tannery effluent. Clean Technol. Envir. 14: 251–256.

Stern, S., R. Azpyrkowicz and I. Rodighiro. 2003. Aerobic treatment of textile dyeing wastewater. Water Sci. Technol. 47(10): 55–59.

Swamy, M. 2000. The role of cyanobacteria in removal of chromium from tannery effluent with special reference to trivalent chromium. National symposium on microbes in bioremediation for eco-friendly environment in the new millennium. University of Madras, Chennai, India, pp. 6–7.

Swartz, C.D., C. Jackson-Moss, R.A. Rowswell, A.B. Mpofu and P.J. Welz. 2017. Water and wastewater management in the tanning and leather finishing industry: NATSURV 10 (2nd Edition). Water Research Commission, South Africa.

Szpyrkowicz, L., S.N. Kaul and R.N. Neti. 2005. Tannery wastewater treatment by electrooxidation coupled with a biological process. J. Appl. Electrochem. 35: 381–390.

Tadesse, I., F.B. Green and J.A. Puhakka. 2004. Seasonal and diurnal variations of temperature, pH and dissolved oxygen in advanced integrated wastewater pond systems treating tannery effluent. Water Res. 38: 645–654.

Talaro, K.P. and B. Chess. 2015. Foundations in Microbiology. McGraw-Hill Education, New York.

Tisler, T., J. Zagorc-Koncan, M. Cotman and A. Drolc. 2004. Toxicity potential of disinfection agent in tannery wastewater. Water Res. 38: 3503–3510.

Torres, M.A., M.P. Barros, S.C.G. Campos, E. Pinto, Rajamani, R.T. Sayre, et al. 2008. Biochemical biomarkers in algae and marine pollution: a review. Ecotoxicol. Environ. Saf. 71: 1–15.

Travieso, L., R.O. Canizares, R. Borja, F. Benitez, A.R. Dominguez, R. Dupeyron, et al. 1999. Heavy metal removal by microalgae. Bull. Environ. Cont. Toxicol. 62: 144–151.

Tünay, O., N.I. Kabdasli, D. Orhon and G. Cansever. 1999. Use and minimization of water in leather tanning processes. Water Sci. Technol. 40: 237–244.

Velma, V., S.S. Vutukuru and P.B. Tchounwou. 2009. Ecotoxicology of hexavalent chromium in freshwater fish: a critical review. Rev. Environ. Health 24(2): 129–145.

Volesky, B. and Z.R. Holan. 1995. Biosorption of heavy metals. Biotechnol. Prog. 11(3): 235–250.

Wang, L., M. Min, Y. Li, P. Chen, Y. Chen, Y. Liu, et al. 2010. Cultivation of green algae chlorella sp. in different wastewaters from municipal wastewater treatment plant. Appl. Biochem. Biotechnol. 162: 1174–1186.

Worms, I., D.F. Simon, C.S. Hassler and K.J. Wilkinson. 2006. Bioavailability of trace metals to aquatic microorganisms: importance of chemical, biological and physical processes on biouptake. Biochimie. 88(11): 1721–1731.

Zayed, A., S. Gowthaman and N. Terry. 1998. Phytoaccumulation of trace elements by wetland plants: I. Duckweed. J. Environ. Qual. 27(3): 715–721.

Zayed, A.M. and N. Terry. 2003. Chromium in the environment: factors affecting biological remediation. Plant Soil 249(1): 139–156.

Zhang, E., B. Wang, Q. Wang, S. Zhang and B. Zhao. 2008. Ammonia–nitrogen and orthophosphate removal by immobilized *Scenedesmus* sp. isolated from municipal wastewater for potential use in tertiary treatment. Bioresour. Technol. 99: 3787–3793.

Pulp and Paper Wastewater

Introduction

Pulp and paper industry is one of the three largest wastewater producing industries along with primary metals and petrochemical industries (Thompson et al. 2001, Sumathi and Hung 2006, Savant et al. 2006). It is estimated that pulp-and-paper mills generate 20-100 m^3 of wastewater per ton of air-dried pulp, and further water consumption would depend on the production process and quality of paper produced (WB 1999, Thompson et al. 2001, Pokhrel and Viraraghavan 2004). The pulp and paper industry involves mainly two major steps: pulping and bleaching, which contributes to a large amount of wastewater. For the pulping, most paper mills use chemical pulping like the Kraft process, in which debarked wood is digested in a heated alkaline solution to separate the cellulosic fibers from other unnecessary components of the wood, namely lignins and tannins. There are alternatives to chemical pulping processes like semi-chemical and mechanical pulping processes, but these are used by a smaller number of paper mills. Pulping is followed by a bleaching process and then the bleached pulp undergoes dewatering and a drying process to produce sheets, liberating a stream of wastewater containing byproducts of the pulping process.

The wastewater from the pulp and paper industry is characterized with a high organic content, dark brown color, the presence of Adsorbable Organic Halide (AOX) and large amounts of other toxic chemicals (Sumathi and Hung 2006). There are 500 different chlorinated organic compounds including resin acids, chlorinated hydrocarbons, chloroform, chlorate, phenols, catechols, guaiacols, furans, dioxins, etc., which are very harmful to the environment and are collectively designated as adsorbable organichalides (AOX). The impact of pulp-and-paper mill wastewater depends on two factors: the chemical nature and dark coloration, both which negatively affect aquatic fauna and flora. The detrimental effects of this wastewater to animals could be in the form of respiratory stress, liver damage and geno-toxicity (Ali and Sreekrishnan 2001, Singhal and Thakur 2009). With regards to human health, the consequences could be diarrhea, vomiting, headaches, nausea and irritation of the eyes in children and workers.

There are many treatment methods including physical, physical and biological; but these are not enough for the efficient treatment of pulp and paper wastewater. Biological treatment methods involving the use of fungi, bacteria and microbial enzymes could provide an alternative to physicochemical processes or could be used in combination with physical and chemical methods for better results. Malaviya and Rathore (2007) suggested that microorganisms like bacteria, fungi or even algae either perform biosorption to retain the components or release enzymes for degradation in the treatment of pulp and paper wastewater. In the future, it was also suggested that enzyme-based processes could be cleaner and more efficient in the treatment of pulp and paper wastewater. Some of the advanced processes such as adsorption, membrane filtration and electro-flocculation proved to be more efficient than physical chemical and biological methods. But the major limitations with these processes are the high cost of building and operations, regular maintenance and the need of skilled workers to operate them.

Microalgae-mediated remediation could serve as a potential role in the treatment of pulp and paper wastewater in terms of cost effectiveness, better adaptability and efficient removal of the specific pollutants present in the wastewater (Kumar and Singh 2016, Singh et al. 2016a, b, Kumar et al. 2017, Kumar and Singh 2017). They are very efficient in phenol degradation in end products that are either used for synthesis of organic compounds or assimilated later in the form of inorganic carbon. Although microalgae are not found to be the most suitable for the degradation of lignins and tannins, they also support the growth of lignin degrading heterotrophic microorganisms by providing the essential nutrients and oxygen for respiration (Kumar 2018, Kumar et al. 2018a, b, Singh et al. 2019, Kumar and Singh 2020a, b).

Pulp and Paper Industry: A Brief Description

The history of papermaking goes back to AD 105, where Ts'ai Lun, an official of the imperial court to the Chinese Emperor, who was considered as the inventor of the paper making. Later this ancient Chinese art of paper making led the way for modern manufacture of paper. Although there are many differences in the current products of paper making, compared to its ancestral materials, but even today papermaking has some diverse similarities to the ancient papermaking processes invented by Ts'ai Lun. The wood logs are loaded into a thawing conveyor and the frozen log goes to a drum where warm water or steam is provided. This warm water makes the surface of wood, mainly the bark usually warmer, which leads to weaking the bond strength between the bark and wood; resulting in easier removal of the bark from the wood. Then debarked logs are washed properly and converted into small pieces known as chips by the use of industrial machines i.e., wood chippers.

The wood chip undergoes the pulping process which involves the digestion of wood chips to obtain fibrous constituents. There are three types of pulping processes: (a) chemical pulping, (b) mechanical pulping and (c) semi-chemical pulping; which are used to break the bonds between fibers.

The chemical pulping process involves a significant amount of chemicals to separate out the cellulosic fibers from the wood by the action of heat and pressure. Based on the use of chemicals, it is categorized into (i) Kraft/soda pulping and (ii) sulfite pulping. In Kraft pulping, an alkaline solution (liquor) is made up of sodium sulfide (Na_2S) and sodium hydroxide ($NaOH$) in a 10% solution that is used; while sulfite pulping involves the use of acidic solutions made up of sulfurous acid (H_2SO_3) and bisulfite ion (HSO_3^-) generally from sodium sulfite. The liquor (white liquor in case of Kraft) is added to the wood chips in a digester, which turns in to a black color known as spent liquor, which is then either recovered or disposed off.

The mechanical pulping process uses industrial machinery such as a rotating grindstone which strips of the fibers from the wood logs usually debarked wood logs. Then they are suspended in the water. The semi-chemical pulping process involves the use of chemicals as well as the mechanical disintegration for fiber separation. The cooking liquor made up of sodium sulfite (Na_2SO_3) and sodium carbonate (Na_2CO_3) is applied to the heated pulp in the semi-chemical pulping process. The main difference between the chemical and semi-chemical pulping, is that in semi-chemical pulping lower temperatures are used, more dilute cooking liquor or shorter cooking times, as compared to the chemical pulping process. The selection of a pulping process is primarily influenced by the type and the desired quality of the finished product, but the chemical pulping process is very often applied in most of the industries.

The pulp processing is used to remove impurities mainly uncooked chips, and recovering the residual cooking liquor from the pulp. Some common methods such as screening, de-fibering and de-knotting are applied for the processing of the pulp. Further excess water might be removed to obtain a thick pulp, and further blending could be required to get a uniform pulp product.

Recovery of residual spent cooking liquor from chemical and semi-chemical process done by washing the pulp using rotary vacuum washers, which are performed consecutively in two or four washing units. It is a critical part of the processing of pulp; in which recovery of processing chemicals from the spent cooking liquor takes place for their reuse.

Bleaching is followed to enhance the brightness of the pulp. The extent of bleaching could depend on the lignin content and the type of the pulping process. It is reported that mechanical or semi-chemical pulps have high lignin content; which proves to be difficult to bleach the pulp completely and further requires longer bleaching. Due to their low lignin content (10%), the bleaching of chemical pulps are found to be much easier. There are two methods i.e., Elemental Chlorine Free (ECF) and Total Chlorine Free (TCF), in which the ECF is usually applied for bleaching. ECF is responsible for 95% of the bleached pulp production which may include the use of chlorine dioxide and hypochlorites.

The bleaching process takes place in traditional bleach plants, which include three to five cycles of chemical bleaching and consequently water washing. The bleaching of semi-chemical pulps uses hydrogen peroxide (H_2O_2), while mechanical pulps include the use of hydrogen peroxide (H_2O_2) and/or sodium hydrosulfite (Na_2SO_3) in a bleach tower. The number of bleaching cycles depends on the plant design, the brightness of prepared pulp stock and demand of the considered product and the whiteness desired.

After the final washing, the pulp undergoes up to 90% drying in a container, it could either in a bleached or unbleached state. For efficient drying, jet dryers of flash dryers are often used for the final drying of the pulp in several layers laying one above the other. This process involves the use of ventilators, cyclone separators and high-performance suspension towers. Finally the dried flakes of pulp are conveyed to a bale press, which produces paper bales with a specific weight of about 0.5 g/cm^3.

Pulp and Paper Industry Wastewater: Characteristics and Specific Pollutants

Pulp and paper wastewater is characterized with color, high BOD, high COD, biochemical and a significant amount of suspended solids and toxic AOX compounds. During the pulping process, various chemicals such as sodium sulfide (Na_2S), sodium hydroxide ($NaOH$), sulfurous acid (H_2SO_3) and bisulfite ion (HSO_3^-) are used; while in the bleaching process, chlorine dioxide, hypochlorites and hydrogen peroxide are added to get a whiter color. Apart from these, other chemicals like organic fillers (starch, latex), colors and aluminum sulfate are also used to obtain the desired product. So pulp and paper wastewater contains a wide variety of chemical residues ranging from chlorine or chlorine dioxide to sodium silicates, sodium carbonate, fatty acids or non ion detergents and also emissions of chlorine, peroxides, oxygen and ozone.

Lignin and its derivatives

Lignin is the collective term that is used for a large group of rigid and impermeable aromatic polymers which are formed from the oxidative coupling of 4-hydroxyphenylpropanoids (Vanholme et al. 2010). They are predominantly found in woody plants. It is suggested that approximately 5-10% of the lignin present in woods cannot be removed, which is mainly responsible for the dark color of the pulp and paper wastewater. Minu et al. (2012) observed that lignin is found be very difficult in chemical or biological degradation, because of the presence of recalcitrant and not-hydrolysable carbon-carbon linkages and aryl ether bonds.

Volatile compounds

The major volatile organic compounds that are present in pulp and paper wastewater: water vapors. Additionally, particulates, nitrogenoxides,

Table 8.1: Characteristics of pulp and paper wastewater

Water quality parameter	Mehmood et al. 2019	Usha et al. 2016	Ginni et al. 2014	Kesalkar et al. 2012	Zwain et al. 2013	Devi et al. 2011	Nagasathya and Thajuddin 2008	Avşar and Demirer 2008	Kirkwood et al. 2005
pH	7.61	5.41	7.20	6.9	6.2-7.8	6.9	5.9	6.5	7.7
EC (ms cm^{-1})	1.348	3.932				2.050	3.290		
TS (mg L^{-1})	3494	6446			3530-6163				126
TDS (mg L^{-1})	2710	3580		910	1630-3025	38.12		1241	
TSS (mg L^{-1})	784	2866	1295	443				1197	
BOD (mg L^{-1})	975	2944	60	202	1650-2565	45.2	3.4	3791	203
COD (mg L^{-1})	2820	3000.15	2133	892	3380-4930	316.2	248		
TN/ NH$_4^+$-N (mg L^{-1})		9.932					3.0		<0.1
PO$_4^{3-}$/ Phosphates (mg L^{-1})		30.25					16.5		0.7
TOC (mg L^{-1})		198							
Sulfate/Sulfides (mg L^{-1})		86.48				5.2	38.5		
Chlorides (mg L^{-1})		204				30.5	1290.7		
Alkalinity (mg L^{-1})		228			300-385	206.3	60		

Note: TS: Total Solids; TDS: Total Dissolved Solids; TSS: Total Suspended Solids; BOD: Biological Oxygen Demand; COD: Chemical Oxygen Demand; TN: Total Nitrogen; NH$_4^+$-N: Ammonical Nitrogen; NO$_3^-$-N: Nitrate Nitrogen; TP: Total Phosphorous; TOC: Total Organic Content.

Table 8.2: Specific pollutants in pulp and paper wastewater

Process	Pollutants
Debarking, Chipping	Suspended solids including bark particles, resin, fiber pigments, dirt, grit
Pulping and Pulp processing	Color, bark particles, soluble wood materials, resin acids, fatty acids, AOX, VOCs, and Dissolved inorganics
Chemical recovery	-
Bleaching	Dissolved lignin, color, carbohydrate, inorganic chlorines, AOX, EOX, VOCs, chlorophenols, and halogenated hydrocarbons
Dewatering Pressing and drying	Particulate wastes, organic and inorganic compounds
Finishing	-

Volatile Organic Compounds (VOCs), sulfur oxides and Total Reduced Sulfur compounds (TRS). Some of the VOCs are terpenes, alcohols, phenols, methanol, acetone, chloroform, methyl ethyl ketonemethylene chloride, chloromethane and trichloroethane; these are more often found in pulp and paper wastewater.

Adsorbable organic halides (AOXs)

Adsorbable organic halides (AOX) are chlorinated organic compounds that are synthesized as a result of the reaction of residual lignin from wood fibers with chlorine/chlorine compounds that are used in the bleaching process. Some common AOX are chlorinated benzenes, phenols, chlorophenols, dioxins, dibenzofurans, chloroform, carbon tetra chloride, epoxystearic acid and dichloromethane.

Impacts on Environment and Public Health

Pulp and paper wastewater have high organic loads and a large amount of chemicals especially chlorinated compounds, which deleteriously impact the aquatic ecosystem in various ways: oxygen depletion in large areas, damage to the benthic communities and numerous negative impacts in fish related to their reproduction and physiology (Springer 2000, Lehtinen 2004). Munkittrick et al. (1997) observed that pulp and paper wastewater might interrupt sexual maturation, leading to the development of smaller gonads in animals. It also influences reproduction in fishes and might hinder the secondary sexual characteristics in fish species.

Pulp and paper wastewater contains chlorinated organic compounds, which are primarily produced during chlorine bleaching operations. The presence of chlorinated organic compounds in wastewater is measured as adsorbable organic halogen (AOX). It is observed that AOX compounds are found to be persistent, recalcitrant and highly toxic; resulted in the high resistance to biological degradation. Savant et al. (2006) described the toxic effects of AOX; it could be from a range of carcinogenic, mutagenic to acute and chronic toxicity.

There is a sharp decrease in the presence of adsorbable organic halogen in pulp and paper wastewater as a gradual replacement of elemental chlorine with chlorine dioxide in bleaching operations; until it is completely substituted with chlorine dioxide. The environmental control authority of many countries set strict restrictions related to discharge limits of these chlorinated organics into the aquatic environment. Several environmental agencies of some countries like Sweden (1992) and Finland (1995) have also limited chlorinated organics emissions to 1.5 kg and 1.4 kg AOX/t of pulp respectively (Lehtinen 2004). Springer (2000), Rodgers and Thomas (2004) and EU 2003 suggested that instead of chlorine, chlorine dioxide could be a better alternative, which helps in reduction of the AOX levels and also reduces the amount of the dioxins, chloroform and polychlorinated compounds in pulp and paper wastewater.

Table 8.3: Toxic effects of major AOX compounds (Savant et al. 2006)

Compounds	Toxic effect
Chlorophenols	Liver and kidney damage, loss in weight, general fatigue and low appetite. In fish, impaired function of liver, enzyme system, metabolic cycle, increase in the incidence of spinal deformities and reduced gonad development
Chlorocatechols	Strong mutagens
Chloroguaiacols	Tetra- and tri-chloroguaiacols are known to bioaccumulate in fish
Chlorobenzenes	Drowsiness, headache, eye irritation, sore throat.
Chlorinated dibenzodioxins and dibenzofurans	Highly toxic, teratogenic. Acute exposures cause severe skin rash, changes in skin color, hyperpigmentation, polyneuropathies in arms and legs. Act as endocrine disrupting factors by interfering production, release, transport, metabolism, binding action or elimination of natural hormones in the body weight.

Current Treatment and Its Challenges

Conventional wastewater treatment involves various Physicochemical process such as screening flotation, sedimentation, coagulation, flocculation and ozonation; which is applied to remove color, suspended solids and toxic compounds from pulp and paper wastewater (Hogenkamp 1999, Bhattacharjee et al. 2007, El-Ashtoukhy et al. 2009, Kishimoto et al. 2010). Besides this, biological processes like activated sludge are preferably applied to remove organic contaminants, because of relatively low capital and operating costs (Pokhrel and Viraraghavan 2004, Buzzini et al. 2005, Habets and Driessen 2007).

The choice of a particular physicochemical process for the treatment primarily depends on the composition and quantity of pollutants in the wastewater and the further need of the desired level of environmental specifications. The conventional physicochemical and biological processes have proved to be costly in building, maintenance and are also responsible for the generation of a significant amount of sludge (Patel and Suresh 2008). Along with conventional physicochemical and biological processes, some advanced processes have proved to be very effective in color reduction, degradation and mineralization of pollutants (Catalkaya and Kargi 2007, Hassan and Hawkyard 2002, Helmy et al. 2003, Daneshvar et al. 2004, Catalkaya and Kargi 2008, Ugurlu and Karaoglu 2009). But these processes are not sustainable and cost effective for industries and agencies due to their high cost and ineffectiveness in the removal of some pollutants. There are some advanced treatments including adsorption, ultra-filtration and electro-flocculation that are applied to remove residual contaminants even after physicochemical and biological processes.

In the adsorption process, activated carbon is used for adsorption of the pollutants present in pulp and paper wastewater. Due to large surface area ($450–1800 \, m^2/g$), activated carbon proves to be an excellent adsorbent. There are many factors such as particle size, carbon pore size, molecular weight of the compound and the pH of wastewater which influence the removal efficiency of activated carbon.

The membrane filtration like ultra filtration and reverse osmosis are processes which are also used for the treatment of pulp and paper wastewater. These techniques rely on the principle of separation of pollutants on the basis of molecular size; the wastewater is conveyed through a semi permeable membrane, which filters the different pollutants and cleans the water generated.

The electro-flocculation process involves the use of an electric current to release metal ions from Al, Fe, Mg metal electrodes that have coagulating properties, which coagulate the pollutants from wastewater. The electric current also produces gas bubbles that capture the coagulated pollutants, resulting in the flotation of most of the pollutant on the surface.

These advanced processes are found to be effective in the removal of pollutants from the pulp and paper wastewater. But high installation and operating costs of these make them unaffordable for large scale operations.

Microalgal Bioremediation

Kirkwood et al. (2001) investigated the presence of microalgae or cyanobacterial communities in pulp and paper waste-treatment systems. They observed that some microalgal especially cyanobacterial genera *Phormidium, Geitlerinema, Pseudanabaena* and *Chroococcus* dominated the pulp and paper waste-treatment systems in Brazil and New Zealand. Despite having different geographical locations, all the microalgal communities were found to be similar in all sites; which could possibly be due to similar temperature conditions in the range of 28°C–33°C at all the sites. So it was suggested that the same cyanobacterial genera could be used for pulp and paper wastewater in regions that have similar temperature conditions.

Pulp and paper wastewater contains a significant amount of phenolic compounds (Sadhu et al. 2013); which are very toxic in nature. Some microalgal strains like *Chlamydomonas ulvaensis*, *Ochromonas danica*, *Chlorella pyrenoidosa*, *Anabaena cylindrica* and *Scenedesmus basiliensis* are able to remove phenol from wastewater (Ellis 1977, Semple and Cain 1996). There are also reports of microalgae species *Phormidium valderianum*, *Synechococcus* PCC 7002, which are able to grow and degrade phenol; and some are also use phenol for their growth (Shashirekha et al. 1997, Pinto et al. 2002, Wurster et al. 2003, Petroutsos et al. 2007).

Phenol and Their Derivatives Degradation

Phenols and their derivatives such as halophenols in low concentrations are not found to be toxic to the microalgal communities and can be used as an alternative carbon source. It is observed that microalgal degradation of phenols takes place only in aerobic conditions. Many microalgal species are mixotrophic in nature; and could be grown in both photoautotrophic and heterotrophic conditions. Tikoo et al. (1997) observed that microalgae are quite sensitive to phenolic compounds either when they are grown in phototrophic or heterotrophic conditions. To remove this limitation, mixotrophic growth conditions can improve the microalgal ability to have an enhanced capacity to mineralize phenolic compounds (Tikoo et al. 1997). Therefore the ability of microalgae to survive in many conditions such low light or shortage of carbon dioxide ; could be a very good approach for the treatment of phenol-containing wastewaters (Table 8.4) (Pinto et al. 2002).

Table 8.4: Microalgae in phenol degradation

Microalgae	*Phenol tolerance/ removal*	*References*
Chamydomonas reinhardtii	Up to 100	Sanchez-Aponte et al. 2019, Ellis 1977; Samanthakamani and Thangaraju 2015
Chlamydomonas sp.	-	Udaiyappan et al. 2017 Al-Fawwaz et al. 2016
Desmodesmus sp.		Al-Fawwaz et al. 2016
Spirulina maxima	97.5 %	Lee et al. 2015
Microalgae	-	Nazos and Kokarakis 2017
Chlamydomonas mexicana	-	Ji et al. 2014
Chlorella vulgaris	-	Ji et al. 2014

Semple et al. (1999) observed the mechanism of phenol degradation by microalgae including cyanobacteria. It was suggested that the presence of molecular oxygen is crucial for microalgae to initiate the enzymatic attack on the aromatic ring of the phenol and this process only takes place only under aerobic conditions. Phenol is initially hydroxylated to form catechol, which further degrades through ortho- or meta- oxidation and final metabolites of this transformation are CO_2 and simple organic molecules, mainly pyruvate (Fig. 8.1) (Semple and Cain 1996, Semple et al. 1999, Papazi and Kotzabasis 2007). These end products of phenol degradation are either directly used for the production of organic compounds, or indirectly in the form of inorganic carbon that are available for autotrophic nutrition. In case of halophenols, first dehalogenation i.e., a split of halogen substituent takes place, and then it goes to the pathway of phenol degradation. Papazi and Kotzabasis (2007) observed that among different halophenols, the degradation of chlorophenol

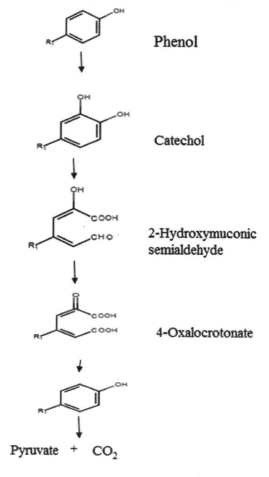

Figure 8.1: Phenol degradation by microalgae (Semple and Cain 1996).

needs maximum energy, and then it is followed by bromophenol and iodophenol. Further the position of halogen on the phenol ring influences the degradation process; the metaposition needs more energy than the ortho- and the para-positions (Papazi and Kotzabasis 2007).

Wurster et al. (2003) investigated the capacity of *Synechococcus* PCC 7002 for phenol degradation under dark conditions, and reported the transformation of phenol in to cis, cis-muconic acid; which is found to be similar as reported in ortho-fission of phenolic compounds by heterotrophic bacteria and yeasts. Wurster et al. (2003) further observed that Synechococcus PCC 7002 was not able to use phenol for growth. In relation to enzymatic degradation, Shashirekha et al. (1997) presented the first report which observed the phenol degradation through activities of polyphenol oxidase and laccase enzymes in *Phormidium valderianum*. Shashirekha et al. (1997) noted that marine microalgae *Phormidium valderianum* successfully tolerate and remove phenol within a retention period of 7 days.

Role of Light in Phenol Degradation

It has been confirmed that microalgae could be grown on phenols using the carbon source only in the light conditions (Semple and Cain 1996, Nakai et al. 2001, Papazi and Kotzabasis 2007, Petroutsos et al. 2007, Lika and Papadakis 2009). Papazi and Kotzabasis (2007) observed the growth of *Scenedesmus obliquus* in light conditions, which exhibited different specific growth rates for different phenolic compounds. Further this study also revealed that phenol degradation grows with the increase in light intensity without supplementing the external carbon. Petroutsos et al. (2007) studied the ability of *Tetraselmis marina* to remove chlorophenols under phototrophic conditions. Further Petroutsos et al. (2007) suggested that the p-chlorophenol removal ability of *Tetraselmismarina* could be increased by extending the duration of light period or the NaHCO$_3$ concentration (Petroutsos et al. 2007).

Petroutsos et al. (2007) emphasized that the position of chloro group also affects its degradation efficiency; o-chlorophenol has no effect on growth while m- and p-chlorophenol could be responsible for decrease in the growth of *T. marina* in comparison of the control. In comparison of p-chlorophenol, m-chlorophenol shows maximum decrease in growth, this is explained by Papazi and Kotzabasis (2007) that meta-position of the halogen requires higher bond dissociation energy compared with those of the ortho- and para-positions. From the findings of Papazi and Kotzabasis (2007), Petroutsos et al. (2007) and Lika and Papadakis (2009), it was established that along with light conditions, oxygen supply also enhances the phenol degradation process. There are very few reports regarding the complete metabolization of phenol by microalgae. Semple and Cain (1996) observed that microalgae *Ochromonas danica* was able to degrade phenol even in dark conditions, and further it also showed the ability of complete metabolization of phenol, from which 20% of the carbon had been used by the microalgae for their biomass.

AOX Removal

Due to the toxic nature of AOX, it becomes a grave concern in the treatment of pulp and paper wastewater (La Fond 1991, Hall and Randle 1992, Ferguson 1994, GoÈkcÈay et al. 1996, Francis et al. 1997). There are various physicochemical and biological treatment methods that have investigated the metabolization of adsorbable organic halides (AOX). GoÈkcÈay et al (1996) claimed that removal of AOX does not always take place through metabolization; it could be possible for Old sludge, but in case of new sludge, it would be adsorption of activated sludge.

A number of studies are available for microalgae mediated treatment of pulp and paper wastewater, but there are no or hardly any studies related to the use of microalgae for the treatment of AOX in particular. It was observed that the removal of AOX can be directly proportional to the initial color value of the pulp wastewater, but color removal was found to be independent of the initial color of the pulp and paper wastewater (Dilek et al. 1999). Lee et al. (1978) investigated the use of algae or microalgae in removal of color and AOX, and reported that 80-95% removal efficiency in pulp, paper and textile wastewater. But in relation to mechanisms of color and AOX removal, there are two opinions because of the metabolic transformation of colored compounds to non colored compounds (Lee et al. 1978) or it only occurs with the adsorption of colored molecules.

Dilek et al. (1999) observed that color removal efficiency of microalgae reached up to 80% for pulp and paper wastewater in 30 days under continuous lighting conditions, while it was reported to be up to 60% in 60 days under simulated field lighting conditions. Further microalgae was able to reduce the color of the pulping process wastewater with low initial color values than pulp wastewater with high-initial-color values. In comparison with the pulping process wastewater, microalgae takes less time for the exact color removal from pulp and paper wastewater including bleaching wastewater, which underlines the fact that the bleaching process wastewater might stimulate the ability of microalgae to treat pulp and paper wastewater effectively. This finding is significant in the field of treatment of pulp and paper wastewater, as GoÈkcÈay and Dilek (1994), Dilek and GoÈkcÈay (1994) reported the inability of physicochemical or biological processes (white-rot fungi) to remove the color and AOX components from the bleaching process wastewater and/or pulp and paper wastewater containing bleaching process wastewater.

Conclusion

The pulp and paper industry involves processes that discharge large amounts of wastewater, which creates significant pollution to the environment (air, water courses and soil). In the last decades, there are strict environment regulations and various technological solutions such as the dry debarking of wood, the introduction of extended cooking and oxygen delignification, the reuse of condensates, improvement in washing efficiency and especially the total substitution of chlorine, which ultimately reduces the large flow of wastewater and load of chlorinated and organic loads in pulp and paper wastewater.

The environmental impact of pulp and paper wastewater and their effective treatment could be determined through estimating their chemical composition and identifying the pollutants especially chlorinated compounds present in pulp and paper wastewater. There are various physical, chemical and biological treatment processes that are applied for the treatment of pulp and paper wastewater. The usual biological treatment for pulp and paper wastewater is activated sludge systems, which are inefficient for treatment beyond a certain level; it is primarily due to the high stability of lignin. Although white-rot fungi showed partial decolorization with removable efficiencies to a great extent, but the need of high glucose feed for the fungi makes this treatment economically unfeasible. There are many advanced oxidation processes like photocatalysis, ozonation and peroxidation, which are found to be efficient in color reduction; and also in detoxification and mineralization of toxic compounds especially chlorinated compounds. But these advanced processes are not industrially feasible due to their high cost .

Microalgae mediated remediation could be a potential approach in decolorizing pulp and paper wastewater. It is also suggested that color removal efficiency by microalgal mainly depends on the incubation period, which could be accelerated as the incubation period increases. Microalgae are found to be very effective in the phenol degradation and detoxification of chlorinated compounds present in pulp and paper wastewater. It is also suggested that more light conditions enhances the phenol degradation. So having the capability to survive in extreme conditions and autotrophic nature, microalgae mediated remediation would provide us a cost effective, sustainable and eco-friendly approach in the treatment of pulp and paper wastewater.

References

Al-Fawwaz, A.T., J.H. Jacob and T.E. Al-Wahishe. 2016. Bioremoval capacity of phenol by green microalgal and fungal species isolated from dry environment. Int. J. Sci. Technol. Res. 5: 155–160.

Ali, M. and T.R. Sreekrishnan. 2001. Aquatic toxicity from pulp and paper mill effluents: a review. Adv. Environ. Res. 5: 175–196.

Avşar, E. and G.N. Demirer. 2008. Cleaner production opportunity assessment study in SEKA Balikesir pulp and paper mill. J. Clean. Prod. 16(4): 422–431.

Bhattacharjee, S., S. Datta and C. Bhattacharjee. 2007. Improvement of wastewater quality parameters by sedimentation followed by tertiary treatments. Desalination 212: 92–102.

Buzzini, A.P., E.P. Gianotti and E.C. Pires. 2005. UASB performance for bleached and unbleached kraft pulp synthetic wastewater treatment. Chemosphere 59: 55–61.

Catalkaya, E.C. and F. Kargi. 2007. Color, TOC and AOX removals from pulp mill effluent by advanced oxidation processes: a comparative study. J. Hazard. Mater. B. 139: 244–253.

Catalkaya, E.C. and F. Kargi. 2008. Advanced oxidation of diuron by Photo-Fenton treatment as a function of operating parameters. J. Environ. Eng. 134(12): 1006–1013.

Daneshvar, N., D. Salari and A.R. Khataee. 2004. Photocatalytic degradation of azo dye acid red 14 in water on ZnO as an alternative catalyst to TiO_2. J. Photochem. Photobiol. A. Chem. 162: 317–322.

Devi, N.L., I.C. Yadav, Q.I. Shihua, S. Singh and S.L. Belagali. 2011. Physicochemical characteristics of paper industry effluents—a case study of South India Paper Mill (SIPM). Environ. Monit. Assess. 177(1–4): 23–33.

Dilek, F.B. and C.F. GoÈkcÈay. 1994. Treatment of effluents from hemp-based pulp and paper industry. I. Waste characterisation and physico-chemical treatability. Water Sci. Technol. 29(9): 165–168.

Dilek, F.B., H.M. Taplamacioglu and E. Tarlan. 1999. Colour and AOX removal from pulping effluents by algae. Appl. Microbiol. Biotechnol. 52: 585–591.

Duan, W., F. Meng, Y. Lin and G. Wang. 2017. Toxicological effects of phenol on four marine microalgae. Environ. Toxicol. Pharmacol. 52: 170–176.

El-Ashtoukhy, E.S.Z., N.K. Amin and O. Abdelwahab. 2009. Treatment of paper mill effluents in a batch-stirred electrochemical tank reactor. Chem. Eng. J. 146: 205–210.

Ellis, B.E. 1977. Degradation of phenolic compounds by fresh-water algae. Plant Sci. Lett. 8: 213–216.

European Commission (EC). 2003. Integrated Pollution Prevention and Control (IPPC)-Reference document on best available techniques for the tanning of hides and skins. European IPPC Bureau. Seville, Spain.

Ferguson, J.F. 1994. Anaerobic and aerobic treatment for AOX removal. Water Sci. Technol. 29(5–6): 149–162.

Francis, D.W., P.A. Turner and J.T. Wearing. 1997. AOX reduction of kraft bleach plant effluent by chemical treatment. Water Res. 31(10): 2397–2404.

Ginni, G., S. Adishkumar, J.R. Banu and N. Yogalakshmi. 2014. Treatment of pulp and papermill wastewater by solar photo Fenton process. Desalination Water Treatment 52(13–15): 2457–2464.

GoÈkcÈay, C.F. and F.B. Dilek. 1994. Treatment of effluents from hemp-based pulp and paper industry. II. Biological treatability of pulping effluents. Water Sci. Technol. 29(9): 161–163.

GoÈkcÈay, C.F., U. Yetis and F.B. Dilek. 1996. Final report of characterization and biological treatability project for Turkish State Pulp and Paper Industry SEKA chlorinated wastewaters. Environmental Engineering Department, Middle East Technical University, Ankara.

Habets, L. and W. Driessen. 2007. Anaerobic treatment of pulp and paper mill effluents-status quo and new developments. Water Sci. Technol. 55(6): 223–230.

Hall, E.R. and W.G. Randle. 1992. AOX removal from bleached kraft mill wastewater: a comparison of three biological treatment processes. Water Sci. Tech. 26(1–2): 387–396.

Hassan, M.M. and C.J. Hawkyard. 2002. Decolourisation of aqueous dyes by sequential oxidation treatment with ozone and Fenton's reagent. J. Chem. Tech. Biot. 77(7): 834–841.

Helmy, S.M., S. El Rafie and M.Y. Ghaly. 2003. Bioremediation post-photo-oxidation and coagulation for black liquor effluent treatment. Desalination 158(13): 331–339.

Hogenkamp, H. 1999. Flotation: the solution in handling effluent discharge. Pap. Asia 15: 16–18.

Ji, M.K., A.N. Kabra, J. Choi, J.H. Hwang, J.R. Kim, R.A.I. Abou-Shanab, et al. 2014. Biodegradation of bisphenol A by the freshwater microalgae *Chlamydomonas mexicana* and *Chlorella vulgaris*. Ecol. Eng. 73: 260–269.

Kesalkar, V.P., I.P. Khedikar and A.M. Sudame. 2012. Physico-chemical characteristics of wastewater from Paper Industry. Int. J. Eng. Res. Appl. 2(4): 137–143.

Kirkwood, A.E., C. Nalewajko and R.R. Fulthorpe. 2001. The occurrence of cyanobacteria in pulp and paper waste-treatment systems. Can. J. Microbiol. 47: 761–766.

Kirkwood, A.E., C. Nalewajko and R.R. Fulthorpe. 2005. The impacts of cyanobacteria on pulp-and-paper wastewater toxicity and biodegradation of wastewater contaminants. Can. J. Microbiol. 51: 531–540.

Kishimoto, N., T. Nakagawa, H. Okada and H. Mizutani. 2010. Treatment of paper and pulp mill wastewater by ozonation combined with electrolysis. J. Water Environ. Technol. 8: 99–109.

Kumar, A. 2018. Assessment of cyanobacterial diversity in paddy fields and their capability to degrade the pesticides. Babasahaeb Bhimrao Ambedkar University, Lucknow, India.

Kumar, A. and J.S. Singh. 2016. Microalgae and cyanobacteria biofuels: a sustainable alternate to crop-based fuels. pp. 1–20. *In*: J.S. Singh, D.P. Singh (eds.). Microbes and Environmental Management. Studium Press Pvt. Ltd. New Delhi, India.

Kumar, A. and J.S. Singh. 2017. Cyanoremediation: a green-clean tool for decontamination of synthetic pesticides from agro- and aquatic-ecosystems. pp. 59–83. *In*: J.S. Singh, G. Seneviratne (eds.). Agro-environmental Sustainability, Vol. II: Managing environment pollution. Springer Int., Cham, Switzerland.

Kumar, A. and J.S. Singh. 2020. Biochar coupled rehabilitation of cyanobacterial soil crusts: a sustainable approach in stabilization of arid and semiarid soils. pp. 167–191. *In*: J.S. Singh, C. Singh (eds.). Biochar Applications in Agriculture and Environment Management. Springer Int., Cham, Switzerland.

Kumar, A. and J.S. Singh. 2020. Microalgal bio-fertilizers. *In*: E. Jacob-Lopes, M.M. Maroneze, M.I. Queiroz, L.Q. Zepka (eds.). Handbook of Microalgae-based Processes and Products. Academic Press, Cambridge, US, In Press.

Kumar, A., S. Kaushal, S.A. Saraf and J.S. Singh. 2017. Cyanobacterial biotechnology: an opportunity for sustainable industrial production. Clim. Chang. Environ. Sustain. 5(1): 97–110.

Kumar, A., S. Kaushal, S.A. Saraf and J.S. Singh. 2018. Microbial bio-fuels: a solution to carbon emissions and energy crisis. Front. Biosci. (Landmark) 23: 1789–1802.

Kumar, A., S. Kaushal, S.A. Saraf and J.S. Singh. 2018. Screening of Chlorpyrifos (CPF) tolerant cyanobacteria from paddy field soil of Lucknow, India. Int. J. Appl. Adv. Sci. Res. 3(1): 100–105.

La Fond, R.A. 1991. Removal of adsorbable organic halogens during anaerobic and aerobic treatment of dilute kraft bleaching effluents. M.Sc. Thesis, University of Washington, US.

Lee, E.G.H., J.C. Mueller and C.C. Walden. 1978. Decolorization of bleached kraft mill effluents by algae. TAPPI J. 61(7): 59–62.

Lee, H.C., M. Lee and W. Den. 2015. Phenol tolerance and biodegradation by *Spirulina maxima*. Water, Air, Soil Pollut. 226(395): 1–11.

Lehtinen, K. 2004. Relationship of the technical development of pulping and bleaching to effluent quality and aquatic toxicity. pp. 273–293. *In*: D.L. Borton, T. Hall, R.

Fisher, J. Thomas (eds.). Pulp & Paper Mill Effluent Environmental Fate & Effects. Destech Publications Inc., Lancaster, Pennsylvania, U.S.

Lika, K. and I.A. Papadakis. 2009. Modeling the biodegradation of phenolic compounds by microalgae. J. Sea Res. 62: 135–146.

Malaviya, P. and V.S. Rathore. 2007. Bioremediation of pulp and paper mill effluent by a novel fungal consortium isolated from polluted soil. Bioresour. Technol. 98: 3647–3651.

Mehmood, K., S.K. Ur Rehman, J. Wang, F. Farooq, Q. Mahmood, A.M. Jadoon, et al. 2019. Treatment of pulp and paper industrial effluent using physicochemical process for recycling. Water 11(2393): 1–15.

Minu, K., K.K. Jiby and V.V.N. Kishore. 2012. Isolation and purification of lignin and silica from the black liquor generated during the production of bioethanol from rice straw. Biomass Bioenergy 39: 210–217.

Munkittrick, K.R., M.R. Servos, J.H. Carey and G.J. Van Der Kraak. 1997. Environmental impacts of pulp and paper wastewater: evidence for a reduction in environmental effects at North American pulp mills since 1992. Water Sci. Technol. 35(2–3): 329–338.

Nagasathya, A. and N. Thajuddin. 2008. Decolourization of paper mill effluent using hyper saline cyanobacterium. Res. J. Environ. Sci. 2: 408–414.

Nakai, S., I. Yutaca and H. Masaaki. 2001. Algal growth inhibition effects and inducement modes by plant-production phenols. Water Res. 35: 1855–1859.

Nazos, T.T., E.J. Kokarakis and D.F. Ghanotakis. 2017. Metabolism of xenobiotics by *Chlamydomonas reinhardtii*: phenol degradation under conditions affecting photosynthesis. Photosynth. Res. 131: 31–40.

Papazi, A. and K. Kotzabasis. 2007. Bioenergetic strategy of microalgae for the biodegradation of phenolic compounds-exogenously supplied energy and carbon sources adjust the level of biodegradation. J. Biotechnol. 129: 706–716.

Patel, U.D. and S. Suresh. 2008. Electrochemical treatment of pentachlorophenol in water and pulp bleaching effluent. Sep. Purif. Technol. 61: 115–122.

Petroutsos, D., P. Katapodis, P. Christakopoulos and D. Kekos. 2007. Removal of p-chlorophenol by the marine microalga *Tetraselmis marina*. J. Appl. Phycol. 19: 485–490.

Pinto, G., A. Pollio, L. Previtera and F. Temussi. 2002. Biodegradation of phenols by microalgae. Biotechnol. Lett. 24: 2047–2051.

Pokhrel, D. and T. Viraraghavan. 2004. Treatment of pulp and paper mill wastewater – a review. Sci. Total Environ. 333: 37–58.

Rodgers, J.H. and J.F. Thomas. 2004. Evaluations of the fate and effects of pulp and paper mill effluents from a watershed multistressor perspective: progree to date and future opportunities. pp. 135–146. *In*: D.L. Borton, T. Hall, R.P. Fisher and J.F. Thomas (eds.). Pulp & Paper Mill Effluent Environmental Fates & Effects. DEStech Publications Inc., Lancaster, Pennsylvania, U.S.

Sadhu, K., A. Mukherjee, K.S. Shukla, K. Adhikari and S. Dutta. 2013. Adsorptive removal of phenol from coke-oven wastewater using Gondwana shale, India: experiment, modeling and optimization. Desalin. Water Treat. 52: 6492–6504.

Samanthakamani, D. and N. Thangaraju. 2015. Potential of freshwater microalgae for degradation of phenol. Ind. J. Sci. Res. Tech. 3: 9–12.

Sanchez-Aponte, J., I. Baldiris-Navarro, M. Torres-Virviescas and C. Bohorquez. 2019. Bioremediation of phenolic waters using the microalgae *Chlamydomonas reinhardtii*. Orient J. Chem. 35(4): 1274-1278.

Savant, D.V., R. Abdul-Rahman and D.R. Ranade. 2006. Anaerobic degradation of

Adsorbable Organic Halides (AOX) from pulp and paper industry wastewater. Biores. Tech. 97: 1092–1104.

Semple, K.T. and R.B. Cain. 1996. Biodegradation of phenolic by *Ochromonas danica*. Appl. Environ. Microbiol. 62: 1265–1273.

Semple, K.T., R.B. Cain and S. Schmidt. 1999. Biodegradation of aromatic compounds by microalgae. FEMS Microbiol. Lett. 170: 291–300.

Shashirekha, S., L. Uma and G. Subramanian. 1997. Phenol degradation by the marine cyanobacterium *Phormidium valderianum* BDU 30501. J. Ind. Microbiol. Biotechnol. 19: 130–133.

Singh, J.S., A. Kumar, A.N. Rai and D.P. Singh. 2016. Cyanobacteria: a precious bio-resource in agriculture, ecosystem, and environmental sustainability. Front. Microbiol.7: 529.

Singh, J.S., S. Koushal, A. Kumar, S.R. Vimal and V.K. Gupta. 2016. Book review: Microbial inoculants in sustainable agricultural productivity, Vol. II: Functional application. Front. Microbiol. 7: 2105.

Singh, J.S., A. Kumar and M. Singh. 2019. Cyanobacteria: a sustainable and commercial bio-resource in production of bio-fertilizer and bio-fuel from wastewaters. Environ. Sustain. Indic. 3: 100008.

Singhal, A. and I.S. Thakur. 2009. Decolorization and detoxification of pulp and paper mill effluent by *Cryptococcus* sp. Biochem. Eng. J. 46: 21–27.

Springer, A.M. 2000. Industrial Environmental Control: Pulp and Paper Industry. 3rd Edition. TAPPI Press, Atlanta, US.

Sumathi, S. and Y.T. Hung. 2006. Treatment of pulp and paper mill wastes. pp. 453–497. In: L.K. Wang, Y.T. Hung, H.H. Lo and C. Yapijakis (eds.). Waste Treatment in the Process Industries. CRC Press, Taylor & Francis, USA.

Thompson, G., J. Swain, M. Kay and C.F. Forster. 2001. The treatment of pulp and paper mill effluent: a review. Bioresour. Technol. 77(3): 275–286.

Tikoo, V., A.H. Scragg and S.W. Shales. 1997. Degradation of pentachlorophenol by microalgae. J. Chem. Technol. Biotechnol. 68: 425–431.

Udaiyappan, A.F.M., H.A. Hasan, M.S. Takriff and S.R.S. Abdullah. 2017. A review of the potentials, challenges and current status of microalgae biomass applications in industrial wastewater treatment. J. Water Proc. Eng. 20: 8–21.

Uğurlu, M. and M.H. Karaoğlu. 2009. Removal of AOX, total nitrogen and chlorinated lignin from bleached Kraft mill effluents by UV oxidation in the presence of hydrogen peroxide utilizing TiO(2) as photocatalyst. Environ. Sci. Pollut. Res. Int. 16(3): 265–273.

Usha, M.T., T.S. Chandra, R. Sarada and V.S. Chauhan. 2016. Removal of nutrients and organic pollution load from pulp and paper mill effluent by microalgae in outdoor open pond. Bioresour. Technol. 214: 856–860.

Vanholme, R., B. Demedts, K. Morreel, J. Ralph and W. Boerjan. 2010. Lignin biosynthesis and structure. Plant Physiol. 153: 895–905.

WB. 1999. Pollution prevention and abatement handbook, 1998: toward cleaner. The International Bank for Reconstruction and Development (Washington D.C., United States).

Wurster, M., S. Mundt, E. Hammer, F. Schauer and U. Lindequist. 2003. Extracellular degradation of phenol by the cyanobacterium *Synechococcus* PCC 7002. J. Appl. Phycol. 15: 171–176.

Zwain, H.M., S.R. Hassan, N.Q. Zaman, H.A. Aziz and I. Dahlan. 2013. The start-up performance of modified anaerobic baffled reactor (MABR) for the treatment of recycled paper mill wastewater. J. Environ. Chem. Eng. 1(1–2): 61–64.

Textile Wastewater

Introduction

The textile industry alone is responsible for 20% of all industrial wastewater production, and it needs 9 billion m^3 of water as the annual global demand of textiles is between 60 and 70 million tons (Rani et al. 2013, World Bank 2014, Holkar et al. 2016). The water footprint of wet textile manufacturing could range from 100-300 m^3 per ton of textiles which mainly depends on the processes and type of fabric manufactured. Bleaching and finishing account for 38%, dyeing for 16% and printing for 8% of the total water demand (Shaikh 2009, Alam and Hossain 2018). Currently the textile industry uses more than 10,000 different dyes and pigments for dyeing and printing purposes; and a large amount of dyes mainly synthetic dyes are produced worldwide which is estimated more than 7×10^5 tons/year of synthetic dyes that are produced worldwide (Bharathi and Ramesh 2013, Khandare and Govindwar 2015, Seow and Lim 2016). Many synthetic dyes especially azo dyes are very toxic and persist in the environment.

Textile wastewater considered is very toxic and hazardous, as it has high pH, a strong color, residual dyes (azo, sulfur, naphthol, vat dyes), bleaching agents, nitrates, acetic acid, soaps and detergents, finishing materials; and the presence of heavy metals like chromium, copper, arsenic, cadmium, mercury, nickel, cobalt (Chung and Chen 2009, Singh and Arora 2011, Seow and Lim 2016). The colloidal materials are abundant in textile wastewater, which along with colors and oily scum leads to enhancing turbidity; resulting in bad aesthetic and a foul smell to the receiving water bodies. This affects the amount of required sunlight and the penetration to Lower layers, which badly impacts the photosynthetic activity of all the submerged and benthic flora of the water bodies (Keharia and Madamwar 2003). Due to the presence of high organic matter that depletes the dissolved oxygen present in the water which is essential for aquatic life (Ghaly et al. 2014). If textile wastewater is used for irrigation of agricultural fields, it might reduce the soil porosity and affect soil texture, leading to soil hardening, that results in less penetration of plant roots, which ultimately decreases soil productivity.

Conventional treatment approaches which can be categorized into physical, chemical and biological processes but they are inefficient to remove dyes and toxic substances, and in addition produce large amounts of colored and toxic sludge. Hence many advanced methods such as membrane filtration, nanofiltration, electrooxidation (sonophotocatalysis), reverse osmosis, electrodialysis and ion exchange; which could be followed after or combined with conventional treatment. These advanced methods have proved to be efficient in removing the residual dyes and toxic chemicals (Chen and Burns 2006) from textile wastewater. But it requires high investment in building or upgrading existing treatment facilities, and further needs extensive energy and skilled labor to operate, which makes them unsustainable for middle and low income countries.

Microalgae mediated discoloration and degradation of dyes especially azo dyes provides us an eco-friendly, economical and sustainable approach (Kumar and Singh 2016; Singh et al. 2016a, b; Kumar et al. 2017; Kumar and Singh 2017) to treat textile wastewater without producing large quantities of sludge (Saratale et al. 2011). A number of studies reported that microalgae not only survived well in textile wastewater but were also able to degrade the recalcitrant dyes. Phang and Chu (2004), Lim et al. (2010) reported the ability of *Chlorella vulgaris* to grow in textile wastewater. Although *Chlorella vulgaris* is able to grow in 100% wastewater but if 20–80% dilution of the textile wastewater follows, it definitely promotes better growth of microalgal cells. Olguín (2003), Ruiz et al. (2011) suggested that microalgae used either bioadsorption or biodegradation processes for the removal of dyes; that degrade dyes primarily for requirement of nitrogen for their growth (Kumar 2018, Kumar et al. 2018a, b, Singh et al. 2019, Kumar and Singh 2020a, b).

This chapter describes textile processes, the characteristics of textile wastewater and their impacts on ecosystems and public health. It also briefly accounts different treatment methods and their demerits in terms of removal efficiency, cost and sustainability. In order to solve this, microalgae-mediated wastewater remediation provides us an efficient, sustainable and cost effective approach which not only discolors textile wastewater but also helps in reduction of the overall toxicity of the wastewater that is not possible with other biological treatments.

Textile Industry: A Brief Description

The textile industry involves the development of yarn from fibers which are then transformed into fabrics. Fibers can be natural or synthetic. This process of producing textiles from raw fiber is a quite long and complex procedure; including pre-treatment, bleaching, dyeing, printing and finishing (Verma et al. 2012). Further, the textile industry could be categorized into: cotton, woollen and synthetic fabrics as per the type of raw fiber being processed. In case of cellulosic fibers especially cotton, a base reduction process known as mercerization also follows before dyeing and printing.

The natural and synthetic fibers have some impurities naturally and some are also added during spinning, knitting and weaving processes. These impurities could cause problems in the processes of dyeing and printing;

which are removed by the pretreatment process. Pre-treatment includes washing, degumming, desizing, scouring, singeing and others. Bleaching and mercerization are processes, in which large amounts of water and chemicals are used to remove the undesired features.

In bleaching, sodium chlorite, sodium hypochlorite or hydrogen peroxide is used for the removal of any undesired coloring from the fabrics, as the desired color is obtained after dyeing or printing. These bleaching agents are generally used in a diluted form or with water, as water facilitates the better adsorption with fabric. Under acidic conditions, sodium chlorite releases chlorine dioxide, which acts as a strong oxidizing agent and is responsible for bleaching of fabrics. After bleaching, mercerization is another important process before dyeing and printing, in which fabrics mainly cottons are treated with a strong caustic alkaline solution to improve the luster, smoothness, strength and affinity to dyestuffs.

When fabric is free from all impurities and undesired colors, the most important and also a polluted one process dyeing is applied to attain the desired color to the required fabric. These are known as colored fabrics. For this, a fabric undergoes a 'wet processing' method; the dye is first dissolved in water and then applied to the fabric to provide the required color and also the durability. Although printing is also one of the dyeing processes, but the main difference between both the processes is the application and the region where are they applied. In the dyeing process, a solution of dyes are used and applied to the whole fabric; while in printing, dyes are used in the form of thick paste and applied on a particular portion that is responsible for a particular design.

Colored fabrics then undergo the finishing process. It is necessary to retain specific properties to the fabric such as softness, strength and durability. This is carried out through applying several finishing processes like heat, combing or agents like starch, resins etc., for softening, creating cross-links within the fabric and waterproofing.

Textile Dyes: Classification and Characteristics

The history of synthetic dye started in 1856 with a chance discovery; a young student William Perkin trying to synthesize quinine, a malaria curing compound. But it led to the synthesis of the first synthetic dye mauveine, a purple solution that could color silk. It was soon considered an important finding and the accident invention made him famous and also paved the way for the growth and development of the textile and printing industry. There are around 10,000 different dyes and pigments that are currently applied in the textile industry for dyeing and printing purposes; and a high amount of dyes (mainly synthetic) are extracted or manufactured. It is estimated that more than 7×10^5 tons/year of synthetic dyes are produced worldwide (Bharathi and Ramesh 2013, Seow and Lim 2016).

A simpler classification of dyes is natural dyes from (plants, insects or minerals) or synthetic dyes (man-made); it is estimated that the consumption of natural dye is only equivalent to 1% as compared the consumption of synthetic dye in the world. Further dyes are primarily classified on the basis of the chemical structure, industrial applications and the type of fibers on which they are applied (Gupta 2009). Textile dyes are classified into azo dyes, nitro dyes, indigo dyes, anthraquinone dyes, phthalein dyes, triphenyl methyl dyes, nitrated dyes, etc. Dyes are categorized into azo dyes, direct dyes, reactive dyes, acid dyes, basic dyes, mordant dyes, vat dyes, disperse dyes and sulfur dyes.

Dyes possess one or more chromophores, which gives a distinct color to the dye and auxochromes, which are responsible for a color enhancer by the electron substitution either by removing or donating (Christie 2001). Chromophores contain functional groups such as $-N=N-$, $-C=O$, $-NO_2$ and $O = (C_6H_4) = O$ (quinoid assemblies), while auxochromes contain functional groups such as NH_3, $-COOH$, $-OH$ and $-SO_3H$ (Srinivasan and Viraraghavan 2010). It is also reported that sulfonate groups of the auxochromes provide a high solubilizing ability to the dyes. Based on study of different auxochromes, Welham (2000) categorized the auxochromes into different groups of reactants such as acid or base, or astringic or disperse, anionic or in grain form, pigmented, vat and also found insoluble or insoluble form.

Textile Wastewater: Characteristics and Specific Pollutants

The textile industry requires a lot of water for the wet processing of fabrics. Further the amount of water also depends on the shade, method and the chemicals used in a particular textile process; and sometimes its consumption goes from 50 to 240 l of water for the production of 1 kilogram of finished textile (Blomqvist 1996, Kocabas et al. 2009). Textile wastewater is characterized with a high organic content, a dark color, presence of dyes, coloring agents, solvents and other toxic chemicals (Table 9.1). It also contains various inorganic compounds such as sodium hydroxide, sodium chloride, sodium hypochlorite, sodium sulfide and hydrochloric acid. There is also the presence of various enzymes, polyvinyl alcohol, carboxymethyl cellulose and polyacrylic acid, which are applied for the removal of residue of soaps and detergents, the sizing agents and other impurities from the fabrics (O'Neill et al. 1999, Asamudo et al. 2005).

Dyes

It is estimated that as much as 15% of the dyes used in textile processes, are unable to dye fibers, and are lost to textile wastewater (O'Neill et al. 1999). The synthetic dyes especially azo dyes are frequently used in textile processes (Ong et al. 2010); due to their low cost, color persistence, and the variety of colors in comparison with natural dyes. Some dyes such as Congo red, amino or azo groups have benzene rings, which makes them more stable and are not degradable even under the effect of light or ozone. Further their

Table 9.1: Specific pollutants from various processes in textile industry

Water quality parameter	Aleem et al. 2020	Aragaw and Asmare 2018	Brahmbhatt and Jasrai 2016	Punzi 2015	Talukder et al. 2015	David Noel and Rajan 2014	El-Kassas and Mohamed 2014	Imtiazuddin et al. 2012	Jonstrup et al. 2011
pH	9.89	10.5	7.23	10	6.3	7.3	8.05	9.9	10
TS (mg L^{-1})							735		
TDS (mg L^{-1})	2009	2012			5100	4301.6	506	4890.4	
TSS (mg L^{-1})	883						229	1261	
BOD (mg L^{-1})		401			110	104		273.8	363
COD (mg L^{-1})	1256	1838	2640	1714	560	459.3	51.2	400.05	1781
TN/NH$_4^+$-N (mg L^{-1})				50			1.15 %		23
TP/PO$_4^{3-}$/Phosphates (mg L^{-1})				16.8			1.51		17
TOC (mg L^{-1})									
Sulfate/Sulfides (mg L^{-1})	630		708	400	300	256	88.67		1400
Chlorides (ppm)			3480						15867
Alkalinity (mg L^{-1})									
Chromium (mg L^{-1})	0.24		0.75 ppm				6.33	1.39	

Note: TS: Total Solids; TDS: Total Dissolved Solids; TSS: Total Suspended Solids; BOD: Biological Oxygen Demand; COD: Chemical Oxygen Demand; TN: Total Nitrogen; NH$_4^+$-N: Ammonical Nitrogen; NO$_3^-$-N: Nitrate Nitrogen; TP: Total Phosphorous; TOC: Total Organic Content.

Table 9.2: Specific pollutants and their nature from various processes in textile industry (Eswaramoorthi et al. 2008, Ghaly et al. 2014)

Process	Compounds	Nature of effluents
Desizing	Sizes, enzymes, starch, waxes, ammonia	Very small volume, high BOD (30-50% of total), PVA.
Scouring	Disinfectants and insecticides residues, NaOH, soaps, surfactants, fats, waxes, pectin's, oils, spent solvents, enzymes	Very small, strongly alkaline, dark color, high BOD values (30% of total)
Bleaching	H_2O_2, sodium silicate or organic stabilizer, high pH	Small volume, strongly alkaline, low BOD (5% of total)
Mercerizing	High pH, NaOH	Small volume, strongly alkaline, low BOD (Less than 1% of total)
Dyeing	Color, metals, salts, surfactants, organic dyes, sulfide, acidity, alkalinity, formaldehyde	Large volume, strongly colored, fairly high BOD (6% of total)
Printing	Urea, solvents, color, metals	Very small volume, oily appearances, fairly high BOD
Finishing	Resins, waxes, chlorinated compounds, acetate, stearate, spent solvents, softeners	Very small volume, less alkaline, low BOD.

biodegradation is also very difficult due to their synthetic origin and complex aromatic molecular structures (Aksu and Tezer 2005).

Dissolved and Suspended Solids

Sodium chloride and other Glauber salts, which are primarily used for the recovery of unused dyes in textile processes; lead to rapid increase in the amount of dissolved solids and results in an increase of turbidity of wastewater. There are many suspended solid materials such as undissolved raw materials, i.e., cellulose, pulp and fiber scrap; that are released during desizing and scouring processes.

Heavy Metals

There are many sources of heavy metals in textile wastewater: chemicals, salts and dyes which are frequently applied during the bleaching and dyeing textile process. Some bleaching chemicals like caustic soda, sodium carbonate and other salts contain mercury heavy metals as impurities, that could be discharged into wastewater. Some metalized mordant dyes are also responsible for heavy metal discharge in the wastewater stream. As it discussed earlier in the tannery industry, most of the metal complex dyes contain a chromium base; from which chromium could be easily released into wastewater.

Residual chlorine

During the bleaching process, chlorine containing compounds such as sodium hypochlorite are used, which leads to the release of residual chlorine in the textile wastewater stream. This residual chlorine carrying wastewater could reduce dissolved oxygen of water bodies and affects aquatic life. Residual chlorine could further react with other substances in water to form harmful and toxic compounds.

Refractory materials

Textile wastewater also includes insoluble and non-biodegradable organic compounds such as surfactants, polyvinyl alcohol, ethylene glycol and dyestuffs which are also known as refractory materials. Due to the presence of such refractory materials could lead to enhancing the COD level of textile wastewater.

Impacts on the Environment and Public Health

Due to presence of high soluble salts in textile wastewater, could be harmful for aquatic life. As soluble salts or TDS increases in the water, it would become more turbid, which affects the light penetration to the lower layer (Tholoana 2007). Some inorganic compounds such as sodium hypochlorite, sodium hydroxide, hydrochloric acid and sodium sulfide, which are used in the bleaching process, have proved to be very dangerous for the entire aquatic life (Blomqvist 1996, Tholoana 2007). The suspended solids in textile wastewater could be combined with the oily scum and affect the oxygen transfer mechanism in the air-water interface.

Besides this, textile wastewater also contains traces of heavy metals like Cr, As and Cu, which are released from chemicals and some dyes, making the freshwater harmful for drinking and agricultural purposes (Eswaramoorthi et al. 2008). The severity and the risk of dye coloration in untreated textile wastewater has been emphasized. This could obstruct light, leading to the photosynthetic process of aquatic plants (Murugesan and Kalaichelvan 2003, Nese et al. 2007). It is also reported that textile dyes might be responsible for the reduction of plant protein, carbohydrates and chlorophyll content of the plants (Sweeney 2015).

Mathur et al. (2005), De Lima et al. (2007) and (Golka et al. 2008) detected benzidine, a known carcinogen from benzidine based azo dyes like disperse orange 37, disperse blue 373 and disperse violet 93 dyes. Puvaneswari et al. (2006) studied the response of benzidine and benzidine analogue to some model organisms like canines, hamsters and mice, and further reported the possibility of aromatic amines and N-acetylated derivative in their urine; this could cause tumor formation in these model organisms. Mathur et al. (2005) and Kumar et al. (2005) reported that exposure to textile dyes could cause headaches, nausea, lung and skin irritations and congenital malformations in humans. The prolonged exposure of textile dyes to workers could cause nasal problems, asthma, dermatitis, rhinitis and cancer in the kidney, liver and urinary bladder (Morikawa et al. 1997). It was observed that azo dyes also showed the evidence of cancer in the bladder, spleen, liver in model organisms and chromosomal deformities in mammalian cells.

Current Treatment and Challenges

Textile wastewaters have very a complex composition and in very large amounts, which could be a difficult and tedious job to handle, manage and treat. Conventional treatment systems comprises of physical, chemical and biological processes that are applied either separately or in one process to support each other; but have not been proved to be sufficient to remove pollutants especially dye color from textile wastewater effectively. Although membrane bioreactors are also in trend as their color removal efficiency is much better than convention biological treatment; but these membrane bioreactors alone might not be able to remove sufficient dye color removal, and needs further post-treatments or bioaugmentation technologies (Bhatia et al. 2017). Besides this, the membrane bioreactor requires higher investment and operational costs in comparison to conventional biological processes (Eswaramoorthi et al. 2008).

There are several advanced processes which are applied or combined with conventional treatment processes like nano-filtration, electro-oxidation, reverse osmosis, electro-dialysis and ion exchange; for the removal or degradation of dye residues and other xenobiotic compounds (Eswaramoorthi et al. 2008). Before applying biological treatment, the electro-oxidation process could be performed to break down the strong bonds of the dyes, which make them easier for biodegradation. Reverse osmosis and electro-dialysis are also used to filter out the dissolved solids, ions and larger chemical species from textile wastewater.

Another method of the ion exchange process that can be used as a tertiary treatment, comprises of ion exchange resins and textile wastewater passing through beds of these resins. These resin ions could charge with either cations or anions. When textile wastewater is conveyed through a cationic resin, it removes the cations from the wastewater and replaces it with hydrogen ions making it acidic. Then this acidic solution is conveyed through the Anionic resin, which substitutes the anions hydroxyl ions. But these methods require high investment for the set up, further needs high operation costs and skilled labor for efficient removal of dye residues and other chemicals.

Microalgal Remediation

Due to their large surface area and binding ability, microalgae proved to be very efficient in biosorption and their charged surface which facilitate the greater affinity to contaminants in textile wastewater. Al-Fawwaz and Abdullah (2016) suggested that various types of groups such as –OH, RCOO⁻, -NH$_2$, and PO$_4^{-3}$ are associated with contaminants in textile wastewater and they are easily adsorbed on the surface of the microalgae. Olguín (2003) and Ruiz et al. (2011) also observed that microalgae generally used the dyes for a nitrogen source. Other advantages of using microalgae in the treatment of textile wastewater, is that they could grow at a rapid pace and, in extreme conditions (Mata et al. 2010, 2011).

Dye Discoloration and Degradation

It was first reported by Jinqi and Houtian (1992) that microalgal species could be used for the dye decolorization. Jinqi and Houtian (1992) observed that *Oscillatoria* sp. and *Chlorella* sp. were able to degrade aniline completely and convert it in to CO_2. It was also reported that *Phormidium* sp., *Synechococcus* sp. and *Leptolyrigbya* sp. degraded the indigo dye more than 50% (Silva-Stenico et al. 2012). Queiroz and Stefanelli (2011) observed that *Anabaena* sp. is able to degrade the Blue Drin dye and reported 81% color removal. Shah et al. (2001) reported that *Phormidium valderianum* was able to decolorize the Acid Red, Acid Red 119 and Direct Black 155 dyes up to 90%. Further there is an increase in discoloration of dyes, if the pH was found to be very alkaline; this condition is very common textile wastewater due to the frequent use of NaOH during the dyeing process. Parikh and Madamwar (2005) found that *Phormidium ceylanicum* is able to degrade Acid Red 97 and FF Sky Blue dyes up to 80% in a month.

Dellamatrice et al. (2017) studied the discoloration of indigo dye and sulfur black by two cyanobacterial species *Phormidium autumnale* UTEX 1580 and *Anabaena flos-aquae* UTCC64; and observed that *Phormidium autumnale* UTEX 1580 is more efficient in the degradation of indigo dye, while *Anabaena flos-aquae* UTCC64 is more efficient in sulfur black dye. Further Dellamatrice et al. (2017) also suggested that filamentous species as *Phormidium autumnale* UTEX 1580 is more efficient in degradation of dyes, as the filaments provides more contact between the microalgae and the dyes; leading to better dye penetration in to the filaments. *Phormidium* sp. proved to be more able in the degradation of different dyes into different chemical classes. It was also suggested that *Anabaenaflos-aquae* UTCC64 could also be used for textile wastewater's sludge discoloration and detoxification.

Jinqi and Houtian (1992), Nguyen and Juang (2013) and Padmanaban et al. (2013) emphasized that dye discoloration or decolorization significantly depends upon the dye structure (Table 9.3). Some dyes having nitro and sulfur groups such as sulfur black, Remazol Brilliant Blue which proved to be very recalcitrant and have greater resistance to degradation.

Mechanism of Dye Degradation

Microalgae have proved to be very efficient in decolonization of dyes through bioadsorption, biodegradation and bioconversion. It is also suggested that dyes undergo decolorization to biosorption which is followed by bioconversion and biocoagulation (Mohan et al. 2002). For decolonization of

Table 9.3: Microalgae in textile dye tolerance and degradation

Dye	Microalgae	References
Methylene Blue	*Desmodesmus* sp.	Al-Fawwaz and Abdullah 2016
Malachite Green	*Desmodesmus* sp.	Al-Fawwaz and Abdullah 2016
	Cosmarium sp.	Daneshvar et al. 2007
	Pithophora sp.	Kumar et al. 2005
Acid Red 274	*Spirogyra rhizopus*	Ozer et al. 2006
Acid blue 9	*Spirulina platensis*	Dotto et al. 2012
Reactive red	*Synechocystis* sp.	Karacakaya et al. 2009
RGB-Red dye	*Nostoc muscorum*	Sinha et al. 2015
Triphenylmethane dye	*Cosmarium* sp.	Daneshvar et al. 2007

dyes, microalgae can be used by two of three different mechanisms (Alvarez et al. 2015, Daneshvar et al. 2007):

(a) Utilization of chromophores for the production of microalgal biomass, CO_2 and H_2O;
(b) Transformation of colored molecules to non-colored molecules;
(c) Adsorption of chromophores on algal biomass.

The process of dye removal starts with accumulation of dye ions on the surface of microalgal EPS matrix and further dye molecules are diffused into a solid phase of the EPS matrix from an aqueous phase. From this, enzymes carry out the process of decolorization and degradation (Ozer et al. 2006). Peroxidase enzymes are primarily involved in the degradation process; other enzymes such as malate dehydrogenase, catalase, bromo and chloroperoxidase are also produced by microalgae. This was also observed in the findings of Murphy et al. (2000), in which the extracts from marine microalgal species i.e., *Porphyridium purpureum*, *Phaeodactylum tricornutum* and *Dunaliella teriolecta* showed the presence of peroxidase enzymes.

Azo dyes

Yan and Pan (2004) and Hernandez-Zamora et al. 2015 reported that *Chlorella pyrenoidosa*, *Chlorella vulgaris* and *Oscillateria tenuis* are able to degrade and decolorize more than 30 azo compounds, in which azo dyes are converted into simpler aromatic amines (Fig. 9.1). Other microalgal species such as *Nostoc muscorum*, *Cosmarium* sp., *Pithophora* sp., *Ulva lactuca*, *Sargassum*, *Desmodesmus* sp. proved to successfully degrade the azo dyes into aromatic amines and further catabolized into organic compounds. Waqas et al. (2015) reported that microalgal species can exploit the azo dyes as a carbon and nitrogen source for their growth. Omar (2008), El-Sheekh et al. (2009), Pathak et al. (2015), Thirumagal and Panneerselvam (2016) showed that microalgae *C. vulgaris* produced the azoreductase enzyme for decolorization of azo dyes.

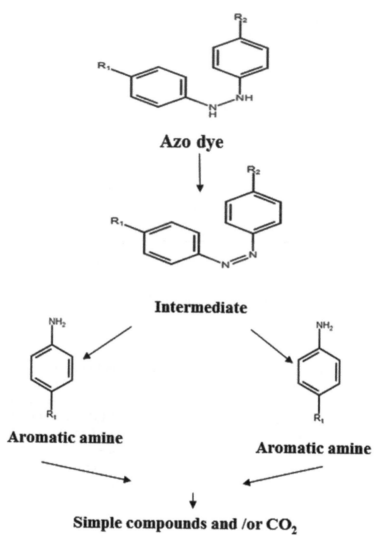

Figure 9.1: Azo dye degradation by microalgae (Jinqi and Houtian 1992).

Indigo dyes

Indigo dyes are extensively applied for denim manufacturing, leading to a recalcitrant substance in the textile wastewater (Balan and Monteiro 2001, Harazono and Nakamura 2005). Dellamatrice et al. (2017) reported that *Phormidium autumnale* UTEX 1580 successfully degraded the indigo dye and transformed it into two metabolites i.e., isatin and anthranilic acid; because of this the color of the dye disappears (Fig. 9.2). Although these metabolites do not increase toxicity, the degradation process is not still complete.

Dellamatrice et al. (2017) also suggested that pure culture of *Phormidium autumnale* UTEX 1580 is not able to degrade in these metabolites; and further

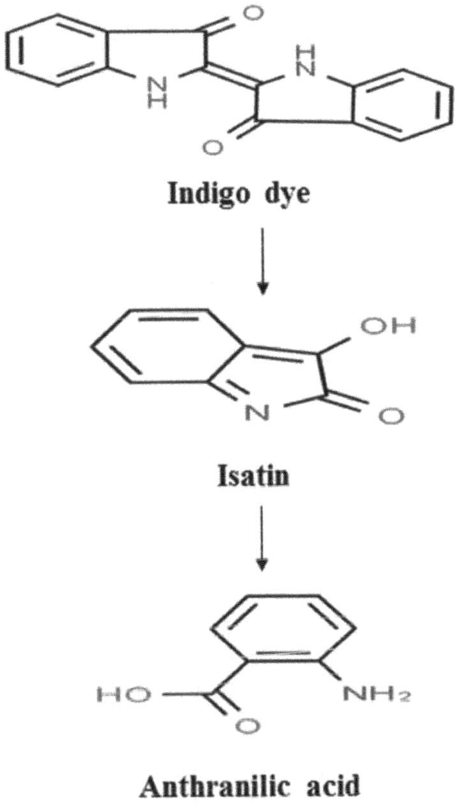

Indigo dye

Isatin

Anthranilic acid

Figure 9.2: Indigo dye degradation by microalgae (Dellamatrice et al. 2017).

consortia of microalgal or microalgal-bacterial cultures could be used for the degradation of metabolites. The end products or metabolites are simpler than the parental indigo dye molecule; and their microbial degradation might occur more easily in the environment.

Triarylmethane/Triphenylmethane dyes

Rhodamine B dye contain phenolic-OH groups, these are acidic in nature; which resonate to its phenoxide ion under normal conditions. But under alkaline conditions, more and more phenolic-OH groups resonate to its sphenoxide ion and maximum dye molecules now exist in a phenoxide ion structure having a negative charge. Shah et al. (2001) suggested that unlike most microbes, the microalgal cell wall possesses a net negative charge on the surface due to the anionic nature of the functional groups that are present on the cell wall. So to facilitate dye removal, divalent metal ions that exist in wastewater act as a bridge between phenoxide oxygen and the cell wall of microalgal cells. This phenomenon of being bridged is further enhanced under alkaline conditions, because the protons detach from the metal-binding sites on the cell wall; which are readily replaced by divalent metal ions that exist in wastewater (Inthorn et al. 1996, Chen et al. 1996). Further alkaline conditions could rupture the microalgal cells and be responsible for the release of additional functional groups (Nagase et al. 1997), that enhance the removal of dyes from wastewater.

Daneshvar et al. (2007) investigated the ability of the *Cosmarium* sp. to decolorize malachite green dye, and it was revealed that microalgae *Cosmarium* sp. decolorizes the dye very efficiently. Further it was also suggested that decolorization of malachite green dye primarily depends on the pH, temperature, dye concentration and microalgal biomass. Daneshvar et al. (2007) also observed that there was a trend of escalating the decolorization rate with increasing temperature in the range of 5–45°C; the pH 9.0 was found to be optimum for the decolorization of malachite green dye.

Conclusion

The textile industry involves wet processing for fabric production, so there is a need of design-related or technological interventions such as bio-scouring, solvent or airflow dyeing to reduce water and dye consumption. Second is the excessive use of synthetic dyes that are very complex in nature and responsible for reducing water transparency and aesthetic deterioration; these can be minimized by the use of natural dyes or other safe alternatives. Third regarding the treatment of wastewater, it is necessary to treat maximum textile wastewater through efficient, cost effective and eco-friendly technologies.

There are many physical, chemical or biological processes, which are conventionally favored for the treatment of textile wastewater. These processes not only generate a huge amount of sludge, but are also unable to provide effective treatment in relation to some dyes; these two factors limit the use of these processes. While photo-degradation and ozonation was found to be quite effective and fairly rapid in dye color removal, it is not necessary that these conventional methods could be satisfactory for the all types of dyes especially for some dispersed dyes.

In agreement with the integrated approach like membrane bioreactor (MBR) which involves membrane filtration in place of the sedimentation process and then biological treatment; could be a suitable method not only for municipal wastewater but also for textile or other industrial wastewater. The MBR method has advantages such as lower space requirement, minimal sludge production, shorter operation time and also disinfection capability; but the cost of building and the operation of such efficiency makes them unsuitable for poor and lower middle income countries. In such countries, the high cost of wastewater treatment is considered overburdening at the time of a race between emerging countries to produce the cheapest textile production. They need cheaper and sustainable treatment methods that provide effective treatment of textile wastewater.

There are many microalgal genera, especially cyanobacterial genera found in textile wastewater and successfully investigated in the treatment of textile wastewater. Due to the large survival potential and autotrophic nature, microalgae-mediated remediation acts as cost effective, and a sustainable and easier operation facility for bioremediation of textile wastewater. It is quite simplistic to suggest the use of one microalgal species and its removal efficiency which could be different in various regions. There is a need of further field scale research to identify the potential microalgal genera and how their effectiveness can be enhanced through improving the physiological, biochemical and genetic aspects of microalgae application. Therefore microalgae being a photosynthetic organism can be applied as a secondary treatment for textile wastewater:

1. Bioremediation approaches involving naturally occurring microalgae or cyanobacteria prove to be eco-friendly, economical and sustainable in relation of wastewater treatment.
2. The potential of microalgae for decolorization and degradation of different dyes provide us with valuable data, which can be used to design or plan an optimum condition to achieve rapid degradation and complete decolorization of azo dyes in textile wastewater.
3. Further there is need of more and extensive research to commercialize these bioremediation approaches through facilitating research from the laboratory scale to experimental studies in the field.

References

Aksu, Z. and S. Tezer. 2005. Biosorption of reactive dyes on the green algae *Chlorella vulgaris*. Process Biochem. 40: 1347–1361.

Alam, F.A. and M.A. Hossain. 2018. Conservation of water resource in textile and apparel industries. IOSR J. Polymer Textile Eng. 5(5): 11–14.

Aleem, M., J. Cao, C. Li, H. Rashid, Y. Wu, M.I. Nawaz, et al. 2020. Coagulation- and adsorption-based environmental impact assessment and textile effluent treatment. Water, Air, Soil Pollut. 231(45): 1–8.

Al-Fawwaz, A.T. and M. Abdullah. 2016. Decolorization of methylene blue and malachite green by immobilized *Desmodesmus* sp. isolated from North Jordan. Int. J. Environ. Sci. Dev. 7(2): 95–99.

Alvarez, M.S., A. Rodriguez, M.A. Sanroman and F.J. Deive. 2015. Microbial adaptation to ionic liquids. RSC. Adv. 5: 17379–17382.

Aragawa, T.A. and A.M. Asmare. 2018. Phycoremediation of textile wastewater using indigenous microalgae. Water Pract. Technol. 13(2): 274–284.

Asamudo, N.U., A.S. Daba and O.U. Ezeronye. 2005. Bioremediation of textile effluent using *Phanerochaete chrysosporium*. Afr. J. Biotech. 4(13): 1548–1553.

Balan, D.S.L. and R.T.R. Monteiro. 2001. Decolorization of textile indigo dye by ligninolytic fungi. J. Biotechnol. 89: 141–145.

Bharathi, K.S. and S.T. Ramesh. 2013. Removal of dyes using agricultural waste as low-cost adsorbents: a review. Appl. Water Sci. 3: 773–790.

Bhatia, D., N.R. Sharma, J. Singh and R.S. Kanwar. 2017. Biological methods for textile dye removal from wastewater: a review. Critical Rev. Environ. Sci. Technol. 47(19): 1836–1876.

Blomqvist, A. 1996. Food and Fashion. Water management and collective action among irrigation farmers and textile industrialists in South India. TEMA Research, Linköping, Sweden.

Brahmbhatt, N.H. and R.T. Jasrai. 2016. The role of algae in bioremediation of textile effluent. Int. J. Eng. Res. Gen. Sci. 4(1): 443–453.

Chen, H.L. and L.D. Burns. 2006. Environmental analysis of textile products. Clothing. Text. Res. J. 24: 248–261.

Chen, K.C., W.T. Huang, J.Y. Wu and J.Y. Houng. 1996. Microbial decolorization of azo dyes by *Proteus mirabilis*. J. Ind. Microbiol. Biotechnol. 23: 686–690.

Christie, R.M. 2001. Color Chemistry. Royal Society of Chemistry, UK.

Chung, Y.C. and C.Y. Chen. 2009. Degradation of azo dye reactive violet 5 by TiO_2 photocatalysis. Environ. Chem. Lett. 7: 347.

Daneshvar, N., M. Ayazloo, A.R. Khataee and M. Pourhassan. 2007. Biological decolorization of dye solution containing Malachite Green by microalgae *Cosmarium* sp. Bioresour. Technol. 98: 1176–1182.

David Noel, S. and M.R. Rajan. 2014. Cyanobacteria as a potential source of phycoremediation from textile industry effluent. J. Bioremed. Biodeg. 5(7): 1–4.

De Lima, R.O.A., A.P. Bazo, D.M.F. Salvadori, C.M. Rech, D.P. Oliveira and G.A. Umbuzeiro. 2007. Mutagenic and carcinogenic potential of a textile azo dye processing plant effluent that impacts a drinking water source. Mutat. Res. Genet. Toxicol. Environ. Mutagen. 626: 53–60.

Dellamatrice, P.M., M.E. Silva-Stenico, L.A.B. de Moraes, M. Fiore and R.T.R. Monteiro. 2017. Degradation of textile dyes by cyanobacteria. Braz. J. Microbiol. 48: 25–31.

Dotto, G.L., E.C. Lima and L.A.A. Pinto. 2012. Biosorption of food dyes onto *Spirulina platensis* nanoparticles: equilibrium isotherm and thermodynamic analysis. Bioresour. Technol. 103: 123–130.

El-Kassas, H.Y. and L.A. Mohamed. 2014. Bioremediation of the textile waste effluent by *Chlorella vulgaris*. Egypt. J. Aquat. Res. 40: 301–308.

El-Sheekh,, M., M. Gharieb and G. Abou-El-Souod. 2009. Biodegradation of dyes by some green algae and cyanobacteria. Int. Biodeter. Biodegr. 63: 699–704.

Eswaramoorthi, S., K. Dhanapal and D. Chauhan. 2008. Advanced in textile waste water treatment: the case for UV-Ozonation and membrane bioreactor for common effluent treatment plants in Tirupur, Tamil Nadu, India. Environment with people's involvement and coordination in India, No. 198, Coimbatore, India.

Ghaly, A.E., R. Ananthashankar, M. Alhattab and V.V. Ramakrishnan. 2014. Production, characterization and treatment of textile effluents: a critical review. J. Chem. Eng. Process Technol. 5(1): 1–16.

Golka, K., P. Heitmann, F. Gieseler, J. Hodzic, N. Masche, H.M. Bolt, et al. 2008. Elevated bladder cancer risk due to colorants – a statewide case-control study in North Rhine-Westphalia, Germany. J. Toxicol. Environ. Health A 71: 851–855.

Gupta, V.K. 2009. Application of low-cost adsorbents for dye removal – a review. J. Environ. Manage. 90: 2313–2342.

Harazono, K. and K. Nakamura. 2005. Decolourization of mixtures of different reactive textile dyes by the white-rot basidomycete *Phanerochaete sordaria* and inhibitory effect of polyvinyl alcohol. Chemosphere 59: 63–68.

Hernández-Zamora, M., E. Cristiani-Urbina, F. Martínez-Jerónimo, H.V. Perales-Vela, T. Ponce-Noyola, M.C. Montes-Horcasitas, et al. 2015. Bioremoval of the azo dye Congo Red by the microalga *Chlorella vulgaris*. Environ. Sci. Pollut. Res. 22(14): 10811–10823.

Holkar, C.R., A.J. Jadhav, D.V. Pinjari, N.M. Mahamuni and A.B. Pandit. 2016. A critical review on textile wastewater treatments: possible approaches. J. Environ. Manage. 182: 351–366.

Imtiazuddin, S.M., M. Mumtaz and K.A. Mallick. 2012. Pollutants of wastewater characteristics in textile industries. J. Basic Appl. Sci. 8: 554–556.

Inthorn, D., H. Nagase, Y. Isaji, K. Hirata and K. Miyamoto. 1996. Removal of cadmium from aqueous solution by the filamentous cyanobacterium *Tolypothrix tenuis*. J. Ferment. Bioeng. 82: 580–584.

Jinqi, L. and L. Houtian. 1992. Degradation of azo dyes by algae. Environ. Pollut. 75: 273–278.

Jonstrup, M., N. Kumar, M. Murtoand and B. Mattiasson. 2011. Sequential anaerobic-aerobic treatment of azo dyes: decolourisation and amine degradability. Desalination 280(1–3): 339–346.

Karacakaya, P., N.K. Kiliç, E. Duygu and G. Donmez. 2009. Stimulation of reactive dye removal by cyanobacteria in media containing triacontanol hormone. J. Hazard. Mater. 172: 1635–1639.

Keharia, H. and D. Madamwar. 2003. Bioremediation concepts for treatment of dye containing wastewater: a review. Ind. J. Exp. Biol. 41: 1068–1075.

Khandare, R.V. and S.P. Govindwar. 2015. Phytoremediation of textile dyes and effluents: current scenario and future prospects. Biotech. Adv. 33: 1697–1714.

Kocabas, A.M., H. Yukseler, F.B. Dilek and U. Yetis. 2009. Adoption of European Union's IPPC Directive to a textile mill: Analysis of water and energy consumption. J. Environ. Manage. 91(1): 102–113.

Kumar, A. 2018. Assessment of cyanobacterial diversity in paddy fields and their capability to degrade the pesticides. Babasaheb Bhimrao Ambedkar University, Lucknow, India.

Kumar, A. and J.S. Singh. 2016. Microalgae and cyanobacteria biofuels: a sustainable alternate to crop-based fuels. pp. 1–20. *In*: J.S. Singh, D.P. Singh (eds.). Microbes and Environmental Management. Studium Press Pvt. Ltd. New Delhi, India.

Kumar, A. and J.S. Singh. 2017. Cyanoremediation: a green-clean tool for decontamination of synthetic pesticides from agro- and aquatic-ecosystems. pp. 59–83. *In*: J.S. Singh, G. Seneviratne (eds.). Agro-Environmental Sustainability, Vol. II: Managing Environment Pollution. Springer Int., Cham, Switzerland.

Kumar, A. and J.S. Singh. 2020. Biochar coupled rehabilitation of cyanobacterial soil crusts: a sustainable approach in stabilization of arid and semiarid soils. pp. 167–191. *In*: J.S. Singh, C. Singh (eds.). Biochar Applications in Agriculture and Environment Management. Springer Int., Cham, Switzerland.

Kumar, A. and J.S. Singh. 2020. Microalgal bio-fertilizers. *In*: E. Jacob-Lopes, M.M. Maroneze, M.I. Queiroz, L.Q. Zepka (eds.). Handbook of Microalgae-based Processes and Products. Academic Press, Cambridge, US, In Press.

Kumar, A., S. Kaushal, S.A. Saraf and J.S Singh. 2017. Cyanobacterial biotechnology: an opportunity for sustainable industrial production. Clim. Chang. Environ. Sustain. 5(1): 97–110.

Kumar, A., S. Kaushal, S.A. Saraf and J.S Singh. 2018. Microbial bio-fuels: a solution to carbon emissions and energy crisis. Front. Biosci. (Landmark) 23: 1789–1802.

Kumar, A., S. Kaushal, S.A. Saraf and J.S Singh. 2018. Screening of Chlorpyrifos (CPF) tolerant cyanobacteria from paddy field soil of Lucknow, India. Int. J. Appl. Adv. Sci. Res. 3(1): 100–105.

Kumar, K.V., S. Sivanesan and V. Ramamurthi. 2005. Adsorption of malachite green onto *Pithophora* sp., a fresh water algae: equilibrium and kinetic modeling. Process Biochem. 40: 2865–2872.

Lim, S., W. Chu and S. Phang. 2010. Use of *Chlorella vulgaris* for bioremediation of textile wastewater. Bioresour. Technol. 101: 7314–7322.

Mata, T.M., A.A. Martins and N.S. Caetano. 2010. Microalgae for biodiesel production and other applications. Renew. Sust. Energ. Rev. 14: 217–232.

Mata, T.M., A.A. Martins, S. Sikdar and C.A.V. Costa. 2011. Sustainability considerations of biodiesel based on supply chain analysis. Clean Technol. Environ. Policy 13: 655–671.

Mathur, N., P. Bhatnagar and P. Bakre. 2005. Assessing Mutagenicity of Textile Dyes from Pali (Rajasthan) using AMES Bioassay. Dept of Zoology, University of Rajasthan, Jaipur, India.

Mohan, S.V., C.N. Roa, K.K. Prasad and J. Karthikeyan. 2002. Treatment of simulated reactive yellow 22 (Azo) dye effluents using Spirogyra species. Waste Manage. 22: 575–582.

Morikawa, Y., K. Shiomi, Y. Ishihara and N. Matsuura. 1997. Triple primary cancers involving kidney, urinary bladder, and liver in a dye worker. Am. J. Ind. Med. 31: 44–49.

Murphy, C.D., M.M. Robert and R.L. White. 2000. Peroxidases from marine microalgae. J. Appl. Phycol. 12: 507–513.

Murugesan, K. and P.T. Kalaichelvan. 2003. Synthetic dye decolourization by white rot fungi. Ind. J. Exp. Biol. 41: 1076–1087.

Nagase, H., D. Inthron, Y. Isaji, A. Oda, K. Hirata and K. Miyamoto. 1997. Selective cadmium removal from hard water using NaOH-treated cells of the cyanobacterium *Tolypothrix tenuis*. J. Ferment. Bioeng. 84: 151–154.

Nese, T., N. Sivri and I. Toroz. 2007. Pollutants of textile industry wastewater and assessment of its discharge limits by water quality standards. Turk. J. Fish. Aquat. Sci. 7: 97–103.

Nguyen, T.A. and R.S. Juang. 2013. Treatment of waters and wastewaters containing sulfur dyes: a review. Chem. Eng. J. 219: 109–117.

O'Neill, C., F.R. Hawkes, D.L. Hawkes, N.D. Lourenco, H.M. Pinheiro and W. Delle. 1999. Colour in textile effluents-sources, measurement, discharge contents and stimulation: a review. J. Chem. Technol. Biotechnol. 74: 1009–1018.

Olguín, E.J. 2003. Phycoremediation: key issues for cost-effective nutrient removal process. Biotechnol. Adv. 22: 1–91.

Omar, H.H. 2008. Algal decolorization and degradation of monoazo and diazo dyes. Pak. J. Biol. Sci. 11: 1310–1316.

Ong, S., K. Uchiyama, D. Inadama, Y. Ishida and K. Yamagiwa. 2010. Treatment of azo dye Acid Orange 7 containing wastewater using up-flow constructed wetland with and without supplementary aeration. Bioresour. Technol. 101: 9049–9057.

Ozer, A., G. Akkaya and M. Turabik. 2006. The removal of Acid Red 274 from wastewater: combined biosorption and biocoagulation with *Spirogyra rhizopus*. Dyes. Pigm. 71: 83–89.

Padmanaban, V.C., S.S. Prakash, P. Sherildar, J.P. Jacob and K. Nelliparambil. 2013. Biodegradation of antraquinone based compounds: review. Int. J. Adv. Res. Eng. Technol. 4: 74–83.

Parikh, A. and D. Madamwar. 2005. Textile dye decolorization using cyanobacteria. Biotechnol Lett. 27: 323–326.

Pathak, V.V., R. Kothari, A.K. Chopra and D.P. Singh. 2015. Experimental and kinetic studies for phycoremediation and dye removal by *Chlorella pyrenoidosa* from textile wastewater. J. Environ. Manage. 163: 270–277.

Phang, S.M. and W.L. Chu. 2004. The University of Malaya Algae Culture Collection (UMACC) and potential applications of aunique *Chlorella* from the collection. Jpn. J. Phycol. 52: 221–224.

Punzi, M. 2015. Treatment of textile wastewater by combining biological processes and advanced oxidation. Ph.D Thesis. Lund University, Sweden.

Puvaneswari, N., J. Muthukrishnan and P. Gunasekaran. 2006. Toxicity assessment and microbial degradation of azo dyes. Ind. J. Exp. Biol. 44: 618.

Queiroz, B.P.V. and T. Stefanelli. 2011. Biodegradation de corantes Têxteispor *Anabaena flos-aquae*. Eng. Amb. 8: 26–35.

Rani, B., R. Maheshwari, R.K. Yadav, D. Pareek and A. Sharma. 2013. Resolution to provide safe drinking water for sustainability of future perspectives. Res. J. Chem. Environ. Sci. 1: 50–54.

Ruiz, J., P. Alvarez, Z. Arbib, C. Garrido, J. Barragan and J.A. Perales. 2011. Effect of nitrogen and phosphorus concentration on their removal kinetic in treated urban wastewater by *Chlorella vulgaris*. Int. J. Phytorem. 13: 884–896.

Saratale, R.G., G.D. Saratale, J.S. Chang and S.P. Govindwar. 2011. Bacterial decolorization and degradation of azo dyes: a review. J. Taiwan Inst. Chem. Eng. 42: 138–157.

Seow, T.W. and C.K. Lim. 2016. Removal of dye by adsorption: a review. Int. J. Appl. Eng. Res. 11: 2675–2679.

Shah, V., N. Garg and D. Madamwar. 2001. An integrated process of textile dye removal and hydrogen evolution using cyanobacterium, *Phormidium valderianum*. World J. Microbiol. Biotechnol. 17: 499–504.

Shaikh, M.A. 2009. Water conservation in textile industry. Pak. Textile J. Nov. 2009: 48–51.

Silva-Stenico, M.E., F.D.P. Vieira, D.B. Genuário, C.S.P. Silva, L.A.B. Moraes and M.F. Fiore. 2012. Decolorization of textile dyes by cyanobacteria. J. Braz. Chem. Soc. 23: 1863–1870.

Singh, J.S., A. Kumar, A.N. Rai and D.P. Singh. 2016. Cyanobacteria: a precious bio-resource in agriculture, ecosystem, and environmental sustainability. Front. Microbiol. 7: 529.

Singh, J.S., S. Koushal, A. Kumar, S.R. Vimal and V.K. Gupta. 2016. Book review: microbial inoculants in sustainable agricultural productivity, Vol. II: Functional application. Front. Microbiol. 7: 2105.

Singh, J.S., A. Kumar and M. Singh. 2019. Cyanobacteria: a sustainable and commercial bio-resource in production of bio-fertilizer and bio-fuel from wastewaters. Environ. Sustain. Indic. 3: 100008.

Singh, K. and S. Arora. 2011. Removal of synthetic textile dyes from wastewaters: a critical review on present treatment technologies. Crit. Rev. Environ. Sci. Tech. 41: 807–878.

Sinha, S., S. Nigam and R. Singh. 2015. Potential of *Nostoc muscorum* for the decolorisation of textiles dye RGB-Red. Int. J. Pharm. Bio. Sci. 6: 1092–1100.

Srinivasan, A. and T. Viraraghavan. 2010. Decolorization of dye wastewaters by biosorbents: a review. J. Environ. Manage. 91: 1915–1929.

Sweeny, G. 2015. Fast fashion is the second dirtiest industry in the world, next to big oil. Eco Watch. http://ecowatch. com/2015/08/17/fast-fashion-seconddirtiest-industry.

Talukder, A.H., S. Mahmud, S.A. Lira and M.A. Aziz. 2015. Phycoremediation of textile industry effluent by cyanobacteria (*Nostoc muscorum* and *Anabaena variabilis*). Biores. Comm. 1(2): 124–127.

Thirumagal, J. and A. Panneerselvam. 2016. Isolation of azo reductase enzyme in its various forms from *Chlorella pyrenoidosa* and its immobilization efficiency for treatment of water. Int. J. Sci. Res. 5: 2133–2138.

Tholoana, M. 2007. Water Management at a Textile Industry: A Case Study in Lesotho. University of Pretoria, South Africa.

Verma, A.K., R.R. Dash and P. Bhunia. 2012. A review on chemical coagulation/ flocculation technologies for removal of colour from textile wastewaters. J. Environ. Manage. 93: 154–168.

Waqas, R., M. Arshad, H.N. Asghar and M. Asghar. 2015. Optimization of factors for enhanced phycoremediation of reactive blue azo dye. Int. J. Agric. Biol. 17(4): 803–808.

Welham, A. 2000. The theory of dyeing (and the secret of life). J. Soc. Dyers Colour. 116: 140–143.

World Bank. 2014. The Bangladesh Responsible Sourcing Initiative: A New Model for Green Growth. South Asia Environment and Water Resources Unit, World Bank.

Yan, H. and G. Pan. 2004. Increase in biodegradation of dimethyl phthalate by *Closterium lunula* using inorganic carbon. Chemosphere 55: 1281–1285.

Food Processing Wastewater

Introduction

Food processing industries are known to use tons of freshwater for their major operations with water like washing, boiling and making the ingredients that are used in the processing of foods (Bustillo-Lecompte and Mehrvar 2015). Food processing industries are divided in to three categories: (a) Dairy industry involving milk processing and production of milk products; (b) Meat and fish processing; (c) Vegetable and fruits processing. There is an estimate that the dairy industry is responsible for the discharge of wastewater of about 2.5 times more volume than the milk processed (Ramasamy et al. 2004, Kushwaha et al. 2011, Kothari et al. 2012). COWI (2000) and Bustillo Lecompte and Mehrvar (2015) estimated that for a metric ton of meat processing, about 2.5 to 40 kL of freshwater is consumed in the meat processing plants (slaughter houses). As compared to other industrial wastewaters, food processing wastewater contains no major toxic components but comprises of a huge load of organic matter containing sugars, fats and protein entities from processed material. The complexity and nature of food processing wastewater depends on amount of organic contents, which can fluctuate according to the processed material and their seasonal availability (in case of fruits and vegetables) (Onet, 2010).

On discharging, this untreated wastewater with a high nutrient content could lead to eutrophic conditions in the receiving water bodies and streams; resulting in excess growth of blooms of the undesired algae or phytoplankton (Cai et al. 2013). In conventional methods, the activated sludge process is popularly used for the treatment of food processing wastewater. The activated sludge process involves physicochemical and biological operations (use of aerobic bacteria for degradation of organic matter). In this process, mechanical aeration is usually provided to the aerobic bacteria for the degradation of organic compounds. But these processes are inefficient in removing the inorganic nutrients (N and P) and further a huge amount of sludge is also generated. There are advanced processes such as filtration and reverse osmosis that could be used for effective treatment, but high investment and the need of skilled workers makes the wastewater treatment unaffordable and unviable for food processing industries.

Microalgal remediation could be a potential and effective approach for the treatment of food processing wastewater (Kumar and Singh 2016, Singh et al. 2016a, b, Kumar et al. 2017, Kumar and Singh 2017, Kumar 2018). It can also provide essential oxygen for bacterial respiration, reducing the cost of mechanical aeration (Olguin et al. 2003, Suad and Gu 2014).They are very effective in sequestrating organic and inorganic of nitrogen and phosphorous (Olguin et al. 2003, Mohamed et al. 2016). Further due to their minimal growth needs and less doubling time, they can grow at a very fast speed in the shortest possible time to attained biomass (Singh and Gu 2010). Therefore, using microalgae for wastewater treatment could be a low cost alternative of the conventional treatment of food processing wastewater (Kumar et al. 2018a, b, Singh et al. 2019, Kumar and Singh 2020a, b).

This chapter provides a brief account of the food processing industry and their categories, wastewater generation, their composition and specific pollutants. It also focuses on the challenges in the current treatment of wastewater and how microalgal remediation could be a sustainable and effective approach in the treatment of food processing wastewater.

Food Processing Industry: Types and Their Brief Description

Dairy industry

Raw milk reaches the processing plant in milk containing trucks, which undergo preliminary analytical tests, this milk is then pumped into bulk storage tanks called milk silos (which have the capacity of as much as up to or more than 300,000 liter). During pumping of milk into the silos, it is cooled simultaneously to 4–6°C with a heat plate exchanger. After storing milk into milk silos, their processing could be different according their needs and processing of the milk processing plant.

In most of cases, a process of milk clarification is carried out to remove sand or dust particles. In addition to dirt particles, microbial load of the milk is also reduced, microbial load could be of two types: (a) unwanted bacterial populations, (b) excess populations of essential bacteria; which can be reduced through centrifugation and microfiltration techniques (Gésan-Guiziou 2010). The explanation of bacterial load not only helps in increasing the efficiency of the following processes but also enhances the shelf life and organoleptic properties of dairy products; and reduces the risk of interference in aging of dairy products especially cheese. The clarifiers are like milk separators (centrifugal separators), but they are used for different functions, and for the purpose of clarification, only such centrifugal separators are used which centrifuges with a high hydraulic capacity. The clarifier could operate either in cold conditions (below 8°C) or hot milk (50–60°C).

The next step in dairy processing is milk separation, in which milk separators are used to centrifuge the hot milk. In this process, fat (cream) is separated from the whole milk (serum); the product is called skimmed milk and the process is designated as skimming. It takes place generally at a temperature of 50–60°C (122–140°F). The fat proportion in the cream can be maintained in the range of 20 to 70%.

This is followed by the pasteurization process and simultaneously with an in-line fat standardization process for both milk and cream separately. Pasteurization requires a temperature of 71.7°C (161°F) for 15 seconds, and this is the current official standard for milk pasteurization (Pexara et al. 2018); while the standard vat pasteurization is 63°C (145°F) for 30 minutes. It is used to kill dangerous bacteria like *Salmonella* sp., *Escherichia coli*, and *Listeria monocytogenes* (Holsomger et al. 1997, Gorman and Adley 2004). Standardization of milk involves raising or lowering the fat and Solids-Not-Fat (SNF) levels up to desired ranges. It is typically followed by maintaining a uniform fat content in the final dairy product (Bird 1993).

Milk primarily comprises of 85–90% water; from which water is removed up to the milk's weight to 12% w/v, to make milk powder. Milk powders could be of many types, like whole milk powder, skim milk powder, Fat included milk powder, infant formula and milk protein concentrate (85% pure milk protein). They are primarily developed by the spray-drying method, which involves a falling film evaporator to concentrate the milk from ~13% Total Solids (TS) to a target of up to 52%. Other methods like freeze-drying are also used to remove water from the milk (Sánchez et al. 2011). These milk powders are used in confectionary, bakery, ice cream and in the formation of yogurt-like products.

Meat and fish processing

There are two types of meat processing plants: (a) the plants involve slaughtering and very limited processing of byproducts. The products from these plants include mainly fresh meat in the form of whole, half or quarter carcasses or in smaller meat cuts; (b) the plants involve slaughtering and very extensive processing of byproducts. These plants have some additional operations like rendering, paunch and viscera handling, blood processing and hide and hair processing. The products from these plants include processing meat into many categories such as canned, smoked and cured meats.

Before slaughtering, animals should be healthy, physiologically normal and provided with adequate rest. The first step of slaughtering involves the stunning process which ensures that the animal is unconscious and insensitive to pain before being bled out at the slaughter. There are some stunning methods such as captive-bolt stunning, electrical stunning, use of CO_2 gas, which are commonly followed. In captive-bolt stunning, several equipments are used as a blow to the animal's skull with a sufficient force that makes the animal unconscious immediately. Electrical stunning involves the use of an electrical current particularly to the animal's head, which interrupts the normal brain activities; resulting in the animal being unconscious immediately. In another stunning method, CO_2 gas is released in a chamber provided to the animal where it is inhaled by it, leading to a decrease in the level of blood oxygen; this results in the loss of brain activities to cause eventual brain death. It is required that persons involved in the stunning process, must be adequately trained and capable of their assigned task to make sure that each animal is humanely stunned.

After sufficient stunning, the unconscious animal undergoes bleeding through a sharp cut in the main blood vessels of the neck. Then animals are suspended to an overhead rail by the hind legs, where the carcass is allowed to drain most of the blood; resulting in the animal's death from cerebral anoxia. In order to make a perfect and precise cut, the bleeding knife should be sharpened from time to time. Drained blood collects in the trough, and could be discharged or processed. Hides or skins of cattle or hogs are removed and transported to tanneries to make leather.

Now the carcass of the animal undergoes different operations such as decapitation, carcass opening by cutting, inspection of the carcass, evisceration (removal of intestines and internal organs), splitting and cutting of the carcass; and finally chilling or freezing. There are many large meat processing plants that deliver whole graded carcasses to retail markets, while others perform on-site processing to split the carcass in to retail cuts. After this, the meat pieces undergo a range of operations including grinding, additives mixing, curing, pickling, smoking, cooking and canning.

The viscera, intestines and internal organs are separated from the carcass, from these some edible parts like the heart and liver or inedible parts like lungs can be recovered for rendering or processing. For the intestines, a de-sliming process is carried out prior to thorough washing and then followed by the rendering process. Rendering involves heating waste materials of meat processing to detach fat from water and protein residues to produce edible lards and dried protein residues. It is classified in to two different categories: (a) high temperature rendering which involves cooking or steam application; (b) low temperature rendering involves a temperature of around 80°C which is slightly above the fat melting point. By this process, the following products like lard, meat meal, bone meal and meat-cum-bone meal are developed

The last step is packing the meat; there are two types of packaging methods that are generally followed: (a) low-processing packing, which only involves slaughtering and their processing in to fresh meat in the form of canned, smoked, cured and other meat products; (b) high-processing packing which includes both in-house processing (slaughtering and their processing) and out-house house processing (which also involves processing of imported meat from outside). Beside this, high-process packing plants have amenities for carrying operations for the tanning industry.

Vegetables and fruits processing

The first step of food processing is the handling and storage of the raw materials such as fruits, vegetable and cereals. Due to season-based availability, the raw materials are handled usually by the seasonal industry such as canning, sugar refining, grains processing and brewing. But in large scale industries, they are handled either manually or through mechanized operations which involve a number of conveyer belts and automated operations. Then raw materials are stored usually in bins, tanks, silos, cellars or cold stores and it is necessary that the premises must have specific conditions for handling and preserving which are required by the respective raw materials.

The first step in food processing might involve basic operations like pounding, grinding, crushing, drying, filtration, heat extraction (either directly or indirectly) and solvent extraction; to develop or make a desired food commodity from the raw materials like cereals, pulses and oil seeds, fruits and vegetables. The methods like pounding, grinding, crushing usually act as preparatory operations for some raw materials—for example, crushing of cocoa beans and slicing of sugar beet; while for the others it might be the actual extraction process, as in milling of flour. A heat extraction method could be applied directly as dry heat in roasting (of cocoa, coffee and chicory beans) or moist heat in the form of steam for the extraction of edible oils; or indirectly (mainly steam) for extraction of sweet juice from thin slices of beet.

Then some general procedures are followed: fermentation, cooking, dehydration and distillation. It can vary according to the industry and specific operations needed for the desired product. Fermentation is generally followed in making cheese, wine and spirits (discussed in the distillery industry). For fruits and vegetables, the cooking process usually follows before the canning and preserving operations in bakeries, biscuit making; while in many cases cooking is finished in a vacuum-sealed container, which is followed to make a concentrated product like in tomato- and chili-paste or sauce production. Dehydration is the method that involves simply drying in the sun (production of dry fruits), passing hot air through fixed dryers or drying tunnels, contacting steam to the drying drum, vacuum drying and freeze drying (lyophilization).

To prevent the food product from any deterioration, there are some basic procedures for food preservation: radiation, dehydration, sterilization and refrigeration. Radiation involves drying the product by direct heat in hot air, superheated steam, in vacuum and in inert gas. Dehydration involves the removal of moisture from the food by transferring heat into the food products, leading to water vaporization. Sterilization is usually used for canned products, in which steam is applied to the canned product to make the product microbes free; this is followed generally in a closed container. Refrigeration or low-temperature preservation includes freezing and deep-freezing procedures, where the food product is allowed to freeze and then placed in a sealed container; resulting in the preservation of the food product in their naturally fresh state.

For packaging of food products, some methods that are used include: canning, aseptic packaging and frozen packaging. Canning is a simple and the most followed method and for this several million tons of tinplate is used in the canning industry; while a significant proportion also involves glass jars for packing of preserved food. In canning, usually cleaned food, raw or partly cooked (but not intentionally sterilized) food products are packed into a tinplate can or glass jar with a lid sealing. Then the tinplate can or glass jar is heated by steam with a certain pressure to attain a specific temperature for a particular time, which facilitates heat penetration at the center of the can; resulting in the sterilization of the container. After completing sterilization, the can is cooled in air or chlorinated water and properly labeled and packed. Aseptic packaging is quite different from conventional canning, in which the food container is first sterilized; and then filling and closing of the container takes place in a sterile atmosphere. Currently food processing industries use deep-freezing procedures for fresh food at temperatures below their freezing point, which gives more time for transportation over long distances and storing for processing and/or sale when the demand arises, and seasonal products can be available at all times.

Food Industry Wastewater: Characteristics and Specific Pollutants

Dairy wastewater

Dairy wastewater is basically biodegradable, but characterized with undesirable color and odor, variable pH, high BOD (in the range of 40 to 48,000 mg/L), high COD (in the range of 80 to 95,000 mg/l) (Table 10.1) (Kothari et al. 2011, Kushwaha et al. 2011). It also comprises of a high amount of nitrogen (in the range of 14 to 830 mg/l) and phosphorous (in the range of 9 to 280 mg/l), which is responsible for excessive biological growth in the receiving water bodies (Rico Gutierrez et al. 1991, Gavala et al. 1999, Demirel et al. 2005). There is also the presence of a significant amount of Na, Cl, K, Ca, Mg, Fe, CO, Ni and Mn in dairy wastewater. The use of a large amount of alkaline cleaners in dairy plants might be another reason for the presence of high Na and Cl in dairy wastewater (Table 10.2) (Prakash et al. 2011).

Meat and fish processing wastewater

Meat slaughtering and processing involves some processes like bleeding, washing, deboning, sterilization and rendering, in which a lot of blood, fat and portions of internal viscera, bones, skin and hair make up the meat processing wastewater (Table 10.1). Due to this, meat processing wastewater is characterized with a strong color, odor, high BOD, rich in soluble and insoluble organics. It also contains high amounts of nitrogen and phosphorus, suspended solids and fats, oil and grease (Table 2) (COWI 2000).

Table 10.1: Characteristics of food processing wastewater

Water quality parameter	Dairy processing				Meat & Fish processing			Vegetable & fruit processing
	Kothari et al. 2012	Daneshvar et al. 2019	Quin et al. 2014	Guruvaiah et al. 2012	Maizatul et al. 2017	Posadas et al. 2014	Lattifi et al. 2016	Posadas et al. 2014
pH	6.2	6.3	9.31	7.46-8.26	-	7.7	8.61	7.1
Conductivity (ms cm⁻¹)	-	-	-	1.83-2.28	-	-	-	-
TS (mg L⁻¹)	975	-	-	-	4528	-	-	-
TDS (mg L⁻¹)	900	-	-	124-160	1753	-	-	-
TSS (mg L⁻¹)	75	-	560	602	5.6	-	500	-
BOD (mg L⁻¹)	-	-	-	8950	326	-	760	-
COD (mg L⁻¹)	-	2128	2128	19497	872	-	2180	-
TN (mg L⁻¹)	-	86	146	295	-	69	113	69
NH₄⁺-N (mg L⁻¹)	-	-	-	-	-	9	-	-
NO₃-N (mg L⁻¹)	66.4	-	121	-	49	-	-	13
TP (mg L⁻¹)	21	9.82	31.6	279	3.90	-	39.53	6
TOC (mg L⁻¹)	-	770.11	-	29434	-	381	-	327

Table 10.2: Specific pollutants in food processing wastewater

Food	Processes	Components
Dairy	Homogenization Condensation Centrifugation Fermentation Coagulation	Protein, nitrogen, phosphorous, dissolved sugars and nutrients, spoilage microbes
Meat, and fish	Size reduction Cutting/Chopping/ Commuting Mixing/Tumbling Salting/Curing (NaCl, NaNO$_2$) Utilization of spices/additives Stuffing/filling into casings Fermentation and drying Heat treatment Smoking	Blood, tissues, salts, spoilage microbes
Vegetables & fruits	Handling & Storage Extraction or manipulation Roasting or cooking Preservation Packaging	Cellulose, starch, hemicellulose, carbohydrates, inorganics, protein, spoilage microbes

Vegetable and fruits processing wastewater

Vegetable and fruits processing wastewater is also characterized with high BOD, COD, solids and a high load of organic matters, but they could be different in composition and concentration of specific constituents (Table 10.1). In comparison to dairy and meat processing wastewater, vegetable and fruit processing wastewater have a high sugar content which includes cellulose, hemicelluloses, starch, carbohydrates and other organic and inorganic compounds are also found in the wastewater (Table 10.2) (Lara et al. 2002).

Impacts on the Environment and Public Health

Food processing wastewaters including dairy, meat processing and vegetable and fruit processing wastewater comprises of high BOD, COD suspended solids, inorganic nitrogen and phosphorus, fats, oils, and grease (Santos and Robbins 2004, Li et al. 2008, Zulkifli et al. 2011, Rajakumar et al. 2011, Kundu et al. 2013, Omar et al. 2016, Jais et al. 2015, 2017).

Food processing wastewater contain a high amount of organic matter, when discharged leads to consuming the dissolved oxygen that is present in the water. This organic matter mediates oxygen depletion, leading to the death of fishes and other aquatic biota in the receiving water bodies (Mohamed et al. 2016). Besides oxygen depletion, high nutrient load including inorganic nitrogen and phosphorus facilitates eutrophication, in which a robust algal growth covers the surface of the water bodies; further depletion of oxygen levels causes permanent removal of certain species (Bello and Oyedemi 2009, Kundu et al.2013, Al- Gheethi et al. 2015, Atiku et al. 2016, Pahazri et al. 2016).

Current Treatment and Challenges

Conventional treatment of food processing wastewater involves many methods such as activated sludge, trickling filters, aerated lagoons or a combination of these processes in a series of steps. These primarily include physicochemical and biological processes, which significantly improve the wastewater through removing BOD and COD. There is a generation of excess sludge in this process, which further undergoes many treatments such as dewatering and sludge digestion that requires a large proportion of electrical energy (Table 10.3, Table 10.4).

Table 10.3: Typical characteristics of wastewater sludge (Metcalf and Eddy 2003)

Parameters	*Primary*	*Secondary*
pH	5-8	6-8
TS (%)	2.0-7.0	0.5-2.0
VS (% of TS)	60-80	50-60
TN (% of TS)	1.5-4	2-2-5
TP (% of TS)	0.8-3	0.5-0.7
Alkalinity (mgL^{-1}	500-1500	580-1100
Thermal content (kJ/kg, dry mass)	23000-30000	18500-23000

In the case of meat processing wastewater, containing huge concentrations of organics, nutrients and pathogens which are require some additional steps before the conventional physicochemical processes like screening, settling, blood collection and fat separation. Some common processes including Dissolved Air Floatation (DAF) and coagulation/flocculation are applied for the removal of solids and fat content from the food processing wastewater. However treated wastewater still contains color, odor and inorganic nitrogen and phosphorous, which require tertiary treatment or advanced processes such as filtration, adsorption and reverse osmosis (Bustillo-Lecompte and Mehrvar 2015).

These advanced processes have proved to be effective in the improvement of various water quality parameters like color, BOD and COD of food processing wastewater (Sombatsompop et al. 2011). These advanced processes have many limitations such as high installation and operating costs and need of skilled personnels for treatment facilities that makes them unviable for the industries and large scale operations.

Table 10.4: Sludge handling processes (Demirbas et al. 2017)

Treatment process	Description
Sludge degritting	Sludge degritting is applied to ensure that the sand particles are separated from the organic sludge by centrifugal forces in a fluid system
Dewatering	For reducing of the moisture content of the sludge. Dewatering can be improved by chemical conditioning such as the addition of a polymer.
Drying	The water content of sludge lowered to less than 10% by evaporation suitable for processing the sludge or fertilizer.
Drying lagoons	The sludge drying lagoons, which are suitable only for digested sludge treatment, are composed of shallow grounded basins surrounded by dykes
Filtration	It consists of four basic steps: the polymer control zone, the gravity drainage zone for excess water, the low pressure, and the high pressure zones. It is used for dewatering water from raw and digested wastewater sludge.
Stabilization	Sludge stabilization includes lime stabilization, heat treatment, aerobic digestion, anaerobic digestion, and composting.
Blending.	The sludge is blended to form a uniform blend with the downstream operations and process
Thickening	Thickening is an application of increasing the solids content of the sludge by removing some of the liquid content
Conditioning	Conditioning contains the chemical or physical treatment of sludge to improve its dewatering characteristics
Chemical precipitation	Chemical precipitation is the most common technology used to remove metal ions from solutions such as wastewater containing toxic metals.
Heat treatment	The process contains the treatment of sludge by heating in a pressure vessel.
Composting	It involves aerobic degradation of organic matter, to stabilize organic matter, to destroy pathogenic organisms, and to reduce the volume/amount of waste.
Anaerobic digestion	The anaerobic digestion process involves the anaerobic reduction of the organic substance with biological activity in the sludge
Reuse as fertilizer	The solids in the sludge or humus like have high amount of N & P, can be used as fertilizer.
Activated sludge	The activated sludge process is a process using a biological flock of air and bacteria and protozoa. The purpose of the activated sludge process is oxidation of organic substances, oxidation of nitrogenous substances to ammonium and nitrogen, and removal of nutrients

Microalgal Remediation

Microalgal remediation could be a potential approach, which has proved to be cheaper and more effective for removing nutrients and other pollutants from food processing wastewater (Kwarciak-Kozłowska et al. 2014, Sethupathy et al. 2015, Wurochekke et al. 2016). Due to their robust growth and survival in extreme conditions, microalgal cells assimilate a large number of nutrients in to the synthesis of biomolecules for their growth (Kwarciak-Kozłowska et al. 2014, Satpal and Khambete 2016, Xin et al. 2011, Cassidy 2011, Baharuddin et al. 2016).

Microalgae are found to very efficient in nutrient removal from food processing wastewater. In Chapter 4, there is a detailed discussion about nutrients i.e., nitrogen and phosphorous and their various forms that exist in different wastewaters including food processing wastewater. Further the strategies and mechanisms of microalgae for sequestration of nutrients are also described. It has been observed that in terms of specific pollutants in food processing wastewater are almost the same as municipal wastewater but the composition could be different and also in the presence of micro flora of the respective wastewater.

Role of Microalgae Consortia in Nutrient Removal

A number of microalgal strains such as *Botrycoccus* sp., *Chlorella* sp., *Chlorella vulgaris*, *Chlorella zofingiensis*, *Chlorella pyrenoidosa*, *Desmodesmus communis*, *Spirulina platensis* and *Scenedesmus* sp. are investigated for nutrient removal from various wastewaters including food processing wastewater (Kumar et al. 2011, Kothari et al. 2011, Samori et al. 2013). Kothari et al. (2012) investigated the ability of *Chlorella pyrenoidosa* for nutrient removal of dairy wastewater and found that approximately 80–85% phosphorus and 60–80% nitrogen is sequestrated by the microalgae. Similar reports by Huo et al. (2012) and Qin et al. (2014) also observed that microalgae *Chlorella zofingiensis* and *Chlorella vulgaris* were able to remove significant amount of the nutrients from dairy wastewater.

Subashchandrabose et al. (2011), Hernandez et al. (2013) and Gonzalez-Fernandez et al. (2011) studied the successful cultivation of cyanobacteria/microalgae-bacteria consortia in the food processing wastewater. Pires et al. (2013) emphasized the association of microalgae with the heterotrophic bacteria, in which they benefitted each other by providing photosynthesis mediated O_2 by microalgae to bacterial respiration, and in return respiration mediated CO_2 supporting the photoautotrophic growth of microalgae. Samori et al. (2013) investigated the ability of *Desmodesmus communis* for nutrient removal in comparison to natural microalgae consortium that was present in the wastewater; and found that *Desmodesmus communis* showed better growth and nutrient removal than the natural consortium.

In contrast to this, Chinnasamy et al. (2010) investigated the consortium of 15 native microalgal species in nutrient removal and reported that the consortium showed more than 96% nutrient removal and 6.82% fat content removal, which was found to be better than all the 15 individual microalgae culture. Many researchers also suggested that microalgae consortia provides better vitality and remarkable resistance to the microalgal species, which minimizes the risk of failing the microalgal cultivation system in wastewater (Chinnasamy et al. 2010, Renuka et al. 2013, Silva-Benavides and Torzillo 2012).

Qin et al. (2014) reported that microalgal consortia (*Scenedesmus* spp./ *Chlorella zofingiensis*) proved to be more effective in COD reduction up to 62.87% and phosphorous removal than *Chlorella* sp. monoculture (dominant species among the three microalgae involved in the study). It has been observed that nitrogen sequestration by microalgae mainly incorporates in the cell growth of synthesis proteins, while a significant part of the phosphorous could be stored or reserved in the cells for later use when phosphorous is limited (Juneja et al. 2013, Markou et al. 2014, Onet 2010). In the case of meat processing wastewater, microalgae *Botrycoccus* sp. showed an excellent ability to reduce pollutants, and performed BOD and COD reduction above 90%. Further it was also found that *Botrycoccus* sp. was able to sequestrate nitrogen and phosphorous from meat processing wastewater (Gani, 2017).

Thermophilic, Psychrophilic or Acidophilic Microalgae

Many microalgal species *Galdieria sulphuraria*, *Chlamydomonas acidophila*, *Chlorella protothecoides* var. *acidicola*, *Koliella antarctica*, *Chlorella sorokiniana* and some rhodophytes are found to be adaptable to thermophilic, psychrophilic or acidophilic conditions, which could be more useful in the remediation of food processing wastewaters that have such conditions. The thermophilic, psychrophilic or acidophilic conditions are not tolerated by such microalgae, but these conditions are necessary for their metabolic activity (Varshney et al. 2015). Gross and Schnarrenberger (1995) and Schmidt et al. (2005) observed that many microalgal species of Rhodophyta (red algae) are found to be grow mixotrophically and heterotrophically in different sugars and sugar alcohols.

Fernández-Rojas et al. (2014) and Delanka-Pedige et al. (2019) studied the most interesting microalgae *Galdieria sulphuraria* (also known as *Cyanidium caldarium*), due to its remarkable extremophilic growth properties. Earlier studies reported that *Galdieria sulphuraria* grew well in highly acidic environments, down to pH 1.8 (Merola et al. 1981, Enami and Kura-Hotta 1984). Besides its ability to survive in acidophilic nature, *Galdieria sulphuraria* also observed significant growth in thermophilic conditions up to 56°C (Selvaratnam et al. 2014). Due to metabolic versatility, *Galdieria sulphuraria* could be a potential candidate for treatment of food processing wastewater, which is characterized with high COD, acidic or high-temperature (Wan et al. 2016, Sloth et al. 2017, Henkanatte-Gedera et al. 2017).

Chlamydomonas acidophila and *Chlorella protothecoides* var. *acidicola* isolated from an acidic mine water area having pH range of 1.7 and 3.1. *Chlamydomonas acidophila* was found to be growing mixotrophically using different carbon sources like glucose, glycerol, starch at pH 2.5 (Cuaresma et al. 2011); while *Chlorella protothecoides* var. *acidicola* were able to grow well heterotrophically on glycolic acid (Nancucheo and Barrie Johnson 2012), which is very often present in vegetable and fruit processing wastewater. *Chlorella sorokiniana* is another thermophilic microalga, which shows high photoautotrophic growth rates up to 43°C (Varshney et al. 2018); and is also found to be grow well heterotrophically (Kim et al. 2013). The psychrophilic microalgal species such as *Koliella antarctica* were found to grow even below 10°C temperature (Andreoli et al. 1998), making them a potential agent for the treatment of wastewater generated from fresh fruit processing industries.

Benefits of Micro-Algal Activated Sludge Process (MAAS)

The interaction between microalgal photosynthesis and bacterial respiration is well recognized by the researchers in petroleum degradation and advocated for the effective treatment of the various wastewaters including food processing wastewater. The necessary inorganic carbon (CO_2), N and P for microalgal photosynthesis could be provided by the bacterial degradation of organic compounds, and the photosynthetic byproduct O_2 used for bacterial respiration; enhancing the overall nutrient removal from the wastewater. This beneficial interaction becomes the base of the MAAS process (Anbalagan 2016, Nordlander et al. 2017), which facilitates the alternative of mechanical aeration and minimizes the CO_2 emission from wastewater treatment. The MAAS process could be useful in removing many drawbacks of the conventional activated sludge process that are most often used for different wastewaters:

(a) Reduces the use of chemicals such as ferric chloride or ferrous sulfate, which are used and to concentrate the suspended biomass and increase the removal of sludge from treated wastewater;

(b) Minimizes the need of mechanical aeration that contributes to the huge cost in wastewater treatment;

(c) Avoids green house gas emissions mainly CO_2, through the use of CO_2 in the microalgal photosynthesis;

(d) Enhances the stability of the microalgal photosynthesis process through not facilitating the alkaline pH conditions that often support the volatilization of nitrogen in the form of NH_3 and the precipitation of phosphorus.

(e) Integration of microalgae cultivation with the activated sludge process helps in reducing the cost of microalgae biomass cultivation, due to the requirement of a large land area and surface lighting throughout the year.

(f) Enhances the settleability (i.e., gravity sedimentation) of microalgal cells, leading to ease in harvesting of microalgal biomass;

(g) Saving the cost of methods that are used for harvesting the microalgal, such as dissolved air flotation, chemical coagulation, centrifugation and filtration;

Conclusion

Food processing wastewater contains a large amount of biodegradable organic matter which contributes in BOD and COD; further there are seasonal variations in the pH, temperature and composition. It makes the food processing wastewater difficult to treat and achieve the desired environmental standards. There are conventional activated sludge processes, including some physicochemical and biological processes, but these are not enough to provide effective treatment for nutrient removal from food processing wastewater. Further huge sludge generation is the main drawback in the conventional treatment, which requires more resources and increased the cost of wastewater treatment. To remove residual inorganic nutrient and other pollutants, many advanced processes like filtration, reverse osmosis and oxidation processes are used for better treatment, but they add to further high installing and operating costs for wastewater treatment.

Microalgal remediation undoubtedly provides an alternative, effective and affordable treatment approach. They are very effective in sequestrating of nutrients especially inorganic nutrients that are left out in conventional processes. In comparison to individual microalgal species, consortia of microalgal species proved to be much more effective in the treatment of food processing wastewater, which may be due to collective survivability and supporting each other in the wastewater environment. Other factors such as pH, light and temperature of wastewater that affects the microalgal growth and cell content, need more research for improved treatment. There are some microalgal species that survive and grow in thermophilic, psychrophilic or acidophilic conditions, which could be very useful in treatment of such wastewater that has variations in composition, temperature and pH conditions.

The beneficial association of microalgae-bacteria is well known in nutrient sequestration and this integration of microalgal remediation with activated sludge process or Micro-Algal Activated Sludge process (MAAS) could be an innovative and affordable approach in the treatment of wastewaters including food processing wastewater. In the comparison to conventional activated sludge, the sludge produced in this system with microalgal-bacterial communities provides less handling and management and their thermophilic and mesophilic anaerobic digestion more feasible. The microalgal communities presented in the MAAS mediated sludge also enrich the sludge and could be more useful as fertilizers.

References

Al-Gheethi, A.A., R.M.S.R. Mohamed, M. Afaiz Ab. Rahman, J. Mas Rahayu and H.K. Amir. 2015. Treatment of wastewater from car washes using natural coagulation and filtration system. *In*: International Conference on Sustainable Environment and Water Research (ICSEWR2015), 25–26 Oct 2015, Johor Baru.

Anbalagan, A. 2016. Indigenous microalgae-activated sludge cultivation system for wastewater treatment. PhD Thesis. Mälardalen University, Sweden.

Andreoli, C., G.M. Lokhorst, A.M. Mani, L. Scarabel, I. Moro, N. La Rocca et al. 1998. *Koliella antarctica* sp. nov. (Klebsormidiales) a new marine green microalga from the Ross Sea (Antarctica). Algological Studies/ Archiv Für Hydrobiologie. Suppl. 90: 1–8.

Atiku, A., R.M.S.R. Mohamed, A.A. Al-Gheethi, A.A. Wurochekke and H. Kassim Amir. 2016. Harvesting microalgae biomass from the phycoremediation process of greywater. Environ Sci. Poll. Res. 23: 24624–24641.

Baharuddin, N.N., N.S. Azizi, H.N. Sohif, W.A. Karim, J.R. Al-Obaidi and M.N. Basiran. 2016. Marine microalgae flocculation using plant: the case of *Nannochlor opsisoculata* and *Moringa oleifera*. Pak. J. Bot. 48(2): 831–840.

Bello, Y.O. and D.T. Oyedemi. 2009. Impact of abattoir activities and management in residential neighborhoods: a case study of Ogbomoso, Nigeria. J. Soc. Sci. 19: 121–127.

Bird, J. 1993. Milk standardisation. Int. J. Dairy Technol. 46(2): 35–37.

Bustillo-Lecompte, C.F. and M. Mehrvar. 2015. Slaughterhouse wastewater characteristics, treatment, and management in the meat processing industry: a review on trends and advances. J. Environ. Manage. 161: 287–302.

Cai, T., S.Y. Park and Y. Li. 2013. Nutrient recovery from wastewater streams by microalgae: status and prospects. Renew. Sustain. Energ. Rev. 19: 360–369.

Cassidy, K.O. 2011. Evaluating algal growth at different temperatures. M.Sc. Thesis, University of Kentucky.

Chinnasamy, S., A. Bhatnagar, R.W. Hunt and K.C. Das. 2010. Microalgae cultivation in a wastewater dominated by carpet mill effluents for biofuel applications. Bioresour. Technol. 101(9): 3097–3105.

COWI. 2000. Cleaner production assessment in meat processing. 1st ed. Division of Technology, Industry and Economics, UNEP, Copenhagen, Paris.

Cuaresma, M., C. Casal, E. Forján and C. Vílchez. 2011. Productivity and selective accumulation of carotenoids of the novel extremophile microalga *Chlamydomonas acidophila* grown with different carbon sources in batch systems. J. Ind. Microbiol. Biot. 38: 167–177.

Daneshvar, E., M.J. Zarrinmehr, E. Koutra, M. Kornaros, O. Farhadian and A. Bhatnagar. 2019. Sequential cultivation of microalgae in raw and recycled dairy wastewater: Microalgal growth, wastewater treatment and biochemical composition. Bioresour. Technol. 273: 556–564.

Delanka-Pedige, H.M.K., S.P. Munasinghe-Arachchige, J. Cornelius, S.M. Henkanatte-Gedera, D. Tchinda, Y. Zhang et al. 2019. Pathogen reduction in an algal-based wastewater treatment system employing. *Galdieria sulphuraria*. Algal Res. 39: 101423.

Demirbas, A., G. Edris and W.M. Alalayah. 2017. Sludge production from municipal wastewater treatment in sewage treatment plant. Energ. Source Part A 39(10): 1–8.

Demirel, B., O. Yenigun and T.T. Onay. 2005. Anaerobic treatment of dairy wastewaters: a review. Process Biochem. 40: 2583–2595.

Enami, I. and M. Kura-Hotta. 1984. Effect of intracellular ATP levels on the light-induced H⁺ efflux from intact cells of *Cyanidium caldarium* 1. Plant Cell Physiol. 25: 1107–1113.

Fernández-Rojas, B., J. Hernández-Juárez and J. Pedraza-Chaverri. 2014. Nutraceutical properties of phycocyanin. J. Func. Food 11: 375–392.

Gani, P.A. 2017. Phycoremediation using *Botryococcus* sp. as nutrients removal in organic wastewaters coupled with hydrocarbon production. PhD Thesis. Universiti Tun Hussein Onn Malaysia.

Gavala, N., H. Kopsinis, I.V. Skiadas, K. Stamatelatou and G. Lyberatos. 1999. Treatment of dairy wastewater using an upflow anaerobic sludge blanket reactor. J. Agric. Eng. Res. 73: 59–63.

Gésan-Guiziou, G. 2010. Removal of bacteria, spores and somatic cells from milk by centrifugation and microfiltration techniques. pp. 349–372. *In*: M.W. Griffiths (ed.). Improving the Safety and Quality of Milk: Milk Production and Processing. Woodhead Publishing, Cambridge, US.

Gonzalez-Fernandez, C., B. Molinuevo-Salces and M.C. Garcia-Gonzalez. 2011. Nitrogen transformations under different conditions in open ponds by means of microalgae-bacteria consortium treating pig slurry. Bioresour. Technol. 102(2): 960–966.

Gorman, R. and C.C. Adley. 2004. Characterization of *Salmonella enterica* serotype Typhimurium isolates from human, food and animal sources in the Republic of Ireland. J. Clin. Microbiol. 42(5): 2314–2316.

Gross, W. and C. Schnarrenberger. 1995. Heterotrophic growth of two strains of the acido-thermophilic red alga *Galdieria sulphuraria*. Plant Cell Physiol. 36: 633–638.

Guruvaiah, M., D. Shah and E. Shah. 2012. Biomass and lipid accumulation of microalgae grown on dairy wastewater as a possible feedstock for biodiesel production. Int. J. Sci. Res. 3(12): 909–913.

Henkanatte-Gedera, S.M., T. Selvaratnam, M. Karbakhshravari, M. Myint, N. Nirmalakhandan, W.V. Voorhies et al. 2017. Removal of dissolved organic carbon and nutrients from urban wastewaters by *Galdieria sulphuraria*: laboratory to field scale demonstration. Algal Res. 24: 450–456.

Hernandez, D., B. Riano, M. Coca and M.C. Garcia-Gonzalez. 2013. Treatment of agro-industrial wastewater using microalgae bacteria consortium combined with anaerobic digestion of the produced biomass. Bioresour. Technol. 135: 598–603.

Holsomger, V.H., K.T. Rajkowski and J.R. Stabel. 1997. Milk pasteurisation and safety; brief history and update. Review Scientific and Technical Review of the Office International des Epizooties (Paris). 16(2): 441–451.

Huo, S., Z. Wang, S. Zhu, W. Zhou, R. Dong and Z. Yuan. 2012. Cultivation of *Chlorella zofingiensis* in bench-scale outdoor ponds by regulation of pH using dairy wastewater in winter, South China. Bioresour. Technol. 121: 76–82.

Jais, N.M., R.M.S.R. Mohamed, A.A. Al-Gheethi and M.K. Hashim Amir. 2017. Dual role of phycoremediation of wet market wastewater for nutrients and heavy metals removal and microalgae biomass production. Clean Technol Environ Policy 19(1): 37–52.

Jais, N.M., R.M.S.R. Mohamed, W.A.W.M. Apandi and H.M.M. Peralta. 2015. Removal of nutrients and selected heavy metals in wet market wastewater by using microalgae *Scenedesmus* sp. Appl. Mech. Mater. 773–774: 1210–1214.

Juneja, A., R.M. Ceballos and G.S. Murthy. 2013. Effects of environmental factors and nutrient availability on the biochemical composition of algae for biofuels production: a review. Energies 6: 4607-4638.

Kim, S., Y. Lee and S.-J. Hwang. 2013. Removal of nitrogen and phosphorus by *Chlorella sorokiniana* cultured heterotrophically in ammonia and nitrate. Int. Biodeter. Biodeg. 85: 511–516.

Kothari, R., V. Kumar and V.V. Tyagi. 2011. Assessment of waste treatment and energy recovery from dairy industrial waste by anaerobic digestion. IIOAB J. 2(1): 1–6.

Kothari, R., V.V. Pathak, V. Kumar and D.P. Singh. 2012. Experimental study for growth potential of unicellular alga *Chlorella pyrenoidosa* on dairy waste water: an integrated approach for treatment and biofuel production. Bioresour. Technol. 116: 466–470.

Kumar, A. 2018. Assessment of cyanobacterial diversity in paddy fields and their capability to degrade the pesticides. Babasahaeb Bhimrao Ambedkar University, Lucknow, India.

Kumar, A. and J.S. Singh. 2016. Microalgae and cyanobacteria biofuels: a sustainable alternate to crop-based fuels. pp. 1–20. *In*: J.S. Singh, D.P. Singh (eds.). Microbes and Environmental Management. Studium Press Pvt. Ltd. New Delhi, India.

Kumar, A. and J.S. Singh. 2017. Cyanoremediation: a green-clean tool for decontamination of synthetic pesticides from agro- and aquatic-ecosystems. pp. 59–83. *In*: J.S. Singh, G. Seneviratne (eds.). Agro-Environmental Sustainability, Vol. II: Managing Environment Pollution. Springer Int., Cham, Switzerland.

Kumar, A. and J.S. Singh. 2020. Biochar coupled rehabilitation of cyanobacterial soil crusts: a sustainable approach in stabilization of arid and semiarid soils. pp. 167–191. *In*: J.S. Singh, C. Singh (eds.). Biochar Applications in Agriculture and Environment Management. Springer Int., Cham, Switzerland.

Kumar, A. and J.S. Singh. 2020. Microalgal bio-fertilizers. *In*: E. Jacob-Lopes, M.M. Maroneze, M.I. Queiroz, L.Q. Zepka (eds.). Handbook of Microalgae-based Processes and Products. Academic Press, Cambridge, US, In press.

Kumar, A., S. Kaushal, S.A. Saraf and J.S. Singh. 2017. Cyanobacterial biotechnology: an opportunity for sustainable industrial production. Clim. Chang. Environ. Sustain. 5(1): 97–110.

Kumar, A., S. Kaushal, S.A. Saraf and J.S. Singh. 2018. Microbial bio-fuels: a solution to carbon emissions and energy crisis. Front. Biosci. (Landmark) 23: 1789-1802.

Kumar, A., S. Kaushal, S.A. Saraf and J.S. Singh. 2018. Screening of Chlorpyrifos (CPF) tolerant cyanobacteria from paddy field soil of Lucknow, India. Int. J. Appl. Adv. Sci. Res. 3(1): 100–105.

Kundu, P., A. Debsarkar and S. Mukherjee. 2013. Treatment of slaughter house wastewater in a sequencing batch reactor performance evaluation and biodegradation kinetics. Biomed. Res. Int. 2013(134872): 1–11.

Kushwaha, J.P., V.C. Srivastava and I.D. Mall. 2011. An overview of various technologies for the treatment of dairy wastewaters. Crit. Rev. Food Sci. Nutr. 51: 442–452.

Kwarciak-Kozłowska, A., L. Sławik-Dembiczak and B. Bańka. 2014. Phycoremediation of wastewater: heavy metal and nutrient removal processes. Environ. Prot. Nat. Resour. 25(4): 51–54.

Lara, M.A., J. Rodríguez-Malaver, O.J. Rojas, O. Hoimquist, A.M. González, J. Bullón et al. 2002. Black liquor lignin biodegradation by *Trametes elegans*. Int. Biodeter. Biodegr. 52: 167–173.

Latiffi, N.A.A., R.M.S.R. Mohamed, V.A. Shanmugan, N.F. Pahazri, A.H.M. Kassim, H.M. Matias-Peralta et al. 2016. Removal of nutrients from meat food processing industry wastewater by using microalgae *Botryococcus* sp. ARPN J. Engi. Appl. Sci. 11(16): 9863–9867.

Lee, K. and C.-G. Lee. 2002. Nitrogen removal from wastewaters by microalgae without consuming organic carbon sources. J. Microbiol. Biotechnol. 12(6): 979–985.

León-Vaz, A., R. León, E. Díaz-Santos, J. Vigara and S. Roposo. 2019. Using agro-industrial wastes for mixotrophic growth and lipids production by the green microalga *Chlorella sorokiniana*. New Biotechnol. 51: 31–38.

Li, J.P., M.G. Healy, X.M. Zhan and M. Rodgers. 2008. Nutrient removal from slaughterhouse wastewater in an intermittently aerated sequencing batch reactor. Bioresour. Technol. 99(16): 7644–7650.

Maizatul, A.Y., R.M.S.R. Mohamed, A.A. Al-Gheethi and M.K. Amir Hashim. 2017. An overview of the utilisation of microalgae biomass derived from nutrient recycling of wet market wastewater and slaughterhouse wastewater. Int. Aquat. Res. 9: 177–193.

Markou, G., D. Vandamme and K. Muylaer. 2014. Microalgal and cyanobacterial cultivation: the supply of nutrients. Water Res. 65: 186–202.

Merola, A., R. Castaldo, P.D. Luca, R. Gambardella, A. Musacchio and R. Taddei. 1981. Revision of *Cyanidium caldarium*. Three species of acidophilic algae. G. Bot. Ital. 115: 189–195.

Metcalf and Eddy. 2003. Wastewater Engineering: Treatment and Reuse, Fourth ed. McGraw-Hill, Boston, USA.

Mohamed, R.M.S.R., A.A. Al-Gheethi, A.M. Jackson and H.K. Amir. 2016. Multi component filters for domestic greywater treatment in village houses. J. Am. Water Works Assoc. 108(7): 405–414.

Nancucheo, I. and D.B. Johnson. 2012. Acidophilic algae isolated from mine-impacted environments and their roles in sustaining heterotrophic acidophiles. Front. Microbiol. 3(Article 325): 1-8.

Nordlander, E., J. Olsson, E. Thorin and E. Nehrenheim. 2017. Simulation of energy balance and carbon dioxide emission for microalgae introduction in wastewater treatment plants. Algal Res. 24(A): 251–260.

Olguin, E.J. 2003. Phycoremediation: key issues for cost-effective nutrient removal processes. Biotechnol. Adv. 22: 81–91.

Omar, D., M.R. Salim and Salmiati. 2016. Nutrient removal of greywater from wet market using sequence batch reactor. Malaysia J. Anal. Sci. 20(1): 142–148.

Onet, C. 2010. Characterization of the untreated wastewater produced by food industry. Analele University din Orade, Fascicula, Protectia Meduilui XV: 709–714.

Pahazri, F., R.M.S. Mohamed, A.A. Al-Gheethi and H. Amir. 2016. Production and harvesting of microalgae biomass from wastewater: a critical review. Environ. Technol. Rev. 5(1): 39–56.

Pexara, A., N. Solomakos and A. Govaris. 2018. Q fever and prevalence of *Coxiella burnetii* in milk. Trends Food Sci. Technol. 71: 65–72.

Pires, J.C., M.C. Alvim-Ferraz, F.G. Martins and M. Simoes. 2013. Wastewater treatment to enhance the economic viability of microalgae culture. Environ. Sci. Pollut. Res. Int. 20(8): 5096–5105.

Posadas, E., S. Bochon, M. Coca, M.C. García-González, P.A. García-Encina and R. Muñoz. 2014. Microalgae-based agro-industrial wastewater treatment: a preliminary screening of biodegradability. J. Appl. Phycol. 26: 2335–2345.

Prakash, K., C. Vimal and D. Indra. 2011. An overview of various technologies for the treatment of dairy wastewaters. Department of Chemical Engineering, Indian Institute of Technology, India. Crit Rev. Food Sci. Nutr. 51: 442–452.

Qin, L., Q. Shu, Z. Wang, C. Shang, S. Zhu, J. Xu, et al. 2014. Cultivation of *Chlorella vulgaris* in dairy wastewater pretreated by UV irradiation and sodium hypochlorite. Appl. Biochem. Biotechnol. 172: 1121–1130.

Rajakumar, R., T. Meenambal, T.R. Banu and I.T. Yeom. 2011. Treatment of poultry slaughterhouse wastewater in up flow anaerobic filter under low up flow velocity. Int. J. Environ. Sci. Tech. 8(1): 149–158.

Ramasamy, E.V., S. Gajalakshmi, R. Sanjeevi, M.N. Jithesh and S.A. Abbasi. 2004. Feasibility studies on the treatment of dairy wastewaters with up flow anaerobic sludge blanket reactors. Bioresour. Technol. 93: 209–2012.

Ras, M., J.P. Steyer and O. Bernard. 2013. Temperature effect on microalgae: a crucial factor for outdoor production. Rev. Environ. Sci. Bio-Technol. 12(2): 153–164.

Renuka, N., A. Sood, S. Ratha, R. Prasanna and A. Ahluwalia. 2013. Evaluation of microalgal consortia for treatment of primary treated sewage effluent and biomass production. J. Appl. Phycol. 25(5): 1529–1537.

Rico Gutierrez, J.L., P.A. Garcia Encina and F. Fdz-Polanco. 1991. Anaerobic treatment of cheese-production wastewater using a UASB reactor. Bioresour. Technol. 37: 271–276.

Samori, G., C. Samori, F. Guerrini and R. Pistocchi. 2013. Growth and nitrogen removal capacity of *Desmodesmus communis* and of a natural microalgae consortium in a batch culture system in view of urban wastewater treatment: Part I. Water Res. 47(2): 791–801.

Sánchez, J., E. Hernández, J.M. Auleda and M. Raventós. 2011. Freeze concentration technology applied to dairy products. Food Sci. Technol. Int. 17(1): 5–13.

Santos, C.P.E. and D.M.R.S. Robbins. 2004. Low-cost innovative solutions for treating public market wastewater in the Philippines: deploying hybrid anaerobic/aerobic coco peat filtration systems. PADCO, Makati.

Satpal and A.K. Khambete. 2016. Waste water treatment using micro-algae—a review paper. Int J. Eng. Technol. Manag. Appl. Sci. 4(2): 188–192.

Schmidt, R.A., M.G. Wiebe and N.T. Eriksen. 2005. Heterotrophic high cell-density fed-batch cultures of the phycocyanin-producing red alga *Galdieria sulphuraria*. Biotechnol. Bioeng. 90: 77–84.

Selvaratnam, T., A.K. Pegallapati, F. Montelya, G. Rodriguez, N. Nirmalakhandan, W. Van Voorhies, et al. 2014. Evaluation of a thermo-tolerant acidophilic alga, *Galdieria sulphuraria*, for nutrient removal from urban wastewaters. Bioresour. Technol. 156: 395–399.

Sethupathy, A.V., A. Subramanian and R. Manikandan. 2015. Phyco-remediation of sewage waste water by using micro-algal strains. Int. J. Eng. Innov. Res. 4(2): 300–302.

Silva-Benavides, A. and G. Torzillo. 2012. Nitrogen and phosphorus removal through laboratory batch cultures of microalga *Chlorella vulgaris* and cyanobacterium *Planktothrix isothrix* grown as monoalgal and as co-cultures. J. Appl. Phycol. 24(2): 267–276.

Singh, J. and S. Gu. 2010. Commercialization potential of microalgae for biofuels production. Renew. Sustain. Energ. Rev. 14: 2596–2610.

Singh, J.S., A. Kumar and M. Singh. 2019. Cyanobacteria: a sustainable and commercial bio-resource in production of bio-fertilizer and bio-fuel from wastewaters. Environ. Sustain. Indic. 3: 100008.

Singh, J.S., A. Kumar, A.N. Rai and D.P. Singh. 2016. Cyanobacteria: a precious bio-resource in agriculture, ecosystem, and environmental sustainability. Front. Microbiol. 7: 529.

Singh, J.S., S. Koushal, A. Kumar, S.R. Vimal and V.K. Gupta. 2016. Book review: microbial inoculants in sustainable agricultural productivity, Vol. II: functional application. Front. Microbiol. 7: 2105.

Sloth, J.K., H.C. Jensen, D. Pleissner and N.T. Eriksen. 2017. Growth and phycocyanin synthesis in the heterotrophic microalga *Galdieria sulphuraria* on substrates made of food waste from restaurants and bakeries. Bioresour. Technol. 238: 296–305.

Sombatsompop, N., T. Markpin, P. Ratchatahirun, W. Yochai, S. Ittiritmeechai, N. Premkamolnetr, et al. 2011. Research productivity and impact of ASEAN countries and universities in the field of energy and fuel. Malays. J. Libr. Inf. Sci. 16(1), 35–46.

Suad, J. and S. Gu. 2014. Commercialization potential of microalgae for biofuels production. Renew. Sustain. Energ. Rev. 14: 2596–2610.

Subashchandrabose, S.R., B. Ramakrishnan, M. Megharaj, K. Venkateswarlu and R. Naidu. 2011. Consortia of cyanobacteria/microalgae and bacteria: biotechnological potential. Biotechnol. Adv. 29(6): 896–907.

Varshney, P., J. Beardall, S. Bhattacharya and P.P. Wangikar. 2018. Isolation and biochemical characterisation of two thermophilic green algal species – *Asterarcysquadri cellulare* and *Chlorella sorokiniana*, which are tolerant to high levels of carbon dioxide and nitric oxide. Algal Res. 30: 28–37.

Varshney, P., P. Mikulic, A. Vonshak, J. Beardall and P.P. Wangikar. 2015. Extremophilic micro-algae and their potential contribution in biotechnology. Bioresour. Technol. 184: 363–372.

Wan, M., Z. Wang, J. Wang, S. Li, A. Yu and Y. Li. 2016. A novel paradigm for the high-efficient production of phycocyanin from *Galdieria sulphuraria*. Bioresour. Technol. 218: 272–278.

Wurochekke, A.A., R.M.S. Mohamed, A.A. Al-Gheethi, H.M. Amir and H.M. Matias-Peralta. 2016. Household greywater treatment methods using natural materials and their hybrid system. J. Water Health 14: 914–928.

Xin, L., H.H. Ying and Z.Y. Ping. 2011. Growth and lipid accumulation of a freshwater microalga *Scenedesmus* sp. under different cultivation temperature. Bioresour. Technol. 102: 3098–3102.

Zang, C., S. Huang and M. Wu. 2011. Comparison of relationships between pH, dissolved oxygen and chlorophyll-a for aquaculture and non-aquaculture waters. Water, Air, Soil Poll. 219(1–4): 157–174.

Zulkifli, A.R., H. Roshadah and T.F.T. Khalkausar. 2011. Control of water pollution from non-industrial premises. Report No. 1563, Department of Environment, Malaysia.

Integrated Microalgal Wastewater Remediation and Microalgae Cultivation

Introduction

Due to increased urbanization and industrialization, there is a significant rise in wastewater production, which requires efficient processes and methods to provide wastewater within established environmental standards. Conventional treatment mainly involves preliminary-, primary-, secondary- and tertiary- processes with disinfection (unlikely followed most of the times); but they alone are not enough, requiring advanced, engineered treatment systems like Activated Sludge Process (ASP), Sequential Batch Reactor (SBR), Membrane Bio Reactor (MBR), Moving Bed Biofilm Reactor (MBBR) and Up-flow Anaerobic Sludge Blanket reactor (UASB).

Although they are more efficient and appropriate but have major constraints, such as high building, operation and maintenance investments that makes them unsuitable for developing countries or poor countries. In light of these limitations and constraints, microalgal remediation has proved to be an efficient, low cost and sustainable treatment approach that not only discharges the desired quality of wastewater but also could be used to cultivate microalgal biomass in wastewater (Kumar and Singh 2016, Singh et al. 2016a, b, Kumar et al. 2017, Kumar and Singh 2017, Kumar 2018). This could be a game changer approach for the overburdened treatment facilities largely funded by governments, which benefit through reduction in the operation and maintenance costs, and further generate extra funds through value added production of microalgal biomass.

Although microalgae cultivation can be done directly in wastewater, it is more feasible in secondary treated wastewater or could be integrated with conventional or modern treatment methods. There are many methods for microalgae cultivation: (a) suspended cultivation systems, including open ponds, closed photo-bioreactors, hybrid systems; (b) immobilized cultivation systems, including matrix-immobilized, microalgal biofilms and submersible algal cultivation (Kumar et al. 2018a, b, Singh et al. 2019, Kumar and Singh 2020a, b).

To harvest this microalgal biomass, chemical, mechanical, electrical methods or combinations of these methods are followed, but mechanical methods are commonly used that involve e.g., sedimentation, centrifugation, filtration and dissolved air flotation. Then thickening and dewatering processes are followed to make the biomass more valuable and less in volume, sometimes subsequent processing of biomass is also followed; otherwise it is transported to further operations to produce biofuels and other industrial products.

Conventional Wastewater Treatment: A Brief Description

Conventional wastewater treatment involves a series of processes which are primarily categorized in to preliminary, primary, secondary, tertiary and disinfection processes (Fig. 11.1). These processes produce large quantities of different types of solids and sludge that undergo further handling, management and treatments:

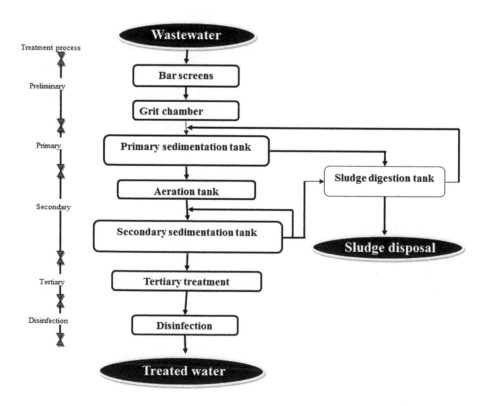

Figure 11.1: A typical scheme of conventional wastewater treatment.

Preliminary treatment

Screening: includes mechanical or manually-cleaned bar screens to remove large solids.

Grit removal: includes tanks that facilitate the removal of grit and scum.

Primary treatment

Mixing and aeration: to make wastewater homogenous and aerated.

Sedimentation (primary): Sludge and scum settles down and collects as primary sludge; the treated wastewater is conveyed to the secondary treatment.

Secondary treatment

Biological treatment: Microorganisms mainly bacteria degrade the organic matter (BOD) in to sludge and scum (suspended solids).

Sedimentation (secondary): The sludge and scum settle and is collect as secondary sludge.

Tertiary treatment

After the primary and secondary treatment, treated wastewater containing color, dissolved solids and toxic compounds are removed through tertiary treatment. Tertiary treatment involves a number of techniques mainly filtration.

Disinfection

Finally treated wastewater undergoes a process such as chlorination, ozonation and that kills the microbes especially pathogenic microbes present in the wastewater.

Sludge-processing

This involves facilities for handling, management and processing of the sludge which take place producing compost, ashes or other end products. The characteristics of the end products primarily depend on the type of sludge (primary sludge, secondary or mixed), their treatment and processes that are used for the procedure. Depending on the wastewater treatment process, sludge could be classified in to: (a) Primary sludge that is produced during primary wastewater settling; (b) Secondary sludge or activated sludge that is produced during secondary biological treatment; (c) Mixed sludge that is produced from mixing of primary and secondary sludge; and (d) Tertiary sludge: produced during tertiary or advanced wastewater treatment.

Sludge Problem and Its Management

Sludge glutinous, humus-like material which are removed through settling from the primary and secondary (activated sludge) or sometimes tertiary treatments. It contains a high amount of nutrients and organic matter which could be used for many applications like fertilizers or bricks etc. Having a high thermal value, it could also be used for heating purposes and the remaining ashes further can be used as a micronutrient source. However it also has some limitations like high moisture content (98-99%) and contains pollutants like heavy metals, synthetic organics, and microbial (Kim and Smith 1997). Sludge components are categorized in to the following groups (Rulkens 2004): (a) moisture up to 99%; (b) nontoxic inorganic substances (e.g., Si, Al, Ca, Mg); (c) nontoxic organic substances that makes up to 60% of dry mass (N and P); (d) toxic substances: Heavy metals (Zn, Pb, Cu, Cr, Ni, Cd, Hg, As) and synthetic organics (PCBs, PAHs, dioxins, pesticides, etc); and (e) pathogenic microorganisms and microbial pollutants.

So the direct discharge of sludge in to the environment might be responsible for many environmental problems like contamination of water bodies (surface and ground water), public health hazards and air pollution. Due to different compositions of wastewater, sludge production also have various pollutant loads and toxic compounds especially heavy metals; which should be considered as an important factor in deciding the proper treatment method for sludge management. Sludge production is only responsible for less than 1% of the total operational cost of wastewater treatment, while its handling and management costs could reach as much as up to 40-50% of the total operational cost of a wastewater treatment facility (Campbell 2000).

For proper handling and management, first there is need to reduce the volume of the main sludge produced and the degradation of their substances. This includes some common procedures like sludge thickening, dewatering and drying. Sludge thickening involves the loss of moisture, which leads to sludge mass reduction up to 70%, and leaving the solid content up to 2-3% of their initial content. For this, many methods have been applied for sludge thickening: (a) gravity belt thickening, (b) thickening by centrifugation, (c) sand drying bed, and (d) belt filter presses.

In dewatering, thickened sludge undergoes an operation which further reduces the water content present in the sludge; leading to an increase in the solid content up to 18-20%. There are many dewatering procedures in which some are similar to the thickening treatments, while others are different. Some of the methods and technologies for dewatering are: (a) vacuum filtering, (b) gravity belt thickening, (c) filter belt press, (d) gravity thickening, (e) centrifugation, (f) membrane press (Dentel 2001, Chen et al. 2002, Wakeman 2007). Besides their efficiency in reducing the total solid content in the range of 5 to 12%, these thickening procedures also have some limitations and demerits like high area requirement, odor incidence, high energy requirement, addition of chemicals and high capital costs. In comparing all the above mentioned methods, belt press methods are found to be the most suitable.

The final step in sludge processing, sludge drying further involves such operations which reduce the water content below 10%. These are the some categories of sludge drying systems: (a) Direct drying systems involves the use of rotating drums, lamps, belt dryers, spray dryers, solar energy drying; (b) Indirect drying systems involves the use of a rotaplate indirect dryer, kneading and self-cleaning disk dryer, porcupine processor, (c) K-S Nara paddle and paddle dryer, (d) Fluidized bed dryers, (e) Combination of drying and incineration.

Conventional or Modern Wastewater Treatment Systems

After a basic scheme of wastewater treatment, there are some technologies or techniques that have been developed for the effective treatment of wastewater (Table 11.1). Their steps could be the same or different from above mentioned basic schemes or some steps could be added, skipped or replaced with the some other modern techniques. Some popular wastewater systems that are followed around the world.

Activated Sludge Process (ASP) or Extended Aeration (EA)

Activated Sludge Process (ASP) or Extended Aeration (EA) is the most adapted treatment method in the world, in which suspended growth microorganisms are facilitated for the treatment in wastewater. It includes provision for aeration and mixing, settling, recovery of activated sludge and removal of solids; these remaining solids collect as waste activated sludge. In this method, raw wastewater reaches into a mix tank with an aeration facility where it is kept for a specific time which would be sufficient for the bacteria to degrade all consumable (most of BOD and some part of COD) present in wastewater. The wastewater is further conveyed to a sedimentation tank where the flocculated materials (with colonies or organism) tend to settle; leading to the flow of clear water.

Sequencing Batch Reactor (SBR)

SBR is a modified form of the activated sludge process, or simply known as fill- and draw- active sludge process. Unlike in the activated sludge process, there is no requirement of clarifiers, because all treatment processes are facilitated in to a reactor tank. It treats the wastewater in to batches; each batch undergoes a sequence of various stages of treatment processes. The SBR method involves two tanks which operate on a fill and draw basis. The wastewater is pumped to the reactor tank, where all the required treatment processes follow. It leads to the settling of solids from the wastewater, from which the clarified water is pumped out from the reactor tank. There are five phases: (a) Fill, (b) React, (c) Settle, (d) Draw, and (e) Idle, that makes

Table 11.1: Major wastewater treatment facilities and their merits and demerits

Treatment process	Merits	Demerits
Activated sludge process (ASP)	Low installation cost Good quality effluent Low land requirement Loss of head is small No risk of fly and odor nuisance High degree of treatment	Not a very flexible method Operation cost is high Sludge disposal is required on a large scale Sensitive to certain industrial wastes Skilled supervision is required
Sequential batch reactor (SBR)	All processes in a single reactor vessel Operating flexibility and control Minimal footprint Potential capital cost savings by eliminating clarifiers and other equipment	Need of higher level of sophistication Higher level of maintenance Potential of discharging floating or settled sludge during the decant phase Potential plugging of aeration devices Potential requirement for equalization
Membrane bio reactor (MBR)	High quality effluent Independent HRT and SRT Small footprint Consistent performance Low sludge production Less sludge dewatering	High capital and operational costs Operational complexity Frequent problem of membrane fouling, clogging and cleaning
Moving bed biofilm reactor (MBBR)	Most adapted for space constraints No such need of skilled plant operators Requires little maintenance Resistant to shock loads works quickly with a low hydraulic retention time	Manual monitoring Skilled experts Insects Escaping carriers
Up-flow anaerobic sludge blanket reactor (UASB)	Low production of sludge Low energy required Methane generation Tolerance to high organic loads Low nutrient consumption	Need of post-treatment Possibility of generation of bad odors Low nutrient and pathogens removal Susceptible to the temperature of the effluent

up a SBR treatment cycle; each phase provides a defined time period (that depends on the aeration and mixing pattern). Further the duration of aeration times depends up on plant size and the volume/composition of the incoming wastewater, but typically lasting for 60 to 90 minutes.

Membrane Bioreactor (MBR)

MBR is a specially suspended growth bioreactor, which involves the use of a perm-selective membrane (known as microfiltration or ultra-filtration) integrated with biological treatment, which replaces the secondary, tertiary and disinfection processes that follow in the traditional activated sludge process. The membrane rejects out the solid materials present in wastewater; resulting in the discharge of clarified and disinfected treated wastewater. These rejected solid materials or sludge undergo a biological process, leading to a microbial degradation like in the activated sludge process. The membrane bioreactor system has large volumetric loading rates, shorter retention time; providing high quality treated and producing less sludge as compared to ASP or SBR systems. MBR also have some demerits which include higher energy costs, a severe membrane fouling problem and high costs burden of regular replacement of the membrane filter.

Moving Bed Biofilm Reactor (MBBR)

The MBBR system involves thousands of polyethylene biofilm carriers which are continuously moving inside an aerated wastewater tank. The polyethylene biofilm carrier operates in a fixed motion, providing a large and protected surface area to facilitate better growth of microorganisms. Moving bed attached microorganisms could be able to degrade organic material very efficiently as compared to Fixed Bed Bio Reactor (FBBR). In comparison to MBR, these systems achieve high density of bacterial populations, which leads to a high rate of biodegradation within the system; resulting in the facilitation of process reliability and ease of operation. Due to minimal maintenance and mobility of biofilms within the system, MBBR could operate in self-maintaining at an optimum level and are also able to respond automatically to load fluctuations. These advantages make the MBBR, a cost effective, flexible and ease of operation wastewater treatment system that is quite suitable for wastewater treatment in current conditions (in terms of volume and characteristics of wastewater); also have the flexibility to expand according to future loads or meet the more strict environmental discharge standards without altering the main design.

Up Flow Anaerobic Sludge Blanket Reactor (UASB)

The UASB reactor involves a single tank to perform all wastewater treatment processes. This single tank includes primary sedimentation, anaerobic digestion, secondary sedimentation including sludge stabilization; that are combined in to a reactor. In the UASB reactor, the aerators are provided for the supply of essential oxygen to decompose the organic matter along with wastewater through facilitating huge surface area; resulting in the further reduction of the BOD load up to 75%. The wastewater is pumped into the reactor through the bottom and then flows upwards. Moving upwards wastewater passes through suspended sludge blanket filters, leading to the formation of millions of small granules of sludge that tends to be suspended in wastewater. This excess sludge is collected and conveyed to the sludge pump house for further processing. Along with nutrient rich sludge, methane enriched biogas are produced as byproducts in the UASB reactor, which could be a major advantage of this system.

Integration of Microalgal Remediation with Wastewater Treatment: An Opportunity for Microalgal Biomass Cultivation

Microalgae mediated-wastewater remediation could be further integrated with conventional or advanced wastewater treatment facilities (Fig. 11.2). This not only provides an efficient alternative to the tertiary treatment but also gives the opportunity for the cultivation of microalgal biomass for value added products. Although microalgae are directly used for the remediation of untreated wastewater, it is not feasible for efficient treatment and better microalgal cultivation as untreated wastewater contains some toxic compounds. Thus secondary-treated wastewater could be the best suitable option as an integrated microalgal remediation, since it requires inorganic nutrients and a less hostile surrounding environment. Further it reduces the need of tertiary treatment and disinfection, which requires the use of

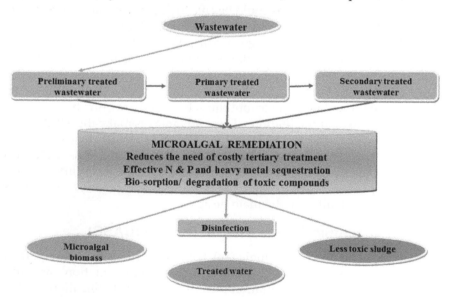

Figure 11.2: Integration of microalgal remediation with wastewater treatment: An opportunity for microalgal biomass cultivation.

expensive techniques and additional building and operational cost; that makes waste water treatment a burden on municipalities and industries.

Integrated microalgae-mediated wastewater remediation includes some components of conventional treatment such as preliminary-, primary- or sometimes secondary- treatment processes, and microalgae are used for efficient removal of excess nutrients and toxic heavy metals, which remain even after the secondary treatment. Further it could improve the overall aesthetic appearance of treated water by removing the color, odor and presence of pathogenic microbes. This approach could provide many benefits including efficient treatment of wastewater, reduction in the building and operating cost of tertiary treatment, the need of fewer disinfectants for pathogenic microbes, increased usability of treated water and minimizing the risk of eutrophication of receiving water bodies.

The primary- or secondary-treated wastewaters have sufficient nutrients and conditions which can support the mass cultivation of microalgal biomass, which could be a value-adding approach for the existing wastewater treatment facilities (Subramanian and Sundaram 1986). This could provide additional benefits to the existing wastewater treatment facilities in two ways: (a) reducing the cost of existing treatment methods; (b) enhancing the profitability and sustainability of wastewater treatment facilities.

Microalgal Biomass Cultivation Systems

Suspended cultivation systems

Open ponds

Since the 1950s, open ponds or raceway ponds are the common cultivation systems that are practiced for large scale production. They are also known as High Rate Algal Ponds (HRAPs). Raceway ponds are open, shallow ponds which are promoted with a paddle wheel for the continuous circulation of the microalgae and nutrients (Fig. 11.3). In theoretical prediction, the productivity of raceway ponds should be of 50– 60 g m^{-2} day^{-1} but practically it is difficult to achieve the productivity of even 10–20 g m^{-2} day^{-1} (Sheehan et al. 1998,Shen et al. 2009).The raceways ponds are relatively cheap low-cost in building and operation, which makes them very convenient for large scale microalgal cultivation. But it has several disadvantages such poor mixing, culture contamination, dark zones and inefficient CO_2 utilization; leading to low cell density (Chisti 2007, Mata et al. 2010). Further high evaporation from raceway ponds is considered as a major limitation, but this could be helpful in temperature regulation of ponds through evaporative cooling (US DOE 2010).

Closed photo-bioreactors (PBRs)

In comparison to raceway ponds, closed photo-bioreactors can facilitate proper mixing, minimum culture contamination and low evaporation, which

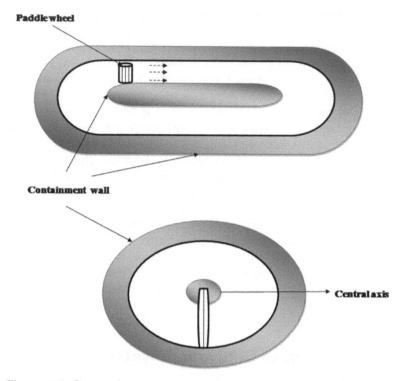

Figure 11.3: Open cultivation systems: Raceway ponds or circular ponds.

ultimately leads to high cell density (Mata et al. 2010). Closed photobioreactors are tubular, plate or fermenter type, but tubular photo-bioreactors are commonly closed systems that are used for large scale production (Fig. 11.4). Tubular photo-bioreactors are vertical, horizontal or helical in designs, but helical designs are preferred due to their ease to scale up (Carvalho et al. 2006, Chisti 2007). The productivity of these photo-bioreactors are reported in the range of 20-40 g m^{-2} day^{-1} (Shen et al., 2009). Besides the advantages and higher productivity, tubular photo-bioreactors also have some problems like overheating, toxic oxygen accumulation, bio-fouling, adverse pH and CO_2 gradients and high cost of material and maintenance (Molina-Grima et al. 1999, Carvalho et al. 2006, Mata et al. 2010).

Hybrid systems

Both open and closed systems have some disadvantages, which limit the performance and efficiency of these systems. To overcome this, hybrid systems are developed which include the advantages of open and closed cultivation systems. There are two stages in hybrid cultivation systems: (a) first stage involves the use of closed photo-bioreactors to culture the microalgal inoculums (b) the second stage uses these microalgal inoculums for the mass cultivation in open ponds. Due to the use closed photo-

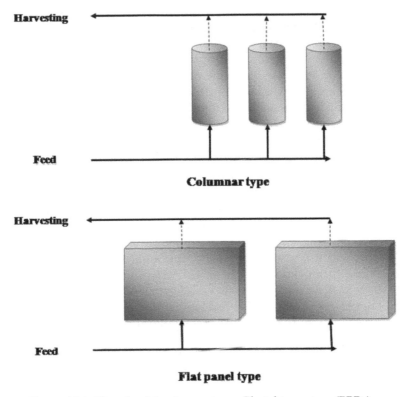

Figure 11.4: Closed cultivation systems: Photobioreactors (PBRs).

bioreactors, microalgal seeds are minimally exposed to contamination before they are supplied to large scale open ponds, where it helps in maintaining the dominance of preferred microalgal species in open ponds. The cost of closed photo-bioreactors in the first stage could be a limitation in large-scale applications.

Immobilized cultivation systems

There are two problems in suspended cultivation systems i.e., low biomass density, high cost of harvesting and downstream processing, which limit the large scale applicability and acceptability of these systems (Posten 2016, Arbib et al. 2013). Immobilization of microalgae could be a potential approach to achieve the following objectives: efficient conversion of wastewater nutrients and cost-effective harvesting and downstream processing (Lam and Lee 2012). There are many immobilization techniques like adsorption, affinity immobilization, covalent coupling, capture behind semi-permeable membrane, confinement in liquid–liquid emulsion and entrapment. Based on these techniques, there are five types of bioreactors developed such as, packed-bed, fluidized-bed, air-lift, parallel-plate and hollow-fiber. Entrapment is the most common immobilization method that is frequently used for microalgae in laboratory experiments (Christenson and Sims 2011). In the entrapment method, microalgal cells are confined in 3D lattice of gel or a polymer, which could be either natural such as agar, cellulose, agarose, alginate, collagen, carrageenan; or synthetic such as acrylamide, polyurethane, polyvinyl. Although microalgal cells are restricted in the polymer, cells are free within their compartments, and their interaction with wastewater nutrients is facilitated through the pores present in the polymer.

Matrix-immobilized microalgae

Matrix-immobilized microalgae involves the immobilization of microalgal cells in carrageenan or alginate matrices, which provide some potential advantages such as efficient nutrient removal and better growth in terms of pigment and lipid production (Chevalier et al. 2000). Although there is no such increase of growth rate reported in immobilized microalgal cultivation systems as in suspended cultivation systems (Hameed and Ebrahim 2007). However it is also reported that immobilization could help in more hydrocarbon production (Bailliez et al. 1985), increase in cellular pigment and enhanced lipid content and its varieties (De-Bashan et al. 2002). Having such benefits, the high cost of immobilization process could be a major limitation that seems to be suitable at the laboratory scale (Hoffmann 1998).

Microalgal biofilms

There are many limitations in harvesting and processing of microalgal biomass in suspended or matrix-immobilized cultivation systems, which could be overcome by microalgal biofilms. Microalgal biofilms can provide efficient wastewater treatment with improved biomass cultivation, and recently they have gained much attention by industries for their large scale application (Woertz et al. 2009). Middlebrooks et al. (1974) suggested that by providing enough surface area, the microalgal biofilm system provides more growth than suspended growth systems. In comparison to suspended cultivation systems, microalgal biofilm systems have some benefits like better integrated production, harvesting, and dewatering operations; leading to a more streamlined process with reduced downstream processing costs. These microalgal biofilms could be integrated into an ongoing wastewater treatment, that could benefit in dual ways by removing inexpensive nutrients and simultaneously providing the treated water. US DOE (1985) reported that surface attached microalgal biofilms can achieve the same increased microalgal cell density at the corresponding cost of the matrix, but requires less land and water as compared to matrix-immobilized systems.

Innovative submersible aquatic algae cultivation technology

In recent years, there have been new and innovative attempts of cultivation of microalgae nutrient-rich polluted water resources such as wastewater lagoons, fish farms, coastal waters, polluted lakes and reservoirs. Bussell (2008) conceptualized these innovative floating ponds which involve microalgae cultivation in a buoyant system suspended in water. The microalgae in floating ponds are cultivated in a controlled space that is separated from the surrounding water, which reduces the risk of uncontrolled microalgal blooms in all the polluted waters. The main advantage of these floating systems is natural mixing by wind and waves to provide better CO_2 transfer rates with the atmosphere, which ultimately reduces the cost of energy that is needed in CO_2 addition and mixing operations. Further efficiency of this system might be enhanced through the adoption of such microalgae species which have high nutrient uptake capability; resulting in maximum removal of nutrients from polluted waters and leaving the treated water more clean. The performance of floating ponds mainly depends on weather conditions i.e., wind, temperature and sunshine. Further there is need of cautionary measures to avoid the risk of escaping microalgae from controlled and enclosed containment system.

Harvesting of Microalgal Biomass

Harvesting of microalgal biomass is considered an energy- and cost-intensive process. There are various chemical, mechanical, electrical methods or a combination of these methods that are used for harvesting (Table 11.2) (Kumar et al. 1981, Bernhardt and Clasen 1991, Danquah et al. 2009). Mechanical methods are found to be more applicable and sustainable in terms of variety and cost reduction of harvesting. Although biological methods are also being observed in order to reduce the cost of harvesting methods. However no single method has been found to be perfect in terms of efficiency, cost and suitability for harvesting of microalgal biomass (Shelef et al. 1984).

Table 11.2: Major biomass harvesting methods and their merits and demerits (Chisti 2007, Amenorfenyo et al. 2019)

Method	Merits	Demerits
Flotation	Large volume of biomass processed	Contamination due to floating agents
Filtration	More suitable for small sized microalgae	High cost, due to species specific, filter clogging/fouling
Centrifugation	Rapid and efficient	High energy and maintenance cost
Gravity sedimentation	Low cost and energy efficient	Time consuming, ineffective for small sized microalgae
Ultrasonication	Feasible for continuous operation	Safety issue, unsuitable for further processing due to disruption of cells
Flocculation	Cost-effective	Unsuitable for further processing due to use of chemical flocculent

Mechanical methods

The common mechanical-based harvesting methods are centrifugation, sedimentation filtration and dissolved air floatation. Centrifugation is the simplest, rapid and reliable method; in which centrifugal forces are used to separate the microalgal biomass based on density differences. This method is found to more effective and sustainable for laboratory scale experiments, but involves high investment and operating costs making this cost-prohibitive for any large scale use (US DOE 2010).

Filtration is another low-cost method that could be used in harvesting of microalgal biomass especially for filamentous microalgal species (Vonshak and Richmond 1988). They can be adjusted in raceway ponds or HRAPs to screen out larger filamentous microalgae cells from smaller non-filamentous microalgae cells (Wood 1987). Danquah et al. (2009) and Uduman et al. (2010) emphasized the risk of membrane fouling and replacement of filters that add significant costs for harvesting. Due to the application in harvesting of filamentous microalgae, Mulbry et al. (2008) considered this method unsuitable for applications in biofuels as filamentous microalgae are known for their low lipid content.

Sedimentation is a relatively slow and low-cost method that could be applied for the harvesting of microalgal biomass. According to Uduman et al. (2010), it is predicted that sedimentation could achieve the biomass concentration of 1.5% solids, but there is a limitation of fluctuating density of microalgal cells, leading to a low consistency of biomass harvesting (Shen et al. 2009). Greenwell et al. (2010) observed the settling rates of 0.1–2.6 cm h^{-1} in the sedimentation process, which makes them a relatively slow harvesting process.

Dissolved air flotation is a process that is primarily applied for sludge removal in conventional wastewater treatment. In biomass harvesting, it is preferred over the sedimentation method due to the greater settling rate and applicability in large scale production (Teixeira and Rosa 2006, Greenwell et al. 2010).

Considering the constraints and limitations of all the above mentioned mechanical methods, attached microalgal biofilms can provide a cheaper, effective and sustainable mechanical harvest method. To harvest the biomass, microalgal turf is removed either through simple scrapping or vacuum application (Adey 1982, 1998, Jensen 1996, Johnson and Wen 2010). Johnson and Wen (2010) achieved the harvest concentration of 6.3% solids in simple scraping harvesting of a *Chlorella* biofilm; which further needs no additional harvesting or concentrating operation.

Chemical methods

Due to the small size of microalgal cells, metal salts or polymers are primarily applied as a pretreatment method for other harvesting methods. These chemicals or polymers help in increasing the particle size, which could further harvest the microalgae through flotation (Bernhardt and Clasen 1991).

Metal salts such as aluminum sulfate and ferric chloride are very commonly used processes in chemical based harvesting, but there is also a need of caution regarding the risk of inhibitory effect on methanogenic activity of some specific bacteria that reside in wastewater sludge (Cabirol et al. 2003). Further application of metal salts containing sludge in agricultural fields could induce more absorption of heavy metals, which might lead to phosphorus deficiencies in crops (Bugbee and Frink 1985).

This concern of secondary pollution is not associated with natural polymers like chitosan and cationic starch, which have not been studied much. Divakaran and Pillai (2002) and Vandamme et al. (2009) investigated the ability of chitosan and cationic starch polymer, and found them to be effective flocculating agents for harvesting of microalgal biomass.

Electrical based

Microalgal cells have a negative charge due to the presence of functional groups on its cell wall. So the application of an electric field or electrophoresis could help in separation of microalgal cells (Kumar et al. 1981). Although it has no requirement of chemicals before the harvesting process which could be a major advantage, the high power requirements and electrode costs makes this method unfeasible for large-scale applications (Uduman et al. 2010).

Biological methods

Certain conditions like specific pH, light, temperature and other conditions induce the ability of microalgae to spontaneously flocculate without any chemical addition; which could significantly help in reduction of harvesting costs (Sukenik and Shelef 1984). This ability of microalgae or the process described by two interchangeable but different processes i.e., auto-flocculation and bio-flocculation.

The autoflocculation phenomenon can be started with more consumption of dissolved CO_2 that develops higher pH conditions, leading to supersaturation of calcium and phosphate ions. Such conditions influence the formation of positively charged calcium phosphate precipitates, which attract and precipitate out the negatively charged microalgal cells (Lavoie and De la Noüe 1987). It was also suggested that autoflocculation might not take place in all type of waters. Sukenik and Shelef (1984) observed the concentration of calcium of 60–100 mg l^{-1} and phosphate of 3.1–6.2 mg l^{-1} that could be optimum for precipitation of calcium phosphate and autoflocculation at a pH of 8.5–9.

In contrast to autoflocculation, the bioflocculation phenomenon is described as the flocculation process facilitated through secretion of Extracellular Polymer Substances (EPS) by microalgal cells. Passow and Alldredge (1995), Staats et al. (1999), Wolfstein and Stal (2002) and Bhaskar and Bhosle (2005) emphasized the importance of increased EPS conditions in enhancing the flocculation or sedimentation of algal blooms. It was also reported that EPS production by algae biofilms enhances the solids flocculation in a trickling filter during the clarifier operation later.

The flocculation of microalgae also facilitates the use of such microbes that could release some flocculent, known as microbial flocculation Lee et al. (2008) and Oh et al. (2001) conducted some studies related to the positive role of microbes in flocculation of microalgal biomass. It was reported that mixing of flocculating microbes with a feed of 0.1 g l^{-1} acetate, glucose or glycerin for a duration of 24 hours, leads the flocculation up to 90% biomass (Lee et al., 2008). Similar findings were reported in a study involving a flocculent originated from soil microbes for *Chlorella vulgaris* biomass, showing better harvesting efficiency than chemical based harvesting aluminum sulfate or polyacrylamide (Oh et al.2001).

Biomass Processing and its Valorization

Biomass processing involves mainly two procedures: (a) thickening, (b) dewatering. The thickening procedure includes gravity sedimentation, coagulation- flocculation, flotation and electrical methods; while dewatering involves filtration and centrifugation methods. Biomass valorization is quite different from processing as it involves only removal of the water from the biomass; which needs specific procedures for a particular demand for extraction or production of a desired product.

Thickening

Thickening procedures commonly include gravity sedimentation, coagulation- flocculation, flotation and electrical or magnetic methods. These methods might be similar as the methods that are used in harvesting but the purpose could be different (Barros et al. 2015). Gravity sedimentation could be a cheap and energy-effective method which is usually used to separate the thickened slurry from the growth medium (Barros et al. 2015). Because of the absence of gas vesicles in microalgae and cyanobacteria, it might be a suitable method for thickening of microalgae biomass if the value of the end product is very low. However in case of low specific gravity of microalgae, it could be a very inefficient method and then a coagulation/flocculation follows d before sedimentation.

Coagulation-flocculation could be done either by the use of chemicals or by the use of biological agents. For industrial purposes, chemical coagulant/flocculent are preferred more where they can be easily applied for different microalgal species and also for large volumes. As an alternative use of chemicals, autoflocculation can be induced through a physical or chemical stimulus. Further bacterial exopolysaccharides can replace the chemical flocculants and facilitate bioflocculation, which is possible through cocultivation of heterotrophic, exopolysaccharide-producing bacteria with the microalgae (Barros et al. 2015).

Flotation facilitates thickening by bubbling the air through the harvested microalgal slurry. It can be used for microalgae species which have low specific gravity due to a small size or the presence of gas vesicles (Barros et al. 2015). It is relatively rapid and can be performed into smaller containers, but it is more costly than sedimentation and chemical flocculation methods that are often used for thickening.

Magnetic micro or nanoparticles are also applied for thickening of harvested microalgal biomass, or they could be used to separate harmful microalgae in the treatment of drinking water (Wang et al. 2015). For this process, naked ferrous oxide (Fe_3O_4) nanoparticles are used as flocculants to attract the microalgal cells through electrical interactions, and the strength of these magnetic particles is increased by a coating with cationic substances. Microalgae attached with magnetic particles can be collected with the use of a magnetic drum and magnetic particles which could be separated from the microalgae with acid treatment and filtration.

Some microalgae secrete Extracellular Polymeric Substances (EPS) like glycoproteins or exopolysaccharides that help the microalgal cells to self agglutinate. The agglutination process might be induced by environmental changes, while the autoflocculation process could be influenced by the increase in pH. It is reported that filamentous microalgae like *Arthrospira* and some diatoms might agglutinate to each other easily to create large flocks or sediments. Due to this, there is no need of added flocculants, autoflocculation is considered as a possible alternative for thickening of microalgal biomass (Kawano et al. 2011).

Dewatering

Dewatering involves the removing of extra water from the microalgal slurry, it could be done by filtration and/or centrifugation and then the dewatered biomass could be finally dried (Barros et al. 2015). Microalgal slurry passes through filters with some pressure and further filters can also be supplemented with vacuum to increase the dewatering efficiency. To avoid clogging, filters must be washed continuously.The choice of a filter depends mainly on the microalgal cell size and filters can also be supplemented with vacuum to increase the dewatering efficiency. Tangential flow filtration has proved to be cheaper than vacuum filtering and it could be able to recover 70-89 % of the microalgal biomass (Pragya et al. 2013). In comparison to tangential flow filtration, vacuum filtering is relatively expensive and has several limitations like repeated replacement of membranes and the risk of damage to microalgal cells.

While comparing the filtration method, centrifugation is considered a rapid but relatively expensive dewatering method which could be preferred for dewatering of high-value end products. It is used for dewatering of biomass slurry of most microalgal species except those species containing gas vesicles. While the centrifugation process usually follows thickening, there is a possibility that centrifuges can handle the microalgal biomass even after the harvesting process (Barros et al. 2015).

Conclusion

Conventional wastewater treatment generates nutrient-rich treated wastewater that could be an ideal and optimum growth medium for microalgal biomass production. This approach of microalgal-mediated wastewater remediation could be helpful in removing the dissolved inorganic nutrients, improving the color, odor and enhancing the oxygenation of treated wastewater. Further integration of this microalgal remediation approach with conventional treatment, could be a game changer approach in relation to efficient treatment and cost-effective microalgal biomass production. However there are many constraints like the design of the treatment method, complex characteristics of wastewater, microalgal culture and the type of cultivation systems; that need to be resolved through modifications and optimization in the existing technologies and systems.

Suspended cultivation systems are more popular and used in microalgal biomass production, in which open ponds are found to be simpler and widely applied, while in the case of production of high-value products, close photo-bioreactors have many advantages like low risk of contamination and better control on pH, temperature and CO_2 concentrations. But they also have some technical constraints like low productivity and high cost of harvesting and downstream processing that makes them less feasible for microalgal biomass production. In order to solve these constraints and limitations, immobilized cultivation systems could be ideal, in which matrix-based are more popular, but currently there is more focus on the microalgal biofilms, because of better nutrient removal and low cost of harvesting and downstream processing.

The harvesting of biomass involves many mechanical, chemical, electrical and biological methods, in which mechanical methods like centrifugation, sedimentation, filtration and dissolved air flotation are frequently used. Further thickening and dewatering of microalgal biomass is followed in order to increase the market value; sometimes many specific processing procedures are followed that are required for the production of specific industrial products. The production cost of microalgal biomass, depends mainly on the microalgal strain, cultivation system, harvesting and downstream processing.

References

Adey, W.H. 1982. Algal turf scrubber. US patent # US4333263A.
Adey, W.H. 1998. Algal turf water purification method. US patent # US5851398.
Amenorfenyo, D.K., X. Huang, Y. Zhang, Q. Zeng, N. Zhang, J. Ren, et al. 2019. Microalgae brewery wastewater treatment: potentials, benefits and the challenges. Int. J. Environ. Res. Public Health 16: 1–19.
Arbib, Z., J. Ruiz, P. Álvarez-Díaz, C. Garrido-Pérez, J. Barragana and J.A. Perales. 2013. Long term outdoor operation of a tubular airlift pilot photobioreactor and

a high rate algal pond as tertiary treatment of urban wastewater. Ecol. Eng. 52: 143–153.

Bailliez, C., C. Largeau and E. Casadevall. 1985. Growth and hydrocarbon production of *Botryococcus braunii* immobilized in calcium alginate gel. Appl. Microbiol. Biotechnol. 23: 99–105.

Barros, A.I., A.L. Gonçalves, M. Simões and J.C.M. Pires. 2015. Harvesting techniques applied to microalgae: a review. Renew. Sustain. Energy Rev. 41: 1489–1500.

Bernhardt, H. and J. Clasen. 1991. Flocculation of micro-organisms. J. Water Supply Res. Technol. Aqua 40: 76–87.

Bhaskar, P.V. and N.B. Bhosle. 2005. Microbial extracellular polymeric substances in marine biogeochemical processes. Curr. Sci. 88: 45–53.

Bugbee, G.J. and C.R. Frink. 1985. Alum sludge as a soil amendment: effects on soil properties and plant growth. Connecticut Agric. Exp. Stn. Bull. 827: 1–7.

Bussell, S. 2010. Submersible aquatic algae cultivation system. US Patent No. 20100287829 A1, Appl. No.12/812532.

Cabirol, N., E.J. Barragán, A. Durán and A. Noyola. 2003. Effect of aluminium and sulphate on anaerobic digestion of sludge from wastewater enhanced primary treatment. Water Sci. Technol. 48: 235–240.

Campbell, W.H. 2000. Sludge management—future issues and trends. Water Sci. Technol. 41(8): 1–8.

Carvalho, A., L. Meireles and F. Malcata. 2006. Microalgal reactors: a review of enclosed system designs and performances. Biotechnol. Prog. 22: 1490–1506.

Chen, G., P.L. Yue and A.S. Mujumdar. 2002. Sludge dewatering and drying. Drying Technol. 20(4–5): 883–916.

Chevalier, P., D. Proulx, P. Lessard, W.F. Vincent and J. De La Noüe. 2000. Nitrogen and phosphorus removal by high latitude mat-forming cyanobacteria for potential use in tertiary wastewater treatment. J. Appl. Phycol. 12: 105–112.

Chisti, Y. 2007. Biodiesel from microalgae. Biotechnol. Adv. 25: 294–306.

Chisti, Y. 2016. Large-scale production of algal biomass: raceway ponds. pp. 21–40. *In*: F. Bux and Y. Chisti (eds.). Algae Biotechnology. Springer Int., Cham, Switzerland.

Christenson, L. and R. Sims. 2011. Production and harvesting of microalgae for wastewater treatment, biofuels, and bioproducts. Biotechnol. Adv. 29: 686–702.

Danquah, M.K., L. Ang, N. Uduman, N. Moheimani and G.M Forde. 2009. Dewatering of microalgal culture for biodiesel production: exploring polymer flocculation and tangential flow filtration. J. Chem. Technol. Biotechnol. 84: 1078–1083.

De-Bashan, L.E., Y. Bashan, M. Moreno, V.K. Lebsky and J.J. Bustillos. 2002. Increased pigment and lipid content, lipid variety, and cell and population size of the microalgae *Chlorella* sp. when co-immobilized in alginate beads with the microalgae-growth-promoting bacterium *Azospirillum brasilense*. Can. J. Microbiol. 48: 514–521.

Dentel, K.S. 2001. Conditioning, thickening, and dewatering: research update/ research needs. Water Sci. Technol. 44(10): 9–18.

Divakaran, R. and V.N.S. Pillai. 2002. Flocculation of algae using chitosan. J. Appl. Phycol. 14: 419–422.

Greenwell, H.C., L.M.L. Laurens, R.J. Shields, R.W. Lovitt and K.J. Flynn. 2010. Placing microalgae on the biofuels priority list: a review of the technological challenges. J. R. Soc. Interface 7: 703–726.

Hameed, M.S. and O.H. Ebrahim. 2007. Biotechnological potential uses of immobilized algae. J. Agric. Biol. 9: 183–192.

Hoffmann, J.P. 1998. Wastewater treatment with suspended and nonsuspended algae. J. Phycol. 34: 757–763.

Jensen, K.R. 1996. Apparatus for water purification by culturing and harvesting attached algal communities. US patent # US 5527456.

Johnson, M.B. and Z. Wen. 2010. Development of an attached microalgal growth system for biofuel production. Appl. Microbiol. Biotechnol. 85: 525–534.

Kawano, Y., T. Saotome, Y. Ochiai, M. Katayama, R. Narikawa and M. Ikeuchi. 2011. Cellulose accumulation and cellulose synthase gene are responsible for cell aggregation in the cyanobacterium *Thermosynocococcus vulcanus* RNK. Plant Cell Physiol. 52: 957–966.

Kim, B.J. and E.D. Smith. 1997. Evaluation of sludge dewatering reed beds: a niche for small systems. Water Sci. Technol. 35(6): 21–28.

Kumar, A. 2018. Assessment of cyanobacterial diversity in paddy fields and their capability to degrade the pesticides. Babasahaeb Bhimrao Ambedkar University, Lucknow, India.

Kumar, A. and J.S. Singh. 2016. Microalgae and cyanobacteria biofuels: a sustainable alternate to crop-based fuels. pp. 1–20. *In*: J.S. Singh, D.P. Singh (eds.). Microbes and Environmental Management. Studium Press Pvt. Ltd. New Delhi, India.

Kumar, A. and J.S. Singh. 2017. Cyanoremediation: a green-clean tool for decontamination of synthetic pesticides from agro- and aquatic-ecosystems. pp. 59–83. *In*: J.S. Singh, G. Seneviratne (eds.). Agro-Environmental Sustainability, Vol. II: Managing Environment Pollution. Springer Int., Cham, Switzerland.

Kumar, A. and J.S. Singh. 2020. Biochar coupled rehabilitation of cyanobacterial soil crusts: a sustainable approach in stabilization of arid and semiarid soils. pp. 167–191. *In*: J.S. Singh, C. Singh (eds.). Biochar Applications in Agriculture and Environment Management. Springer Int., Cham, Switzerland.

Kumar, A. and J.S. Singh. 2020. Microalgal bio-fertilizers. *In*: E. Jacob-Lopes, M.M. Maroneze, M.I. Queiroz, L.Q. Zepka (eds.). Handbook of Microalgae-based Processes and Products. Academic Press, Cambridge, US, In press.

Kumar, A., S. Kaushal, S.A. Saraf and J.S. Singh. 2017. Cyanobacterial biotechnology: an opportunity for sustainable industrial production. Clim. Chang. Environ. Sustain. 5(1): 97–110.

Kumar, A., S. Kaushal, S.A. Saraf and J.S. Singh. 2018. Microbial bio-fuels: a solution to carbon emissions and energy crisis. Front. Biosci. (Landmark) 23: 1789–1802.

Kumar, A., S. Kaushal, S.A. Saraf and J.S. Singh. 2018. Screening of chlorpyrifos (CPF) tolerant cyanobacteria from paddy field soil of Lucknow, India. Int. J. Appl. Adv. Sci. Res. 3(1): 100–105.

Kumar, H., P. Yadava and J. Gaur. 1981. Electrical flocculation of the unicellular green alga *Chlorella vulgaris* Beijerinck. Aquat. Bot. 11: 187–195.

Lam, M.K. and K.T. Lee. 2012. Immobilization as a feasible method to simplify the separation of microalgae from water for biodiesel production. Chem. Eng. J. 191: 263–268.

Lavoie, A. and J. de la Noüe. 1987. Harvesting of *Scenedesmus obliquus* in wastewaters: auto or bioflocculation? Biotechnol. Bioeng. 30: 852–859.

Lee, A.K., D.M. Lewis and P.J. Ashman. 2008. Microbial flocculation, a potentially low-cost harvesting technique for marine microalgae for the production of biodiesel. J. Appl. Phycol. 21: 559–567.

Mata, T.M., A.A. Martins and N.S. Caetano. 2010. Microalgae for biodiesel production and other applications: a review. Renew. Sustain. Energy Rev. 14: 217–232.

Middlebrooks, E.J., D.B. Porcella, R.A. Gearheart, G.R. Marshall, J.H. Reynolds and W.J. Grenney. 1974. Techniques for algae removal from wastewater stabilization ponds. J. Water Pollut. Control Fed. 46: 2676–2695.

Molina-Grima, E., F.G.A. Fernández, F. García-Camacho and Y. Chisti. 1999.

Photobioreactors: light regime, mass transfer, and scale-up. J. Biotechnol. 70: 231–247.

Mulbry, W., S. Kondrad and J. Buyer. 2008. Treatment of dairy and swine manure effluents using freshwater algae: fatty acid content and composition of algal biomass at different manure loading rates. J. Appl. Phycol. 20: 1079–1085.

Oh, H.M., S.J. Lee, M.H. Park, H.S. Kim, H.C. Kim, J.H. Yoon, et al. 2001. Harvesting of *Chlorella vulgaris* using a bioflocculant from *Paenibacillus* sp. AM49. Biotechnol. Lett. 23: 1229–1234.

Pal, S., D. Mal and R. Singh. 2005. Cationic starch: an effective flocculating agent. Carbohydr. Polym. 59: 417–423.

Passow, U. and A.L. Alldredge. 1995. Aggregation of a diatom bloom in a mesocosm: the role of transparent exopolymer particles (TEP). Deep Sea Res Part II: Top Stud Oceanogr. 42: 99–109.

Posten, C. 2016. Aquatische biomasse: verfahrenstechnische grundlagen, pp. 254–272. *In*: M. Kaltschmitt, H. Hartmann, H. Hofbauer (eds.). Energie aus Biomasse: Grundlagen, Techniken und Verfahren. Springer Berlin Heidelberg, Berlin, Heidelberg.

Pragya, N., K.K. Pandey and P.K. Sahoo. 2013. A review on harvesting, oil extraction and biofuels production technologies from microalgae. Renew. Sustain. Energy Rev. 24: 159–171.

Rulkens, W.H. 2004. Sustainable sludge management—what are the challenges for the future? Water Sci. Technol, 49(10): 11–19.

Sheehan, J., T. Dunahay, J. Benemann and P. Roessler. 1998. A look back at the US department of energy's aquatic species program—biodiesel from algae. NREL, USDoE, Report no. # NREL/TP-580-24190.

Shelef, G., A. Sukenik and M. Green. 1984. Microalgae harvesting and processing: a literature review. US TRDF Report No. # SERI/STR-231-2396.

Shen, Y., W. Yuan, Z.J. Pei, Q. Wu and E. Mao. 2009. Microalgae mass production methods. Trans. ASABE 52: 1275–1287.

Singh, J.S., A. Kumar, A.N. Rai and D.P. Singh. 2016. Cyanobacteria: a precious bio-resource in agriculture, ecosystem, and environmental sustainability. Front. Microbiol. 7: 529.

Singh, J.S., S. Koushal, A. Kumar, S.R. Vimal and V.K. Gupta. 2016. Book review: microbial inoculants in sustainable agricultural productivity, Vol. II: Functional application. Front. Microbiol. 7: 2105.

Singh, J.S., A. Kumar and M. Singh. 2019. Cyanobacteria: a sustainable and commercial bio-resource in production of bio-fertilizer and bio-fuel from wastewaters. Environ. Sustain. Indic. 3: 100008.

Staats, N., B. De Winder, L. Stal and L. Mur. 1999. Isolation and characterization of extracellular polysaccharides from the epipelic diatoms *Cylindrotheca closterium* and *Navicula salinarum*. Eur. J. Phycol. 34: 161–169.

Subramanian, G. and S.S. Sundaram. 1986. Induced ammonia release by the nitrogen fixing cyanobacterium *Anabaena*. FEMS Microbiol. Lett. 37: 151–154.

Sukenik, A. and G. Shelef. 1984. Algal autoflocculation-verification and proposed mechanism. Biotechnol. Bioeng. 26: 142–147.

Teixeira, M.R. and M.J. Rosa. 2006. Comparing dissolved air flotation and conventional sedimentation to remove cyanobacterial cells of *Microcystis aeruginosa*: part I: the key operating conditions. Sep. Purif. Technol. 52: 84–94.

Uduman, N., Y. Qi, M.K. Danquah, G.M. Forde and Hoadley. 2010. Dewatering of microalgal cultures: a major bottleneck to algae-based fuels. J. Renew. Sustain. Energy 2: 012701.

US DOE. 1985. Review and evaluation of immobilized algae systems for the production of fuels from microalgae. U.S. DOE Report no.: SERI/STR-231-2798, Alexandria, Virginia (VA).

US DOE. 2010. National algal biofuels technology roadmap. US DOE Report no. # DOE/EE-0332, Washington, D.C.

Vandamme, D., I. Foubert, B. Meesschaert and K. Muylaert. 2009. Flocculation of microalgae using cationic starch. J. Appl. Phycol. 22: 525–530.

Vonshak, A. and A. Richmond. 1988. Mass production of the blue-green alga *Spirulina*: an overview. Biomass 15(4): 233–247.

Wakeman, R.J. 2007. Separation technologies for sludge dewatering. J. Hazard. Mater. 144(3): 614–619.

Wang, S.-K., A.R. Stiles, C. Guo and C.-Z. Liu. 2015. Harvesting microalgae by magnetic separation: a review. Algal Res. 9: 178–185.

Woertz, I., A. Feffer, T. Lundquist and Y. Nelson. 2009. Algae grown on dairy and municipal wastewater for simultaneous nutrient removal and lipid production for biofuel feedstock. J. Environ. Eng. 135: 1115–1122.

Wolfstein, K. and L.J. Stal. 2002. Production of extracellular polymeric substances (EPS) by benthic diatoms: effect of irradiance and temperature. Mar. Ecol. Prog. Ser. 236: 13–22.

Wood, A. 1987. A simple wastewater treatment system incorporating the selective cultivation of a filamentous algae. Water Sci. Technol. 19: 1251–1254.

Microalgal Biomass: An Opportunity for Sustainable Industrial Production

Introduction

Microalgae are a large group of phototrophic organisms that have high production rate, minimum growth requirements and resilience to adverse conditions; which could be an excellent feedstock for the production of various industrial products (Fig. 12.1) (Kumar and Singh 2016, Singh et al. 2016a, b, Kumar et al. 2017). They are very rich in carbohydrates, proteins, lipids, vitamins and minerals; which makes them a potential, cheap and sustainable raw material for the development of human supplements, animal-feed and aquaculture feed-stocks. Microalgae are found to be an excellent and rich source of omega-3 fatty acids and β-carotene, and they could be a better alternative to the traditional source of these compounds. Many Asian countries have known the nutritional importance of these microalgae from centuries and they were cultivated for the nutritional purposes before the modern exploration for microalgae for healthy foods (Kumar and Singh 2017, Kumar 2018, Kumar et al. 2018a, b, Singh et al. 2019, Kumar and Singh 2020a, b).

Microalgae are known to synthesize significant amounts of the polyunsaturated fatty acids such as EPA, DHA, ERA and GLA, precursors of eicosanoids; these are known to signal the molecule in different human metabolic activities. They also produce such compounds which have antioxidant and anti-bacterial properties that could be a new and alternative research field to discover and develop new drugs. Other interesting compounds from the microalgal source are mycosporine-like amino acids that are known to provide protection from harmful UV radiation; which have a potential role in the cosmetics industry.

They are known to produce a wide variety of pigments: chlorophyll (green), carotenoids (yellow, orange, reddish brown) and phycobiliproteins (blue or red). Although they are not a suitable and cost effective source for chlorophyll extraction, they could be excellent feed-stocks for the extraction of carotenoids and phycobiliproteins. Beside pigments, they store the polyhydoxy butyrate (PHBs) granule, which have great thermoplastic

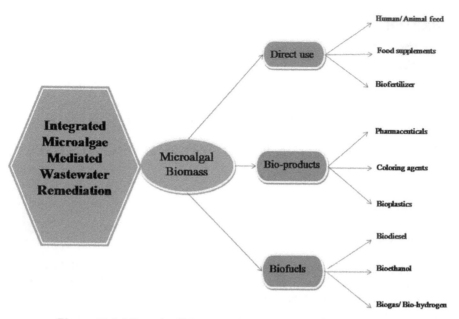

Figure 12.1: Microalgal biomass: An opportunity for sustainable industrial production.

processability, water resistance and complete biodegradability that makes them an eco-friendly alternative to conventional plastics.

Due to the capability to fix atmospheric nitrogen and mobilization of inorganic phosphorous, microalgae could be a better and organic alternative to inorganic fertilizers. Further they have the capacity to secrete plant-growth promoting substances like vitamins, amino acids, hormones and some anti-microbial compounds that provide a new horizon in the field of sustainable and organic agriculture. Having a good amount of lipid content in their cells, they can be used to develop biofuels like biodiesel, bioethanol, biomethane and biohydrogen, which not only reduce the pressure on fossil fuels but also give a more environmentally sound energy source in the wake of fossil fuel mediated risk of severe air pollution, global warming and ozone layer deterioration.

Food Supplements and Animal Feed

Microalgal cells have significant amount of carbohydrates, proteins, lipids, vitamins and minerals that could be an excellent source of food supplements (Table 12.1) (Sathasivam et al. 2017, Becker 2013). Since 2000 years, the Chinese have consumed microalgae *Nostoc* sp.as healthy food; and later other microalgae *Chlorella* and *Spirulina* species became the choice of functional healthy food in Taiwan, Japan and Mexico (Sathasivam et al. 2017). Having a high protein content and nutritional value, microalgal species like *Spirulina plantesis*, *Chlorella* sp., *Dunaliella terticola*, *Dunaliella saline* and *Aphanizomenon*

Table 12.1: Composition of some microalgal genera
(Modified from Tibbetts 2018, Koyande et al. 2019)

Microalgae	Carbohydrate (%)	Lipid (%)	Protein (%)
Anabaena cylindrica	25-30	4-7	43-56
Botryococcus braunii	2	33	40
Chlamydomonas rheinhardii	17	21	48
Chlorella pyrenoidosa	26	2	57
Chlorella vulgaris	12-17	10-22	41-58
Dunaliella bioculata	4	8	49
Dunaliella salina	32	6	57
Euglena gracilis	14-18	14-20	39-61
Porphyridium cruentum	40-57	9-14	28-39
Prymnesium parvum	25-33	22-39	28-45
Scenedesmus dimorphus	21-52	15-40	8-18
Scenedesmus obliquus	10-17	12-14	50-56
Scenedesmus quadricauda	-	19	47
Spirogyra sp.	23-64	11-21	6-20
Spirulina maxima	13-16	6-7	60-71
Spirulina platensis	25-30	4-7	43-56
Synechoccus sp.	15	11	63
Tetraselmis maculata	15	3	52
Crypthecodinium	-	20-56	15-23
Desmodesmus	-	1	21-27
Haematococcus	26-55	7-67	3-48
Isochrysis	13-18	16-53	20-45
Nanochloropsis	8-36	2-68	18-48
Pheodactylum	8-25	7-57	30-49
Schizochytrium	32-39	15-71	12-39

flos-aquae are commercially well exploited as healthy food in many countries including Chile, Mexico, Peru and the Philippines; currently it is available in the form of capsules, tablets, powders and liquids (Pulz and Gross 2004, Liang et al. 2004, Soletto 2005, García et al. 2017). They could be a potential source for vitamins such as vitamin A, B1, B2, B6, B12, C and E and minerals such as potassium, iron, magnesium, calcium and iodine (Kulshreshtha et al. 2008, Prasanna et al. 2010, Becker 2013).

Due to the nutritional benefits of microalgae, they can also be mixed with candies, biscuits, gums, snacks, noodles, breakfast cereals, wine and other beverages (Liang et al. 2004). The World Health organization (WHO) labeled the *Spirulina* sp. as a 'superfood' because of its interesting nutritional and pharmaceutical values (Soletto 2005). It is estimated that a spoonful (approximately 7 g) dried biomass of *Spirulina* comprises of about 4 g of protein, 1 g of fat which includes PUFAs like omega-3 and omega-6 fatty acids; and contains a significant amount of vitamin B_1, B_2 and B_3 up to11, 15 and 4% of Required Daily Allowance (RDA) respectively. It also constitutes copper and iron up to 21 and 11% of RDA and small amounts of other minerals such as magnesium, manganese and potassium respectively (Khan et al. 2005, USDA, 2018). Capelli and Cysewski (2010) reported that *Spirulina* was found to be very nutritive as it contains 670% more protein as compared to tofu, 180% more calcium compared to milk, 5100% more iron as compared to spinach and 3100% more β-carotene as compared to carrots.

Like human food supplements, microalgae could be a significant source of food for aquaculture, pigs, poultry, cattle and other animals. The following microalgal species such as *Spirulina, Scenedesmus, Chlorella, Nannochloropsis, Dunaliella, Nitzchia, Tetraselmis, Isochrysis, Pavlova, Navicula, Haematococcus, Schizochytrium, Chaetoceros* and *Crypthecodinium* are found to be suitable as animal feed for terrestrial and aquatic animals (Yaakob et al. 2014, García et al. 2017, Madeira et al. 2017). Due to its high nutritional qualities i.e., proteins, carbohydrates, vitamins and fatty acids, microalgae could be a better alternative of meal for aquaculture and could be used for aquaculture production of molluscs, crustaceans, shrimp and fish farming (Batista et al. 2013, Chauton et al. 2015). They not only provide valuable nutrition for aquatic organisms but also enhance their immunostimulant and disease resistant properties. Burr et al. (2011) suggested that before selecting the microalgae as an ingredient of fish feed; there should be a correct examination of the nutritional value and digestibility of each microalgal biomass.

Pharmaceuticals

Microalgal lipids contain significant amounts of polyunsaturated fatty acids (PUFAs), which are gaining significance in pharmaceuticals (Koller et al. 2014, Polishchuk et al. 2015). The important microalgal PUFAs are eicosapentaenoic acid (EPA), arachidonic acid (ARA), docosahexaneoic acid (DHA) and Gamma-Linolenic Acid (GLA) which play an important role the human metabolism as they are precursors of the signaling molecules known as eicosanoids. Eicosanoids are known for their function of signaling molecules in mammals which participate in cell growth, immune response, inflammation and blood pressure regulation (Karmali 1996). Although the human body is able to synthesize all the precursors of eicosanoids, but their amounts are found to be sub-optimal, and to make up the total requirement, fish or fish oil are recommended as an excellent source. There are various microalgae species such as *Nannochloropsis* sp. and *Phaeodactylum tricornutum* for EPA (Koller et al. 2014, Polishchuk et al. 2015), *Crypthecodinium cohnii* and *Schizochytrium pavlova* lutheri for DHA (Koller et al. 2014), *Spirulina*

(*Arthrospira*) for GLA and ARA (Mendes et al. 2006) that have been commercially investigated for the production of these fatty acids.

Apart from this, many microalgal strains like *Nostoc, Spirulina, Chlorella, Botryococcus, Dunaliella, Phaeodactylum, Haematococcus* and *Chaetoceros* were investigated for antioxidant properties (Table 12.2). From these studies, it was evident that microalgae could be an alternative for potent antioxidants in which carotenoids are the major ones (Herrero et al. 2006, Li et al. 2007, Ceron et al. 2007, Goh et al. 2010). Some microalgae are also used for the commercial production of carotenoid antioxidants such as astaxanthin from *Haematococcus* and β-carotene from *Dunaliella*; and further they are used as additives in food and feed applications, dietary supplements or used in cosmetics (Pulz and Gross 2004, Spolaore et al. 2006, Takaichi 2011).

Table 12.2: Antioxidant compounds from microalgae

Microalgae	Bioactive compounds	References
Padina gymnospora	Sulfated polysaccharides	De Souza et al. 2007
Bifurcaria bifurcata	Phenols	Zubia M et al. 2009
Dunaliella salina	Carotenoid	Herrero et al. 2006
Spirulina platensis	Antioxidants—carotenoids	Jaime et al. 2005
Haematococcus pluvialis	Nutraceuticalastaxanthin	Minhas et al. 2016
Spirulina platensis	Phycocyanin	Minhas et al. 2016
Chlorococcum humicola	Bioactive compounds	Sanmukh et al. 2014
Chlorella vulgaris	Biomass, Ascorbic acid	Pulz and Gross 2004, Priyadarshani et al. 2012
D. salina	Carotenoid, β carotene	Pulz and Gross 2004, Priyadarshani et al. 2012
H. pluvialis	Carotenoids, astaxanthin	Pulz and Gross 2004, Priyadarshani et al. 2012
D. salina	Lutein	Talero et al. 2015
Synechocystis sp.	Zeaxanthin	Talero et al. 2015
S. platensis	Phycobiliproteins	Talero et al. 2015
C. pyrenoidosa	Peptides	Talero et al. 2015
S. maxima	Phenolic compounds	Talero et al. 2015
Porphydium sp.	Tocopherols	Talero et al. 2015

The problem of multi-drug-resistance of many bacteria appears such as, *Staphylococcus aureus* Enterococci and Enterobacteriaceae to some antibiotics methicillin, vancomycin and AmpC respectively (Reinert et al. 2007); there is an urgent need of research on new antibiotics that treat these multi-drug-resistance bacterial infections. Microalgae including cyanobacteria are a great source of many biologically active compounds such as polysaccharides, fatty acids, phycobilins, phenols, terpenes, halogenated aliphatic compounds and sulfur containing hetero-cyclic compounds; these are identified as potential antimicrobial compounds (Li et al. 2007, Kannan et al. 2010). For the development of new antibiotics, researchers are continuously engaged in screening microalgal extracts for their antibacterial activity (Kreitlow et al. 1999, Skulberg 2000, Biondi et al. 2008), which could be effective against various bacterial infections.

Besides antibacterial, microalgae are found to be very promising in the development of anticancer metabolites. Patterson et al. (1991) isolated the potent anticancer metabolite cryptophycin 1 from *Nostoc* sp. GSV224 in Moore's laboratory. Another anticancer agent Scytonemin obtained from *Stigonema* sp. showed anti-proliferative and anti-inflammatory activities. Foster et al. (1999) successfully investigated the ability of the cell extract of *Calothrix* isolate to inhibit the growth of human cancer cells, and found that Calothrixin A (I) and B (II), pentacyclic metabolites are helpful in the inhibition of their growth in a dose-dependent manner. Some other microalgal metabolites such as Apratoxins, Largazole from *Symploca* sp. Curacin-A from *Lyngbya majuscule* were found to be potential anti-proliferative substances and could be useful in the inhibition of a variety of cancer cells.

There is also the presence of Mycosporine-like Amino Acids (MAAs) in microalgae which are primarily known for providing protection to organisms from solar UV radiation (Stochaj et al. 1994, Carroll and Shick 1996, Bhandarnayake 1998). MAAs are water soluble pigments, containing acyclohexenone or cyclohexeniminechromophore which is conjugated with the nitrogen component of an aminoacid or its imino alcohol. They have an absorption maximum of about 310-360 nm (Kedar et al. 2002, Volkmann and Gorbushina 2006). Some common MAAs are mycosporine Gly, asterina 330, porphyra 334 and shinorine, that are found in many microalgal species. Prasanna et al. (2010) reported that MAAs are very often found in terrestrial cyanobacterial species.

Natural Colorants

Microalgae contain many types of pigments such as chlorophylls, carotenoids and phycobiliproteins, which play a role in photosynthetic activities. It is well known that chlorophylls are green colored pigments, while carotenoids are yellow, orange, reddish or brown colored (Table 12.3). Besides chlorophyll and carotenoids, microalgae especially cyanobacteria and red algae also comprise blue or red phycobilins pigments. Chlorophylls and carotenoids are lipophilic molecules that can be easily extracted as free pigments; while phycobilins are hydrophilic molecules that are covalently bound to the protein, and on heat treatment (i.e., cooking) they lose their color due to denaturation of the protein.

Table 12.3: Pigments extracted from cyanobacteria and microalgae

Microalgae	Pigments	Reference
Spirulina platensis	Phycocyanin, C-phycocyanin	De Morais et al. 2015, Ibanez et al. 2013, Mostafa et al. 2012
Haematococcus pluvialis	Astaxanthin, lutein, zeaxanthin, canthaxanthin, lutein, β-carotene	De Morais et al. 2015, Markou et al. 2013,
Chlorella sp.	Carotenoids	De Morais et al. 2015, Ibanez et al. 2013
C. vulgaris	Canthaxanthin, astaxanthin	Priyadarshani et al. 2012, Mostafa et al. 2012,
C. ellipsoidea	Zeaxanthin, violaxanthin	De Morais et al. 2015, Amaro et al. 2013
C. azofingiensis	Astaxanthin	De Morais et al. 2015, Markou et al. 2013
C. protothecoides	Lutein, zeaxanthin, canthaxanthin	De Morais et al. 2015, Markou et al. 2013
C. pyrenoidosa	Lutein	De Morais et al. 2015, Plaza et al. 2009,
Dunaliella salina	β-Carotene	De Morais et al. 2015, Markou et al. 2013

Microalgal chlorophylls are like the chlorophylls (a and b) as in plants and are relatively similar in chemical nature. Microalgal chlorophyll could be applied in food coloring and cosmetics applications (Koller et al. 2014), but extraction of chlorophyll a and b from microalgae is probably more costly than that extracted from grass and vegetables. Carotenoids are categorized into two groups: (a) carotenes which do not have oxygen moiety, (b) xanthophylls which have oxygen and are primarily yellow, orange, reddish or brown colored. They function as antioxidants in all organisms, which help in removing harmful reactive oxygen species (primarily singlet oxygen) from the cells.

There are more than 1000 different carotenoids in nature, but two carotenoids i.e., β-carotene and astaxanthin are considered the prominent ones. β-carotene (provitamin A) are found in all photosynthetic organisms, while astaxanthin are restricted in their presence typically in algae including microalge. β-carotene are known for their role in the biosynthesis of rhodops in the pigment which is required for the retina and β-carotene lacking in a diet could be a major cause of blindness. Brányiková et al. (2011) observed that halophilic microalgae *Dunaliella salina* contains up to 14% β-carotene content which makes them a potential candidate for the production of β-carotene. Astaxanthin are known as their role in antibody production, anti-cancer therapies and protection from UV radiation-related damages (Koller et al. 2014). For microalgal production of astaxanthin, *Haematococcus pluvialis* is found to be most suitable for industrial production. It could be used as natural food colorants or further added to the feed of salmon fishes to give better color (Wu and Shi 2007).

Phycobiliproteins are water soluble and fluorescent proteins, which are categorized in to three basic groups: (a) phycoerythrin (red pigment), (b) phycocyanin (blue pigment), and (c) allophycocyanin (bluish green pigment).They make up the important part of photosynthetic apparatus in cyanobacteria as accessory pigments along with chlorophyll a. Due to fluorescence and high solubility, phycobiliproteins are used as food colorants, as chemical tags in cosmetics and biochemical research (Koller et al. 2014, Arad and Yaron 1992). Prasanna et al. (2010) observed that the proportion of phycocyanin in all cyanobacterial phycobiliproteins, alone makes about 20% of total dry weight, and is considered as an extensive pigment in bioindustry.

Bio-plastics (Poly-hydroxybutyrate)

Poly-hydroxybutyrate (PHB) is an intracellular storage compound particularly in prokaryotic organisms including cyanobacteria, which could be an ideal biodegradable plastic material because of its complete microbial mineralization into CO_2 and water (Mallick et al. 2007, Abed et al. 2008, Melnicki et al. 2009, Liu et al. 2010). PHB could be applied in biomedical and biopharmaceutical applications (Sudesh 2004, Abed et al. 2008). It could be a potential alternative to conventional plastics due to its properties including thermoplastic processability, water resistance and complete biodegradability; which follows new waste management strategies (Ben Rebah et al. 2007, Chen et al. 2008).

Due to minimal nutritional needs and photoautotrophic nature, cyanobacteria are considered to be a potential alternative machinery for the production of PHBs (Dias et al. 2008, Jyotsana et al. 2010). Currently many microalgal genera such as *Spirulina* sp., *Aphanothece* sp., *Gloeothece* sp., and *Synechococcus* sp. also showed the occurrence of PHB (Fernandez-Nova et al. 2008). For the detection of PHB in different cyanobacterial species, several techniques are applied like electron microscope for *Gleocapsa* sp. and *Nostoc* sp. (Jau et al. 2005, Shrivastav et al. 2010) and gas–liquid chromatography for *Oscillatoria limosa* and *Gloeothece* sp. (Miyake et al.1996). Hein et al. (1998) and Quillaguaman et al. (2010) reported the presence of PHB synthase genes and further studied its characteristics in *Synechocystis* sp. There are very few reports about the presence of PHB in microalgal species.

Recently cyanobacterial based plastics are in trend with the need of bioplastics over the use of traditional petroleum-derived plastics (Castilho et al. 2009, Chen and Li 2008). Due to the high yield and ability to grow in different environments, they could be a potential and sustainable feedstock for bioplastics production (Sudesh et al. 2000, Seiichi and Doi 2004). The use of cyanobacterial plastics could provide an opportunity of minimizing the risk of the plastic problem and simultaneously neutralizing greenhouse gas emissions from their production facilities (Wang et al. 2008, Senthikumar and Prabhakaran 2006, Dias et al. 2006, Lee et al. 2008).

Biofertilizers

Cyanobacteria (prokaryotic microalgae) are known for their ability of nitrogen (N_2) fixation either by free living or symbiotic species. For nitrogen fixation, they have modified cells designated as heterocyst, which are thick-walled cells compared with normal vegetative cells. Heterocyst cells contain nitrogenase enzyme, which act as a catalyst in the conversion of the molecular nitrogen (N_2) into ammonia and other nitrogenous forms. This fixed nitrogen might be available to the soil in the form of free amino acids, polypeptides, vitamins, and auxin like plant-growth promoting substances; either through secretion or microbial degradation on the death of cyanobacterial cells (Subramanian and Sundaram 1986, Venkataraman 1993). It is not necessary that only heterocystous cyanobacterial genera show a nitrogen-fixing ability, there are several non-heterocystous cyanobacterial genera, which also reported nitrogen fixation.

It is predicted that cyanobacteria is able to provide about 20-30 kg N ha^{-1} season^{-1} along with significant organic matter to the soil (Issa et al. 2014). In many Asian countries like China, Vietnam, India, etc., the association of cyanobacteria-Azolla is successfully used as nitrogen fertilizers in tropical rice fields (Singh and Singh 1987). There are a number of studies where cyanobacteria (*in vitro* conditions) was inoculated in the wheat crop in laboratory conditions, and reported the increase in the plant shoot/root length, dry weight and yield (Spiller and Gunasekaran 1990, Obreht et al. 1993, Karthikeyan et al. 2007, 2009), but there are not enough data related to its agronomic efficiency (Gantar et al. 1991b, 1995).

Besides nitrogen fixation, microalgae including cyanobacteria can also be helpful in enhancing the plant bioavailable phosphorus through solubilization and mobilization of the insoluble organic phosphate that occurs in the soil. There are various reports regarding the ability of microalgae including cyanobacteria in solubilization of an insoluble form like $(Ca)_3(PO_4)_2$, $AlPO_4$, $FePO_4$ and $Ca_5(PO_4)_3OH$ in soils and sediments (Cameron and Julian 1988). Regarding the mechanism of phosphate solubilization, two hypotheses are proposed:

(a) The first hypothesis emphasizes the ability of microalgae to produce a chelator for Ca^{2+} ions that influence the dissolution reaction of insoluble calcium phosphates to release the phosphate ions without any change in the pH (Cameron and Julian 1988) as mentioned below:

$$Ca_{10}(OH)_2(PO_4)_6 \longrightarrow Ca^{2+} + 2OH^- + 6PO_4^{3-}$$

The second hypothesis emphasizes that the microalgae tend to release H_2CO_3 and other organic acids that could dissolve insoluble calcium phosphorus (Bose et al. 1971) as given below:

$$Ca_3(PO_4)_2 + 2H_2CO_3 \longrightarrow 2CaHPO_4 + Ca(HCO_3)_2$$

In contradiction to these two hypotheses, a third possibility has also received much attention these days. It was suggested that when inorganic phosphate is solubilized either by the action of chelator or organic acids; the released PO_4^{3-} are sequestrated by the microalgae for their own nutrition needs. On the death of these microalgal cells, this cell locked PO_4^{3-} is released in the soil and is readily available to plants and other organisms after mineralization (Mandal et al. 1999).

Microalgae especially cyanobacteria are well known to secrete plant-growth promoting substances extracellularly; which could be either plant hormones i.e., auxin, gibberellins, cytokinin, abscisic acids (Singh and Trehan1973) or vitamins (particularly vitamin B), amino acids, antibiotics and toxins. There are many studies related to inoculation of microalgae including cyanobacteria related to study the plant-growth promoting ability in paddy crop and it was found that it can enhance seed germination, root and shoot growth in paddy crop (Singh and Trehan 1973, Mishra and Kaushik1989). Obreht et al. (1993) and Gantar et al. (1995) reported that co-inoculation of cyanobacteria in wheat crop helped in improving the root dry weight and chlorophyll content. Due to their natural inhabitation, regional diversity, simple nutritional needs and ability to survive in extreme conditions, they could be exploited commercially as plant-growth promoters for agricultural application (Ruffing 2011).

Besides plant-growth promotion, cyanobacteria are also known for production of various biologically active compounds which have antimicrobial activities that could help in controlling many plant diseases (Teuscher et al. 1992, Dahms et al. 2006). These bioactive compounds could be fatty acids, indoles, amides, alkaloids, lipopeptides and polyketides, (Abarzua et al. 1999, Burja et al. 2001). Kulik (1995) reported that cyanobacterial cell extract could help in controlling the incidence of *Botrytis cinerea* causing gray mold rot in vines and *Erysiphe polygoni* causing powdery mildew and damping off disease. Besides this, *Nostoc muscorum* cell extract could also reduce the growth of saprophyte such as *Cunninghamella blakesleeana*, *Chaetomium globosum*, *Aspergillus oryzae* and plant fungal pathogens such as *Sclerotiana sclerotium* (cause cottony rot of vegetables and flowers, *Rhizoctonia solani* (cause root and stem rots) (Kulik 1995). It is also reported that several compounds from cyanobacteria like Fischerellin from *Fischerella muscicola* were found to be effective against some plant fungal pathogens fungi such as *Erysiphe graminis* (powdery mildew), *Uromyces appendiculatus* (brown rust), *Pyricularia oryzae* (rice blast) and *Phytophthora infestans*, but less effective against *Pseudocercosporella herpotrichoides* (stem break) and *Monilinia fructigena* (brown rot) (Hagmann and Juttner 1996, Papke et al. 1997).

Biodiesel

Biodiesel is primarily synthesized through the transesterification of lipids mainly triglycerides in the presence of an acyl acceptor such as methanol, ethanol or isopropanol; and a catalyst such as NaOH, KOH, CH_3ONa and CH_3OK (Ghadge and Raheman 2006, Meher et al. 2006). Microalgae genera like *Botryococcus, Chlorella, Nannochloropsis, Neochloris, Nitzschia, Scenedesmus,* and *Dunaliella* comprises high amounts of lipids under optimal conditions (Koller et al. 2014). To produce biodiesel from microalgae, there is always the need of neutral lipids like triacylglycerols, which are hydrophobic molecules lacking charged groups. These fatty acids of triacylglycerols undergo transesterification with an acid or alkali to convert it in to biodiesel and glycerol as byproducts. This process of transesterification can be followed directly to convert microalgal biomass in to biodiesel (Johnson and Wen 2009); or carried out in two steps involving extraction of lipids from microalgal biomass and then transesterified (Mulbry et al. 2009). For the extraction of lipids from microalgal biomass, many solvents such as methanol, isopropanol and petroleum ether are used. The production of biodiesel through direct transesterification of microalgal biomass has been found to be a fast and cost-effective process. Microalgal biodiesel could provide a potential and eco-friendly alternative to the fossil fuel-based diesel, but there is need of microalgal strains which have a high growth rate and oil content to increase the efficiency and sustainability of the microalgal biodiesel (Xiao-li et al. 2017).

Bio-ethanol

Microalgae have significant amounts of carbohydrates in their cells that can be used as a carbon source to convert it into ethanol via the microbial fermentation process. In this process, ethanol is produced as the main product, while CO_2 and water as byproducts. There are many microalgal species such as *Synechococcus* sp. PCC 7002, *Chlorella* sp., *Chlamydomonas* sp., *Oscillatoria* sp., *Cyanothece* sp. and *Spirulina platensis*, which are found to promising candidates for ethanol production (Ueno et al. 1998, Koller et al. 2014). Trivedi et al. (2015) reported that microalgal species *Chlorococcum* and *Chlorella vulgaris* have shown good conversion rates in ethanol production. It also investigated the ability of *Synechococcus* sp. PCC 7002 to synthesize carbohydrate content in nitrogen-starved conditions, and found that it might be able to accumulate fermentable carbohydrates up to 60 % of the dry weight.

In the conversion of microalgal biomass in to ethanol involves several steps like cultivation and harvesting of biomass (earlier described in Chapter 11), preparation of biomass for ethanol production, fermentation and extraction of ethanol. To prepare the microalgal biomass for ethanol production, the cell wall of microalgae includes disintegrating through many methods such as mechanical, chemical or enzymatic methods; enhancing the availability of more and more carbohydrate content for the fermentation process. Then the disrupted biomass supplied is with yeast *Saccharonmyces cerevisiae* to carry out the fermentation of the carbohydrate content. In the end, the distillation process is carried out to extract ethanol from fermented biomass (Amin 2009).

There are some limitations in yeast-based fermentation as they provide low yield and are further unable to ferment the cellulosic material in to ethanol (which makes up a significant part of the microalgal cell wall); these limitations might be barriers in the large-scale production of ethanol from microalgal biomass (Koller et al. 2014). There is a need of new technologies or efficient fermentation microorganisms that could first hydrolyze the cellulose and hemicellulose material into sugars and then make them available for fermentation (Hamelinck et al. 2005); this could help in increasing the overall ethanol production from microalgal biomass.

Biomethane or Biogas

Biomethane or biogas is primarily a mixture of the CH_4 55.0–65.0% and CO2 (30–45 %) with a small amount of hydrogen sulfide (H_2S) and water vapor (Cooney et al. 2007, Kapdi et al. 2005, Costa and Morais 2011, Trivedi et al. 2015); may also contain traces of H_2 and CO (Bailey and Ollis 1986). The organic content (i.e., carbohydrates, lipids and proteins) in microalgal biomass are first hydrolyzed and converted into their monomers, then anaerobic digestion is followed to produce biomethane (Table 12.4). As CO_2 makes the second most component of biomethane, it can be separated out to improve the quality of biomethane, leading to the enhancement in the calorific value of the biogas (Hankamer et al. 2007). This purified CH_4 or biomethane is further compressed to use as compressed natural gas in vehicles, providing a more environmental friendly alternative to conventional fossil fuels.

The factors which affect biomethane yield from the microalgal biomass includes biomass digestibility, requirement of pretreatment, different biomethane yield from various microalgal species and other characteristics of the biomass. Compared to biodiesel, biogas is considered as a low-value energy product, but its production from microalgal biomass is simpler than extraction of oil and transesterification of the fatty acids to produce biodiesel. The economic feasibility of biogas production from biomass is not achieved by biogas production alone, it also needs extraction of high-value products.

Further the residual microalgal biomass could be used for anaerobic digestion, which can provide an economical advantage over combustion-based energy production from dry microalgal biomass. The solid residue of the anaerobically digested biomass also contains a high amount of ammonium, potassium, phosphate and other mineral nutrients, which can be used for production of fertilizers. Koller et al. (2014) also suggested that the liquid part of the digestate could be useful as a nutrient supply

Table 12.4: Biofuels from microalgae

Microalgae	Yield	References
Biohydrogen		
Anabaena sp. PCC 7120	2.6 mmol/mg Chl a/h	Masukawa et al. 2001
A. cylindrica IAMM-58	4.2 mmol/mg Chl a/h	Masukawa et al. 2001
A. flosaquae UTEX LB 2558	3.2 mmol/mg Chl a/h	Masukawa et al. 2001
Anabaenopsis circularis IAM M-13	0.31 mmol/mg Chl a/h	Masukawa et al. 2001
Nostoc muscorum IAM M-14	0.60 mmol/mg Chl a/h	Masukawa et al. 2001
A. variabilis AVM13	68 mmol/mg Chl a/h	Happe et al. 2000
A. variabilis PK84	167.6 mmol/mg Chl a/h	Fedorov et al. 2001
A. variabilis ATCC 29413	45.16 mmol/mg Chl a/h	Fedorov et al., 2001
A. variabilis 1403/4B	20 mmol/mg Chl a/h	Moezelaar et al. 1996
A. azollae	38.5 mmol/mg Chl a/h	Fedorov et al. 2001
Synechococcus PCC 602	0.66 mmol/mg Chl a/h	Serebrykova et al. 2000
Gloebacter PCC 7421	1.38 mmol/mg Chl a/h	Serebrykova et al. 2000
Synechocystis PCC 6308	0.13 mmol/mg Chl a/h	Serebrykova et al. 2000
Synechocystis PCC 6714	0.40 mmol/mg Chl a/h	Serebrykova et al. 2000
Gloeocapsa alpicola CALU 743	0.58 mmol/mg protein	Antal & Lindblad 2005
Mycrocystis PCC 7806	11.3 mmol/mg prot/h	Lindberg et al. 2004
Chlamydomonas rheinhardii CC124	102 mL/1.2 L	Cuellar-Bermudez et al. 2014
C. rheinhardii Dang 137MT+	4.5 mmol/L	Cuellar-Bermudez et al. 2014
Chlorella vulgaris MSU 01	26 mL/0.5 L	Cuellar-Bermudez et al. 2014
Scenedesmus obliquus	3.6 mL/µgChla	Cuellar-Bermudez et al. 2014
Platymonas subcordiformis	7.20 mL/h	Cuellar-Bermudez et al. 2014
Biomethane		
Arthrospira maxima	173 mL g^{-1}	Inglesby and Fisher 2012
A. platensis	481 mL g^{-1}	Mussgnug et al. 2010
Microcystis sp.	70.33–153.51 mL	Zeng et al. 2010
Spirulina Leb 18	0.79 g/L	Costa et al. 2008
S. maxima	0.35–0.80 m^3	Samson and Leduy 1986
S. platensis UTEX1926	0.40 m^3 kg	Converti et al. 2009

?

Table 12.4: *(Contd.)*

Microalgae	Yield	References
Dunaliella salina	0.63-0.79 LCH$_4$/gVS	Cuellar-Bermudez et al. 2014
C. vulgaris	0.68 LCH$_4$/gVS	Cuellar-Bermudez et al. 2014
Euglena graciis	0.13 LCH$_4$/gVS	Cuellar-Bermudez et al. 2014
Scenedesmus	140 LCH$_4$/kgVS	Cuellar-Bermudez et al. 2014
Scenedesmus (Lipid free biomass)	212 LCH$_4$/kgVS	Cuellar-Bermudez et al. 2014
Scenedesmus (Protein free biomass)	272 LCH$_4$/kgVS	Cuellar-Bermudez et al. 2014
S. obliquus	0.59-0.69 LCH$_4$/gVS	Cuellar-Bermudez et al. 2014

for microalgae algae. Polishchuk et al. (2015) observed that diluted liquid digestate from anaerobic treated biosludge of the pulp and paper industry could be a suitable media for the cultivation of microalgae.

Biohydrogen

Microalgae including cyanobacteria are able to produce molecular hydrogen (H_2) which can provide another eco-friendly alternative of conventional fossil fuels. H_2 is considered as a potential fuel, as it can provide the heating value of H_2 is 141.65 MJ kg^{-1}, which is the highest amongst known fuels (Ali and Basit 1993). Further it is the cleanest fuel as its only byproduct is water as compared to the emission of greenhouse gases in the case of all known fuels. Thus it could be a game changer in terms of providing maximum energy efficiency without harming the environment. Hydrogen fuel could easily be stored as gas-metal hydride or as a liquid for longer storage and transport. There are number of microalgal genera such as *Anabaena, Calothrix, Oscillatoria, Cyanothece, Nostoc, Synechococcus, Microcystis, Gloeobacter, Aphanocapsa, Chroococcidiopsis* and *Microcoleus*; these are found to produce H_2 gas under many conditions (Masukawa et al. 2001). In microalgal species, molecular H_2 production is achieved through these two mechanisms (Pinzon-Gamez et al. 2005):

(a) It is mediated by the nitrogenase enzyme which catalyzes the following

$$N_2 + 8H^+ + 8e^- + 16ATP \xrightarrow{\text{Nitrogenase}} 2NH_3 + H2 + 16ADP + 16 \text{ pi}$$

(b) It is nearest to the hydrogenases enzymes which catalyzes reversible reaction:

$$H_2 \xrightleftharpoons{\text{Hydrogenases}} 2H+ + 2e^-$$

Although molecular H_2 is a clean and renewable fuel; several methods and processes are also developed for the successful production of H_2 from the microalgae (Table 12.4). However less H_2 production is the major limitation for the commercialization of these fuels (Tiwari and Pandey 2012).

Conclusion

Microalgal biomass could be a potential feedstock for the production of various value-added products. Their use as nutritive human food or food supplement is well known in many countries as it capable to meet the required daily allowance of many essential elements. Besides human consumption, it could be a cheap and healthy feed option for animals and fishes. There are limited microalgal genera like *Spirulina, Chlorella, Dunaliella, Aphanizomenon* that are commercially exploited for these options. Further microalgal biomass have been successfully investigated for antioxidants, antimicrobial, anticancer and anti-UV properties, that would be a new arena of economical and sustainable production of these drugs. But there is a need to investigate potential microalgal strains, their ability to survive in different environments and how their capability is enhanced to meet the required standards.

In relation to microalgal pigments and PHBs, there are some disadvantages like relatively low productivity and high cost of production, that becomes a major limitation in the commercialization of these microalgal pigments and PHBs. Thus these microalgal pigments and PHBs are not able to compete with inexpensive pigments produced from other sources. This problem of high production cost of microalgal pigments can be solved through the use of residual pigment- and PHBs-containing fraction that remains after oil extraction from the microalgal biomass.

Microalgae especially cyanobacteria has proved to an excellent biofertilizers, because to their ability to fix atmospheric nitrogen and secretion of plant-growth promoting substances in the surrounding environment. There have been many studies conducted to investigate the plant-growth promotion ability of microalgal strains in many agricultural crops, and many studies showed promising results which indicate that they could be a better alternative for chemical fertilizers. But most studies are related to laboratory conditions or small field experiments, so there is need of large scale field studies to evaluate the agronomic efficiency of these microalgal biofertilizers.

Biofuel production from microalgal biomass could be an eco-friendly alternative to conventional fossil fuels. Microalgal lipid, sugar and protein content could serve as a feedstock for production of biodiesel, bio-ethanol and bio-methane respectively. Besides these biofuels, some microalgal strains are able to produce H_2, which is a clean and renewable fuel and several approaches have been successfully adapted for the production of H_2 from these strains, but less hydrogen production acts as a major limitation in the commercialization of this fuel. The ability of microalgae to produce the respective content for a particular biofuel can be increased by the use of genetic engineering. Further, there is need of improvement in production techniques and downstream processing to ensure best quality biofuels end products from microalgal biomass.

References

Abarzua, S., S. Jakubowski, S. Eckert and P. Fuchs. 1999. Biotechnological investigation for the prevention of marine biofouling II. Blue-green algae as potential producers of biogenic agents for the growth inhibition of microfouling organisms. Botanica. Mar. 42: 459–465.

Abed, R.M.M., S. Dobretsov and K. Sudesh. 2009. Applications of cyanobacteria in biotechnology. J. Appl. Microbiol. 106(1): 1–12.

Ali, I, and M.A. Basit. 1993. Significance of hydrogen content in fuel combustion. Int. J. Hydrogen Energ. 18(12): 1009–1011.

Amin, S. 2009 Review on biofuel oil and gas production processes from microalgae. Energ. Convers. Manage. 50: 1834–1840.

Antal, T.K. and P. Lindblad. 2005. Production of H_2 by sulphur-deprived cells of the unicellular cyanobacteria *Gloeocapsa alpicola* and *Synechocystis* sp. PCC 6803 during dark incubation with methane or at various extracellular pH. J. Appl. Microbiol. 98: 114–120.

Arad, S. and A. Yaron. 1992. Natural pigments from red microalgae for use in foods and cosmetics. Trends Food Sci. Technol. 3: 92–96.

Bailey, J.E. and D.F. Ollis. 1986. Biochemical Engineering Fundamentals, second ed. McGraw-Hill, Singapore.

Bandaranayake, W.M. 1998. Mycosporines: are they nature's sunscreens? Nat. Prod. Rep. 15(2): 159–172.

Batista, A.P., L. Gouveia, N.M. Bandarra, J.M. Franco and A. Raymundo. 2013. Comparison of microalgal biomass profiles as novel functional ingredient for food products. Algal Res. 2(2): 164–173.

Becker, E.W. 2013. Microalgae for human and animal nutrition, pp. 461–503. *In*: A. Richmond and Q. Hu (eds.). Handbook of Microalgal Culture: Applied Phycology and Biotechnology, Second Edition. John Wiley& Sons, Ltd., New Jersey, US.

Ben-Rebah, F., D. Prevost, A. Yezza and R.D. Tyagi. 2007. Agro-industrial waste materials and wastewater sludge for rhizobial inoculants production: a review. Bioresour. Technol. 98: 3535–3546.

Biondi, N., M.R. Tredici, A. Taton, A. Wilmotte, D.A. Hodgson, D. Losi, et al. 2008. Cyanobacteria from benthic mats of Antarctic lakes as a new source of bioactivities. J. Appl. Microbiol. 105: 105–115.

Bose, P., U.S. Nagpal, G.S. Venkataraman and S.K. Goyal. 1971. Solubilization of tricalcium phosphate by blue-green algae. Curr. Sci. 40: 165–166.

Brányiková, I., B. Maršálková, J. Doucha, T. Brányik, K. Bišová, V. Zachleder, et al. 2011. Microalgae-novel highly efficient starch producers. Biotechnol. Bioeng. 108: 766–776.

Burja, A.M., B. Banaigs, E. Abou-Mansour, J.G. Burgess and P.C. Wright. 2001. Marine cyanobacteria – a prolific source of natural products. Tetrahedron 57: 9347–9377.

Burr, G.S., F.T. Barrows, G. Gaylord and W.R. Wolters. 2011. Apparent digestibility of macro-nutrients and phosphorus in plant-derived ingredients for Atlantic salmon, Salmo salar and Artic charr, *Salvelinus alpinus*. Aquac. Nutr. 17: 570–577.

Cameron, H.J. and G.R. Julian. 1988. Utilisation of hydroxyapatite by cyanobacteria as their sole source of phosphate and calcium. Plant Soil 109: 123–124.

Capelli, B. and G.R. Cysewski. 2010. Potential health benefits of *Spirulina* microalgae. Nutrafoods 9: 19–26.

Carroll, A.K. and J.M. Shick. 1996. Dietary accumulation of UV-absorbing mycosporine-like amino acids (MAAs) by the green sea urchin (*Stongylocentrotrus droebachiensis*). Mar. Biol. 124: 561–569.

Castilho, L.R., D.A. Mitchell and D.M.G. Freire. 2009. Production of polyhydroxyalkanoates (PHAs) from waste materials and by-products by submerged and solid-state fermentation. Bioresour. Technol. 100: 5996–6009.

Ceron, M.C., M.C. Garcia-Malea, J. Rivas, F.G. Acien, J.M. Fernandez, E. Del Rio, et al. 2007. Antioxidant activity of *Haematococcus pluvialis* cells grown in continuous culture as a function of their carotenoid and fatty acid content. Appl. Microbiol. Biot. 74: 1112–1119.

Chauton, M.S., I.R. Kjell, H.N. Niels, T. Ragnar and T.K. Hans. 2015. A techno-economic analysis of industrial production of marine microalgae as a source of EPA and DHA-rich raw material for aquafeed: research challenges and possibilities. Aquacult. 436: 95–103.

Chen, C.Y., W.B. Lu, C.H. Liu and J.S. Chang. 2008. Improved phototrophic H₂ production with *Rhodopseudomonas palustris* WP3-5 using acetate and butyrate as dual carbon substrates. Bioresour. Technol. 99: 3609–3616.

Chen, H. and X. Li. 2008. Effect of static magnetic field on synthesis of polyhydroxyalkanoates from different short-chain fatty acids by activated sludge. Bioresour. Technol. 99: 5538–5544.

Converti, A., R.P. Oliveira, B.R. Torres, A. Lodi and M. Zilli. 2009. Biogas production and valorization by means of a two-step biological process. Bioresour. Technol. 100: 5771–5776.

Cooney, M., M. Maynard, C. Cannizzaro and J. Benemann. 2007. Two-phase anaerobic digestion for production of hydrogen methane mixtures. Bioresour. Tecnhol. 98: 2641–2651.

Costa, J.A.V. and M.G. de Morais. 2011. The role of biochemical engineering in the production of biofuels from microalgae. Bioresour. Technol. 102: 2–9.

Costa, J.A.V., F.B. Santana, M.R. Andrade, M.B. Lima and D.T. Frank. 2008. Microalgae biomass and biomethane production in the south of Brazil. Biotechnol. Lett. 136: 402–403.

Cuellar-Bermudez, S.P., M.A. Romero-Ogawa, B.E. Rittmann and R. Parra-Saldivar. 2104. Algae biofuels production processes, carbon dioxide fixation and biorefinery concept. J. Pet. Environ. Biotechnol. 5(4): 1–8.

Dahms, H.U., Y. Xu and C. Pfeiffer. 2006. Antifouling potential of cyanobacteria: a mini-review. Biofouling 22: 317–327.

De Morais, M.G., B. da Silva Vaz, E.G. de Morais and J.A.V. Costa. 2015. Biologically active metabolites synthesized by microalgae. BioMed Res. Int. 2015(835761): 1–15.

De Souza, M.C.R., C.T. Marques, C.M.G. Dore, F.R.F. da Silva, H.A.O. Rocha and E.L. Leite. 2007. Antioxidant activities of sulfated polysaccharides from brown and red seaweeds. J. Appl. Phycol. 19(2): 153–160.

Dias, J.M.L., A. Oehmen, L.S. Serafim, P.C. Lemos, M.A.M. Reis and R. Oliveira. 2008. Metabolic modelling of polyhydroxyalkanoate copolymers production by mixed microbial cultures. BMC Syst. Biol. 2: 59.

Fedorov, A.S., A.A. Tsygankov, K.K. Rao and D.O. Hall. 2001. Production of hydrogen by an *Anabaena variabilis* mutant in photobioreactor under aerobic outdoor conditions. pp. 223–228. *In*: J. Miyake, T. Matsunaga and A. San Pietro (eds.). BioHydrogen II. Elsevier Science Ltd. Oxford, UK.

Fernandez-Nava, Y., E. Maranon, J. Soons and L. Castrillon. 2008. Denitrification of wastewater containing high nitrate and calcium concentrations. Bioresour. Technol. 99: 7976–7981.

Foster, B.J., M. Fortuna, J. Media, R.A. Wiegand and F.A. Valeriote. 1999. Cryptophycin1 cellular levels and effects in vitro using L1210 cells. Invest. New Drugs 16: 199–204.

Gantar, M., N.W. Kerby, P. Rowell, Z. Obreht and R. Scrimgeour. 1995. Colonization of wheat (*Triticum vulgare* L.) by N2-fixing cyanobacteria. IV. Dark nitrogenase activity and effects of cyanobacteria on natural ^{15}N abundance on plants. New Phytol. 129: 337–343.

Gantar, M., N.W. Kerby, P. Rowell and Z. Obreht. 1991. Colonization of wheat (*Triticum vulgare* L.) by N2-fixing cyanobacteria: a survey of soil cyanobacterial isolates forming associations with roots. New Phytol. 118: 477–483.

Gantar, M., P. Rowell, N.W. Kerby and I.W. Sutherland. 1995. Role of extracellular polysaccharide in the colonization of wheat (*Triticum vulgare* L.) roots by N_2-fixing cyanobacteria. Biol. Fertil. Soils 19: 41–48.

García, J.L., M. de Vicente and B. Galán. 2017. Microalgae, old sustainable food and fashion nutraceuticals. Microb. Biotechnol. 10: 1017–1024.

Ghadge, S. and H. Raheman. 2006. Process optimization for biodiesel production from mahua (*Madhuca indica*) oil using response surface methodology. Bioresour. Technol. 97(3): 379–384.

Goh, S.H., F.M. Yusoff and S.P. Loh. 2010. A comparison of the antioxidant properties and total phenolic content in a diatom, *Chaetoceros* sp. and a green microalga, *Nannochloropsis* sp. J. Agr. Sci. 2: 123–130.

Gouveia, L., A.P. Batista, I. Sousa, A. Raymundo and N.M. Bandarra. 2008. Microalgae in Novel Food Products. pp. 1–37. 2008. *In*: K.N. Papadopoulos (ed.). Food Chemistry Research Developments. Nova Science Publishers, Inc., New York, US.

Hagmann, L. and F. Juttner. 1996. Fischerellin A, a novel photosystem-II-inhibiting allelochemical of the cyanobacterium *Fischerella muscicola* with antifungal and herbicidal activity. Tetrahedron Lett. 37: 6539–6542.

Hamelinck, C.N., G. Van Hooijdonk and A.P.C. Faaij. 2005. Ethanol from lignocellulosic biomass: techno-economic performance in short-, middle- and long-term. Biomass Bioenerg. 28(4): 384–410.

Hankamer, B., F. Lehr, J. Rupprecht, J.H. Mssgnug, C. Posten and O. Kruse. 2007. Photosynthetic biomass and H_2 production by green algae: from bioengineering to bioreactor scale-up. Physiol. Plant. 131: 10–21.

Happe, T., K. Schütz and H. Böhme. 2000. Transcriptional and mutational analysis of the uptake hydrogenase of the filamentous cyanobacterium *Anabaena variabilis* ATCC 29413. J. Bacteriol. 182: 1624–1631.

Hein, S., H. Tran and A. Steinbuchel. 1998. *Synechocystis* sp. PCC6803 possesses a two component polyhydroxyalkanoic acid synthase similar to that of oxygenic purple sulfur bacteria. Arch Microbiol. 170: 162–170.

Herrero, M., L. Jaime, P.J. Martin-Alvarez and A. Cifuentes. 2006. Optimization of the extraction of antioxidants from *Dunaliella salina* microalga by pressurized liquids J. Agric. Food Chem.54: 5597–5603.

Herrero, M., L. Jaime, P.J. Martín-Alvarez, A. Cifuentes and E. Ibanez. 2006. Optimization of the extraction of antioxidants from *Dunaliella salina* microalga by pressurized liquids. J. Agric. Food Chem. 54: 5597–5603.

Ibanez, E. and A. Cifuentes. 2013. Benefits of using algae as natural sources of functional ingredients. J. Sci. Food Agric. 93(4): 703–709.

Inglesby, A.E. and A.C. Fisher. 2012. Enhanced methane yields from anaerobic digestion of Arthrospira maxima biomass in an advanced flow-through reactor with an intergrated recirculation loop microbial fuel cell. Energy Environ. Sci. 5: 7996–8006.

Issa, A.A., M.H. Abd-Alla and T. Ohyama. 2014. Nitrogen fixing cyanobacteria: future prospect. pp. 23–48. *In*: T. Ohyama (ed.). Advances in Biology and Ecology of Nitrogen Fixation. IntechOpen Limited, London, UK.

Jaime, L., J.A. Mendiola, M. Herrero and C. Soler. 2005. Separation and characterization of antioxidants from *Spirulina platensis* microalga combining pressurized liquid extraction, TLC and HPLC-DAD. J. Sep. Sci. 28: 2111–2119.

Jau, M.H., S.P. Yew, P.SY. Toh, S.C.A. Chong, S.M.C. Phang, N. Najimudin, et al. 2005. Biosynthesis and mobilization of poly(3-hydroxybutyrate) P (3HB) by *Spirulina platensis* Int. J. Biol. Macromolecules 36: 144–151.

Johnson, M. and Z. Wen. 2009. Production of biodiesel fuel from the microalga *Schizochytrium limacinum* by direct transesterification of algal biomass. Energy Fuels 23(10): 294–306.

Jyotsana, D., M.P. Sarma, L. Meeta, K.A. Mandal and I. Banwari. 2010. Evaluation of bacterial strains isolated from oil-contaminated soil for production of polyhydroxyalkanoic acids (PHA). Pedobiologia 54: 25–30.

Kannan, R.R.R., R. Arumugam and P. Anantharaman. 2010. In vitro antioxidant activities of ethanol extract from *Enhalusacoroides* (L.F.) Royle. Asian Pac. J. Trop. Med. 3(11): 898–901.

Kapdi, S.S., V.K. Vijay, S.K. Rajesh and R. Prasad. 2005. Biogas scrubbing, compression and storage: perspective and prospectus in Indian context. Renew Energ. 30: 1195–1202.

Karmali, R.A. 1996. Historical perspective and potential use of n-3 fatty acids in therapy of cancer cachexia. Nutr. 12: S2–S4.

Karthikeyan, N., R. Prasanna, A. Sood, P. Jaiswal, S. Nayak and B.D. Kaushik. 2009. Physiological characterization and electron microscopic investigations of cyanobacteria associated with wheat rhizosphere. Folia Microbiol. 54: 43–51.

Karthikeyan, N., R. Prasanna, D.P. Lata and B.D. Kaushik. 2007. Evaluating the potential of plant growth promoting cyanobacteria as inoculants for wheat. Eur. J. Soil Biol. 43: 23–30.

Kedar, L., Y. Kashman and A. Oren. 2002. Mycosprine-2-glycine is the major mycosprine-like amino acid in a unicellular cyanobacterium (*Euhalothece* sp.) isolated from a gypsum crust in a hypersaline saltern pond. FEMS Microbiol. Lett. 208: 233–237.

Khan, Z., P. Bhadouria and P. Bisen. 2005. Nutritional and therapeutic potential of *Spirulina*. Curr. Pharm. Biotechnol. 6: 373–379.

Koller, M., A. Muhr and G. Braunegg. 2014. Microalgae as versatile cellular factories for valued products. Algal Res. 6: 52–63.

Koyande, A.K., K.W. Chew, K. Rambabu, Y. Tao, D.-T. Chud and P.-L. Show. 2019. Microalgae: a potential alternative to health supplementation for humans. Food Sci. Human Wellness 8: 16–24.

Kreitlow, S., S. Mundt and U. Lindequist 1999. Cyanobacteria – a potential source of new biologically active substance. J. Biotechnol. 70: 61–73.

Kulik, M.M. 1995. The potential for using cyanobacteria (blue-green algae) and algae in the cyanobacterial control of plant pathogenic bacteria and fungi. Eur. J. Plant Path. 101: 585–599.

Kulshreshtha, A., J. Zacharia, U. Jarouliya, P. Bhadauriya, G.B.K.S. Prasad and P.S. Bisen. 2008. *Spirulina* in healthcare management. Curr. Pharm. Biotechnol. 9: 400–405.

Kumar, A. 2018. Assessment of cyanobacterial diversity in paddy fields and their capability to degrade the pesticides. Babasahaeb Bhimrao Ambedkar University, Lucknow, India.

Kumar, A. and J.S. Singh. 2016. Microalgae and cyanobacteria biofuels: a sustainable alternate to crop-based fuels. pp. 1–20. *In*: J.S. Singh, D.P. Singh (eds.). Microbes and Environmental Management. Studium Press Pvt. Ltd. New Delhi, India.

Kumar, A. and J.S. Singh. 2017. Cyanoremediation: a green-clean tool for decontamination of synthetic pesticides from agro- and aquatic-ecosystems. pp. 59–83. *In*: J.S. Singh, G. Seneviratne (eds.). Agro-environmental Sustainability, Vol. II: Managing Environment Pollution. Springer Int., Cham, Switzerland.

Kumar, A. and J.S. Singh. 2020. Biochar coupled rehabilitation of cyanobacterial soil crusts: a sustainable approach in stabilization of arid and semiarid soils. pp. 167–191. *In*: J.S. Singh, C. Singh (eds.). Biochar Applications in Agriculture and Environment Management. Springer Int., Cham, Switzerland.

Kumar, A. and J.S. Singh. 2020. Microalgal bio-fertilizers. *In*: E. Jacob-Lopes, M.M. Maroneze, M.I. Queiroz, L.Q. Zepka (eds.). Handbook of Microalgae-based Processes and Products. Academic Press, Cabridge, US, In press.

Kumar, A., S. Kaushal, S.A. Saraf and J.S. Singh. 2017. Cyanobacterial biotechnology: an opportunity for sustainable industrial production. Clim. Chang. Environ. Sustain. 5(1): 97–110.

Kumar, A., S. Kaushal, S.A. Saraf and J.S. Singh. 2018. Microbial bio-fuels: a solution to carbon emissions and energy crisis. Front. Biosci. (Landmark) 23: 1789–1802.

Kumar, A., S. Kaushal, S.A. Saraf and J.S. Singh. 2018. Screening of chlorpyrifos (CPF) tolerant cyanobacteria from paddy field soil of Lucknow, India. Int. J. Appl. Adv. Sci. Res. 3(1): 100–105.

Lee, C., J. Kim, H. Do and S. Hwang. 2008. Monitoring thiocyanate-degrading microbial community in relation to changes in process performance in mixed culture systems near washout. Water Res. 42: 1254–1262.

Li, H.-B., K.-W. Cheng, C.-C. Wong, K.-W. Fan, F. Chen and Y. Jiang. 2007. Evaluation of antioxidant capacity and total phenolic content of different fractions of selected microalgae. Food Chem. 102: 771–776.

Liang, S., X. Liu, F. Chen and Z. Chen. 2004. Current microalgal health food R & D activities in China. pp. 45–48. *In*: P.O. Ang (ed.). Asian Pacific Phycology in 21st Century: Prospect and Challenges. Springer, Dordrecht, Netherlands.

Lindberg, P., P. Lindblad and L. Cournac. 2004. Gas exchange in the filamentous cyanobacterium *Nostoc punctiforme* strain ATCC 29133 and its hydrogenase-deficient mutant strain NHM5. Appl. Environ. Microbiol. 70: 2137–2145.

Liu, B.F., N.Q. Ren, J. Tang, J. Ding, W.Z. Liu, J.F. Xu, et al. 2010. Bio-hydrogen production by mixed culture of photo and dark fermentation bacteria. Int. J. Hydrogen Energ. 35: 2858–2862.

Madeira, M.S., C. Cardoso, P.A. Lopes, D. Coelho, C. Afonso, N.M. Bandarra, et al. 2017. Microalgae as feed ingredients for livestock production and meat quality: a review. Livest. Sci. 205: 111–121.

Mallick, N., L. Sharma and A.K. Singh. 2007. Polyhydroxyalkanoate (PHA) synthesis by *Spirulina subsalsa* from Gujarat coast of India. J. Plant Physiol. 164: 312–317.

Mandal, B., S.C. Das and L.N. Mandal. 1992. Effect of growth and subsequent decomposition of blue-green algae in the transformation of phosphorus in submerged soils. Plant Soil 143: 289–297.

Markou, G. and E. Nerantzis. 2013. Microalgae for high-value compounds and biofuels production: a review with focus on cultivation under stress conditions. Biotechnol. Adv. 31(8): 1532–1542.

Masukawa, H., K. Nakamura, M. Mochimaru and H. Sakurai. 2001. Photobiological hydrogen production and nitrogenase activity in some heterocystous cyanobacteria. pp. 63–66. *In*: J. Miyake, T. Matsunaga and A. San Pietro (eds.). Biohydrogen II. Elsevier Science Ltd. Oxford, UK.

Meher, L.C., S.D. Vidya and S.N. Naik. 2006. Technical aspects of biodiesel production by transesterification—a review. Renew Sust. Energ. Rev. 10: 248–268.

Melnicki. M.R., E. Eroglu and A. Melis. 2009. Changes in hydrogen production and polymer accumulation upon sulfur-deprivation in purple photosynthetic bacteria. Int. J. Hydrogen Energ. 34: 6157–6170.

Mendes, R.L., A.D. Reis and A.F. Palavra. 2006. Supercritical CO_2 extraction of γ-linolenic acid and other lipids from *Arthrospira (Spirulina) maxima*: comparison with organic solvent extraction. Food Chem. 99: 57–63.

Minhas, A.K., P. Hodgson, C.J. Barrow and A. Adholeya. 2016. A review on the assessment of stress conditions for simultaneous production of microalgal lipids and carotenoids. Front. Microbiol. 7: 546.

Misra, S. and B.D. Kaushik. 1989. Growth promoting substances of cyanobacteria. I. Vitamins and their influence on rice plant. Proc. Indian Sci. Acad. B 55: 295–300.

Moezelaar, R., S.M. Bijvank and L.J. Stal. 1996. Fermentation and sulfur reduction in the mat-building cyanobacterium *Microcoleus chtonoplastes*. Appl. Environ. Microbiol. 62: 1752–1758.

Mostafa, S.S.M. 2012. Microalgal biotechnology: prospects and applications. pp. 275–314. *In*: N.K. Dhal and S.C. Sahu (eds.). Plant Science. IntechOpen Limited, London, UK.

Mulbry, W., S. Kondrad, J. Buyer and D. Luthria. 2009. Optimization of an oil extraction process for microalgae from the treatment of manure effluent. J. Am. Oil Chem. Soc. 86: 909–915.

Mussgnug, J.H., V. Klassen, A. Schlüter and O. Kruse. 2010. Microalgae as substrates for fermentative biogas production in a combined biorefinery concept. J. Biotechnol. 150: 51–56.

Obreht, Z., N.W. Kerby, M. Gantar and P. Rowell. 1993. Effects of root associated N_2-fixing cyanobacteria on the growth and nitrogen content of wheat (*Triticum vulgare* L.) seedlings. Biol. Fertil. Soils 15: 68–72.

Papke, U., E.M. Gross and W. Francke. 1997. Isolation, identification and determination of the absolute configuration of Fischerellin B. A new algicide from the freshwater cyanobacterium *Fischerellin muscicola* (Thuret). Tetrahedron Lett. 38: 379–382.

Patterson, G.M.L., C.L. Baldwin, C.M. Bolls, F.R. Caplan, H. Karuso, L.K. Larsen, et al. 1991. Antineoplastic activity of cultured blue-green algae (cyanophyta). J. Phycol. 27: 530–536.

Pinzon-Gamez, N.M., S. Sundaram and L.K. Ju. 2005. Heterocyst differentiation and H_2 production in N_2-fixing cyanobacteria. pp. 8949–8951. *In*: AIChE Annual Meeting, Conference Proceedings, Cincinnati, Ohio, US.

Plaza, M., M. Herrero, A. Cifuentes and E. Ibanez. 2009. Innovative natural functional ingredients from microalgae. J. Agric. Food Chem. 57(16): 7159–7170.

Polishchuk, A., D. Valev, M. Tarvainen, S. Mishra, V. Kinnunen, T. Antal, et al. 2015. Cultivation of *Nannochloropsis* for eicosapentaenoic acid production in wastewaters of pulp and paper industry. Bioresour. Technol. 193: 469–476.

Prasanna, R., A. Sood, P. Jaiswal, S. Nayak, V. Gupta, V. Chaudhary, et al. 2010. Rediscovering cyanobacteria as valuable sources of bioactive compounds (Review). Appl. Biochem. Microbiol. 46(2): 119–134.

Priyadarshani, I. and B. Rath. 2012. Commercial and industrial applications of micro algae: a review. J. Algal Biomass Util. 3(4): 89–100.

Pulz, O. and W. Gross. 2004. Valuable products from biotechnology of microalgae. Appl. Microbiol. Biotechnol. 65: 635–648.

Quillaguaman, J., H. Guzman, D.V. Thuoc and R.H. Kaul. 2010. Appl. Microbiol. Biotechnol. 85: 1687–1696.

Reinert, R., E.L. Donald, F.X. Rosi, C. Watal and M. Dowzicky. 2007. Antimicrobial susceptibility among organisms from the Asia/Pacific Rim, Europe and Latin and North America collected as part of TEST and the in vitro activity of tigecycline. J. Antimicrob. Chemother. 60: 1018–1029.

Ruffing, A.M. 2011. Engineered cyanobacteria: teaching an old bug new tricks. Bioeng. Bugs 2: 136–149.

Samson, R. and A. Leduy. 1986. Detailed study of anaerobic-digestion of *Spirulina maxima* algal biomass. Biotechnol. Bioeng. 28 (7): 1014–1023.

Sanmukh, S., B. Bruno, U. Ramakrishnan, K. Khairnar, S. Swaminathan and W. Paunikar. 2014. Bioactive compounds derived from microalgae showing antimicrobial activities. J. Aquac. Res. Dev. 5(3): 1–4.

Sathasivam, R., R. Radhakrishnan, A. Hashem and E.F. Abd-Allah. 2019. Microalgae metabolites: a rich source for food and medicine. Saudi J. Biol. Sci. 26(4): 709–722.

Seiichi, T. and Y. Doi. 2004. Evolution of polyhydroxyalkanoate (PHA) production system by enzyme evolution: successful case studies of directed evolution. Macromolecular Biosci. 3: 145–156.

Senthilkumar, B. and G. Prabhakaran. 2006. Production of PHB (bio plastics) using bio-effluents as substrates by *Alcaligenseutropha*. Ind. J. Biotechnol., 5: 76–79.

Serebryakova, L.T., M.E. Sheremetieva and P. Lindblad. 2000. H_2-uptake and evolution in the unicellular cyanobacterium *Chroococcidiopsis thermalis* CALU 758. Plant Physiol. Biochem. 38: 525–530.

Shrivastav A, S.K. Mishra, B. Shethia, I. Pancha, D. Jain and S. Mishra. 2010. Isolation of promising bacterial strains from soil and marine environment for polyhydroxyalkanoates (PHAs) production utilizing *Jatropha* biodiesel byproduct. Int. J. Biol. Macromol. 47(2):283–287.

Singh, A.L. and P.K. Singh. 1987. Influence of Azolla management on the growth, yield of rice and soil fertility. II. N and P contents of plants and soil. Plant Soil 102: 49–54.

Singh, J.S., A. Kumar, A.N. Rai and D.P. Singh. 2016. Cyanobacteria: a precious bio-resource in agriculture, ecosystem, and environmental sustainability. Front. Microbiol. 7: 529.

Singh, J.S., S. Koushal, A. Kumar, S.R. Vimal and V.K. Gupta. 2016. Book review: microbial inoculants in sustainable agricultural productivity, Vol. II: Functional application. Front. Microbiol. 7: 2105.

Singh, J.S., A. Kumar and M. Singh. 2019. Cyanobacteria: a sustainable and commercial bio-resource in production of bio-fertilizer and bio-fuel from wastewaters. Environ. Sustain. Indic. 3: 100008.

Singh, V.P. and A. Trehan. 1973. Effect of extracellular products of *Aulosira fertilissima* on the growth of rice seedlings. Plant Soil 38: 457–464.

Skulberg, O.M. 2000. Microalgae as a source of bioactive molecules; experience from cyanophyte research. J. App. Phycol. 12: 341–348.

Soletto, D., L. Binaghi, A. Lodi, J.C.M. Carvalho and A. Converti. 2004. Batch and fed-batch cultivations of *Spirulina platensis* using ammonium sulphate and urea as nitrogen sources. Aquacult. 243: 217–224.

Spiller, H. and M. Gunasekaran. 1990. Ammonia-excreting mutant strain of the cyanobacterium *Anabaena variabilis* supports growth of wheat. Appl. Microbiol. Biotechnol. 33: 477–480.

Spolaore, P., C. Joannis-Cassan, E. Duran and A. Isambert. 2006. Commercial applications of microalgae. J. Biosci. Bioeng. 101: 87–97.

Stochaj, W.R., W.C. Dunlap and J.M. Shick. 1994. Two new UV-absorbing mycosporine-like amino acids from the sea anemome *Anthopleura elegantissima* and the effects of zooxanthellae and spectral irradiance on chemical composition and content. Mar. Biol. 118: 149–156.

Subramanian, G. and S.S. Sundaram. 1986. Induced ammonia release by the nitrogen fixing cyanobacterium *Anabaena*. FEMS Microbiol. Lett. 37: 151–154.

Sudesh, K. 2004. Microbial polyhydroxyalkanoates (PHAs): an emerging biomaterial for tissue engineering and therapeutic applications. Med. J. Malaysia. 59(Suppl B): 55–56.

Sudesh, K., H. Abe and Y. Doi. 2000. Synthesis, structure and properties of polyhydroxyalkanoates: biological polyesters. Progress Polymer Sci. 25: 1503–1555.

Takaichi, S. 2011. Carotenoids in algae: distributions, biosyntheses and functions. Mar. Drugs 9: 1101–1118.

Talero, E., S. Garcia-Maurino, J. Avila-Roman, A. Rodriguez-Luna, A. Alcaide and V. Motilva. 2015. Bioactive compounds isolated from microalgae in chronic inflammation and cancer. Mar. Drugs 13: 6152–6209.

Teuscher, E., U. Lindequist and S.Mundt. 1992. Cyanobakterien, Quellen biogener Wirkstoffe. Pharm. Ztg. Wiss. 137: 57–69.

Tibbetts, S.M. 2018. The potential for 'next-generation', microalgae-based feed ingredients for salmonid aquaculture in context of the blue revolution. pp. 151–175. *In*: E. Jacob-Lopes, L.Q. Zepka and M.I. Queiroz (eds.). Microalgal Biotechnology. IntechOpen Limited, London, UK.

Tiwari, A. and A. Pandey. 2012. Cyanobacterial hydrogen – a step towards clean environment. Int. J. Hydrogen Energy. 37: 139–150.

Trivedi, J., M. Aila, D.P. Bangwal, S. Kaul and M.O. Garg. 2015. Algae based biorefinery—how to make sense? Renew. Sustain. Energ. Rev. 47: 295–307.

USDA. 2018. Spirulina, dried. National Nutrient Database for Standard Reference Basic Report No. 11667.

Venkataraman, G.S. and S. Neelakantan. 1967. Effect of cellular constituents of nitrogen fixing blue-green alga *Cylindrospermum* on root growth of rice plants. J. Gen. Appl. Microbiol. 13: 53–62.

Volkmann, M. and A.A. Gorbushina. 2006. A broadly applicable method for extraction and characterization of mycosporines and mycosporine-like amino acids of terrestrial, marine and freshwater origin. FEMS Microbiol. Lett. 255: 286–295.

Wang, Y.P., Q.B. Li, J.Y. Shi, Q. Lin, X.C. Chen, W.X. Wu, et al. 2008. Assessment of microbial activity and bacterial community composition in the rhizosphere of a copper accumulator and a non-accumulator. Soil Biol. Biochem. 40: 1167–1177.

Wu, Z. and X. Shi. 2007. Optimization for high-density cultivation of heterotrophic Chlorella based on a hybrid neural network model. Lett. Appl. Microbiol. 44: 13–18.

Xiao-li, L., M.T. Kiran, T. Ling, T. Archana and L. Gu. 2017. A novel growth method for diatom algae in aquaculture wastewater for natural food development and nutrient removal. Water Sci. Technol. 75(12): 2777–2783.

Yaakob, Z., E. Ali, A. Zainal, M. Mohamad and M.S. Takriff. 2014. An overview: biomolecules from microalgae for animal feed and aquaculture. J. Biol. Res. (Greece) 21(1): 1–10.

Zeng, S., X. Yuan, X. Shi and Y. Qiu. 2010. Effect of inoculum/substrate ratio on methane yield and orthophosphate release from anaerobic digestion of *Microcystis* spp. J. Hazard. Mater. 178(1–3): 89–93.

Zubia, M., M.S. Fabre, V. Kerjean and K.L. Lann. 2009. Antioxidant and antitumoural activities of some Phaeophyta from Brittany coasts. Food Chem. 116: 693–701.

Index